Charting the Water Regulatory Future

NEW HORIZONS IN ENVIRONMENTAL AND ENERGY LAW

Series Editors: Kurt Deketelaere, *Professor of Law, University of Leuven, Belgium and University of Dundee, Scotland* and Zen Makuch, *Reader in Law, Barrister, Imperial College, London, UK*

Environmental law – including the pressing considerations of energy law and climate change – is an increasingly important area of legal research and practice. Given the growing interdependence of global society and the significant steps being made towards environmental protection and energy efficiency, there are few people untouched by environmental and energy lawmaking processes.

At the same time, environmental and energy law is at a crossroads. The command and control methodology that evolved in the 1960s and 1970s for air, land and water protection may have reached the limit of its environmental protection achievements. New life needs to be injected into our environmental protection regimes – perhaps through the concept of sustainability in its environmental, economic and social forms. The same goes for energy policy and law, where liberalization, environmental protection and security of supply are at the centre of attention. This important series seeks to press forward the boundaries of environmental and energy law through innovative research into environmental and energy law, doctrine and case law. Adopting a wide interpretation of environmental and energy law, it includes contributions from both leading and emerging international scholars.

Titles in the series include:

The Fragmentation of Global Climate Governance
Consequences and Management of Regime Interactions
Harro van Asselt

Renewable Energy Law in the EU
Legal Perspectives on Bottom-up Approaches
Edited by Marjan Peeters and Thomas Schomerus

Environmental Enforcement Networks
Concepts, Implementation and Effectiveness
Edited by Michael Faure, Peter De Smedt and An Stas

Regulation of the Upstream Petroleum Sector
A Comparative Study of Licensing and Concession Systems
Edited by Tina Hunter

Regional Environmental Law
Transregional Comparative Lessons in Pursuit of Sustainable Development
Edited by Werner Scholtz and Jonathan Verschuuren

Earth Governance
Trusteeship of the Global Commons
Klaus Bosselmann

Waste Policy
International Regulation, Comparative and Contextual Perspectives
Alexander Gillespie

Shale Gas and the Future of Energy
Law and Policy for Sustainability
Edited by John C. Dernbach and James R. May

The Privatisation of Biodiversity
New Approaches to Conservation Law
Colin T. Reid and Walters Nsoh

Aquaculture Law and Policy
Global, Regional and National Perspectives
Edited by David L. VanderZwaag, Nigel Bankes and Irene Dahl

Charting the Water Regulatory Future
Issues, Challenges and Directions
Edited by Julien Chaisse

Charting the Water Regulatory Future

Issues, Challenges and Directions

Edited by

Julien Chaisse

Professor, Faculty of Law & Director, Center for Financial Regulation and Economic Development, The Chinese University of Hong Kong

NEW HORIZONS IN ENVIRONMENTAL AND ENERGY LAW

Cheltenham, UK • Northampton, MA, USA

© The Editor and Contributors Severally 2017

All rights reserved. No part of this publication may be reproduced, stored in a retrieval system or transmitted in any form or by any means, electronic, mechanical or photocopying, recording, or otherwise without the prior permission of the publisher.

Published by
Edward Elgar Publishing Limited
The Lypiatts
15 Lansdown Road
Cheltenham
Glos GL50 2JA
UK

Edward Elgar Publishing, Inc.
William Pratt House
9 Dewey Court
Northampton
Massachusetts 01060
USA

A catalogue record for this book
is available from the British Library

Library of Congress Control Number: 2016949977

This book is available electronically in the **Elgar**online
Law subject collection
DOI 10.4337/9781785366727

ISBN 978 1 78536 671 0 (cased)
ISBN 978 1 78536 672 7 (eBook)

Typeset by Servis Filmsetting Ltd, Stockport, Cheshire

Printed and bound in Great Britain by TJ International Ltd, Padstow, Cornwall

Contents

About the editor	vii
List of contributors	ix
Foreword	xvii
Table of cases	xx
Table of legislation	xxiv

1. Introduction 1
   Julien Chaisse

PART I   THE WATER CHALLENGE TO PUBLIC INTERNATIONAL LAW

2. Promoting global water-use efficiency – promises and shortcomings of international trade rules 23
   Manzoor Ahmad

3. The trade in water services – how does GATS apply to the water and sanitation services sector? 36
   Rebecca Bates

4. Virtual water: a global economic solution to a local environmental and political problem? 55
   Paolo Turrini

5. Foreign investment in water: privatization, globalization and the law 76
   Julien Chaisse

6. The right of the host state to regulate water services 91
   Catharine Titi

7. Regulation and protection of water in international law: terrestrial and marine perspectives 105
   Virginie J.M. Tassin

PART II ETHICAL, LEGAL AND SOCIAL ISSUES

8. Is investment arbitration inimical to the human right to water? The re-examination of arbitral decisions on water services 145
    Miharu Hirano and Shotaro Hamamoto

9. The provision and violation of water rights (the case of Pakistan) – a human rights based approach 167
    Sikander Ahmed Shah

10. The human right to clean water and sanitation – a perspective from Nigeria 195
    Cosmas Emeziem

11. Troubled waters: impact of the private sector in implementing the right to water 215
    Preetha Mahadevan

12. Sanitation rights, public law litigation and inequality: a case study from Brazil 236
    Ana Paula de Barcellos

PART III ECONOMIC DRIVERS SHAPING THE FUTURE OF WATER

13. Demand for infrastructure investment for water services: key features and assessment methods 257
    Sacchidananda Mukherjee and Debashis Chakraborty

14. Residential water charges in Ireland: policy objectives and funding models 297
    Thomas McDonnell

15. The role of multinationals in providing water services – are they more efficient? 321
    Tihomir Ancev, Samad Azad and Francesc Hernandez-Sancho

16. Microfinance in water and sanitation services: identifying best practices 343
    Jonatan A. Lassa and Allen Yu-Hung Lai

Bibliography 356
Index 395

About the editor

Julien Chaisse is Professor in the Faculty of Law at the Chinese University of Hong Kong (CUHK). Julien Chaisse is an award-winning specialist in international economic law with particular expertise in the regulation and economics of foreign investment. His research also covers other relevant fields, such as the law of natural resources, energy law and World Trade Organization law. Before joining the Chinese University of Hong Kong (CUHK) Law Faculty in 2009, Julien Chaisse served in the Ministry of Foreign Affairs of France, and started his academic career in Europe. Professor Chaisse is frequently invited as a guest lecturer to many academies and universities around the world, including the Academy of International Investment and Trade Law, Boston University (US), Yokohama National University (Japan), Passau University (Germany) and Melbourne University (Australia) where he is a Senior Fellow to the Law School.

Julien Chaisse has authored a broad body of well-regarded and widely-cited articles on topics such as the rise of sovereign wealth funds, the regulation of foreign investment and decision-making challenges facing the WTO, which have been published in the top refereed journals of international law. In recognition of his outstanding scholarly achievements, Professor Chaisse received the CUHK Research Excellence Award in 2012 and has been Director of the Centre for Financial Regulation and Economic Development of the Faculty since 2013.

In addition to scholarly work, Professor Chaisse has wide experience as a practitioner, and has been engaged as expert, counsel and arbitrator in transnational dispute settlements. As well as being frequently interviewed by local and international media on current events and legal issues, Professor Chaisse is also regularly invited to provide legal advice and training courses on cutting-edge issues of international economic law for international organizations, governments, multinational law firms and private investors, including Ernst & Young, Deloitte US, Maxwell Stamp, the United Nations ESCAP and ITC, World Trade Organization, ASEAN Secretariat, European Commission, Asian Development Bank, and many European countries and ASEAN member states. Professor Chaisse currently advises the E15 Task Force on Investment Policy, sponsored jointly by the World Economic Forum (WEF) and the International Centre for Trade and Sustainable Development (ICTSD).

Contributors

Manzoor Ahmad is an international trade expert. Currently, as well as serving as a Senior Fellow at the International Centre for Trade and Sustainable Development (ICTSD), he is also the Chief Executive of World Trade Advisors, an independent consulting company. From 2002 to 2008, he served as Pakistan's Ambassador to WTO and from 2008 to 2010 as the Director, Food and Agriculture Organization. He has also been a panellist for several WTO trade disputes. He is also working in senior advisory positions on several national and international think tanks.

Tihomir Ancev is an Associate Professor in Environmental and Resource Economics, at the School of Economics, University of Sydney. He has research interests in water economics, economics of air pollution, design of policies to address pollution problems, and economics of regulation. He has published extensively on these topics in top international journals and in edited book volumes. Tiho has participated in, and led a number of national and international research projects focusing on water economics and economics of pollution.

Samad Azad is a Research Fellow at the Tasmanian School of Business and Economics of the University of Tasmania. His particular research interest is to measure productivity and efficiency of enterprises that use natural resources in their production system. Samad earned his PhD in Agricultural Resource Economics from the University of Sydney in 2012. Prior to joining the University of Tasmania in January 2015, Samad worked as a Research Associate at the Department of Agricultural & Resource Economics and the School of Economics of the University of Sydney (2012–14). He has also worked as a Teaching Assistant at the same University between 2008 and 2014. As an Agricultural Economist, Samad worked at the Bangladesh Rice Research Institute from 1999 to 2006. During this period he visited the International Rice Research Institute (IRRI) a number of times in his role as a Research Collaborator in different socio-economic projects. Samad has published his research works in international refereed journals including *Journal of Environmental Management*, *Ecological Economics* and *Water Resources Management*.

Ana Paula de Barcellos, JD, LLM, SJD, is Constitutional Law Professor at the State University of Rio de Janeiro, and was the 2012–13 Takemi Fellow at the Harvard School of Public Health, Boston, MA, USA. She has authored three books and numerous articles, many of them on human dignity, its legal meaning and consequences, human rights public policies in democracies and social and economic rights. She is also a frequent guest lecturer in Brazil and abroad. She has been a member of the Yale Law School SELA (Seminario en LatinoAmérica de Teoría Constitucional y Política) Organizing Committee since 2012. She is the vice chairman of the Committee on Constitutional Law at the Brazilian Bar Association – Rio de Janeiro (2016/2018).

Rebecca Bates is a Lecturer in Environmental Law at Queen Mary University of London. Prior to joining Queen Mary in 2014, she held lectureships at the University of Sydney and Brunel University. Dr Bates completed her doctorate at the University of Sydney in 2009 with a thesis that examined the governance of privatised water markets in the domestic and international context. Dr Bates' research is focused on the areas of natural resources and water law with a particular focus on the economic and social aspects of these areas. She regularly contributes to international publications and conferences and has had publications appear in the *Sydney Law Review*, RECIEL and the *University of Denver Water Law Review*. At Queen Mary, Dr Bates teaches and convenes International Resources Law on the LLM Programme and teaches Tort at undergraduate level. She is currently the Book Review editor for the *International Human Rights Law Review* (Brill). Outside academia, Dr Bates is admitted as a Solicitor of the Supreme Court of New South Wales and has previously worked at a leading Australian law firm in the area of Product Liability.

Debashis Chakraborty is a member of the Faculty of Economics at the Indian Institute of Foreign Trade (IIFT), New Delhi, which is a leading Deemed University in India. Before joining IIFT, he was a Research Associate at Rajiv Gandhi Institute for Contemporary Studies (RGICS), New Delhi, a renowned policy think tank. His area of specialisation includes trade and WTO issues, environmental economics and economic development concerns in India. He received his Doctorate Degree from Jawaharlal Nehru University (JNU), New Delhi. Dr Chakraborty has conducted several research projects on regional trade agreements, India's trade and investment flows, WTO related commitments, topics relating to Indian Economic Development etc. for State and Central Governments in India as well as for multilateral funding agencies. He has also organised a number of academic conferences as well as training programmes for government officials and corporate executives at IIFT over the last decade. He has extensively published research articles in national and international refereed journals. He also has ten volumes (including both co-authored and co-edited ones) to his credit. His co-edited volume on *Environmental Challenges and Governance: Diverse Perspectives from Asia*, was published in 2015. His forthcoming co-edited volume, *Trade, Investment and Economic Development in Asia: Empirical and Policy Issues*, will be published in mid-2016. He has also been a reviewer of several national and international journals. For the last three and a half years, Dr Chakraborty has been Associate Editor of *Foreign Trade Review* (Sage), which is a comprehensive forum for disseminating theoretical and empirical research on international trade and investment related issues.

Cosmas Emeziem, LLB (Nigeria), LLM (Cornell), Candidate for Doctor of the Science of Law (JSD), Senior Research Assistant, Cornell University Law School, Myron Taylor Hall Ithaca New York. Mr Emeziem is also enrolled as a Barrister and Solicitor of the Nigerian Supreme Court. He has attended courses in Public and Private International Law at The Hague Academy of International Law, The Hague, Netherlands. Additionally, he has practical experience in human rights advocacy, litigation, foreign direct investments, and community development services in Nigeria. Presently, he is researching on transitional justice measures and the intersection between human rights, democratic governance, institutional reforms, public policy, rule of law and post-conflict reconstruction.

Shotaro Hamamoto is Professor of the Law of International Organizations, Graduate School of Law, Kyoto University and Senior Fellow, Centre for International Governance Innovation (Ottawa) (since 2015). He has also taught at the Université de Paris I (2009) and Sciences Po de Paris (2012) as *professeur invité*. He was Counsel and Advocate for the Japanese Government in *Whaling in the Antarctica* (ICJ, Australia v. Japan, New Zealand intervening, 2010), *Hoshinmaru* (ITLOS, Japan v. Russia, 2007), and *Tomimaru* (ITLOS, Japan v. Russia, 2007), assistant for the Spanish Government in *Fisheries Jurisdiction* (ICJ, Spain v. Canada, 1997–98). He represented Japan in UNCITRAL WG II (Arbitration/Conciliation) (2010–15) and in the OECD Investment Committee (2011). He is an arbitrator at the Japan Sports Arbitration Agency (2008). His research subjects include the theory of juridical acts in international law, international investment law, the law of the sea and sports law. His recent publications include: 'Domestic Review of Treaty-Based International Investment Awards: Effects of the Metalclad Judgment of the British Columbia Supreme Court', *in* Machiko Kanetake and André Nollkaemper (eds), *The Rule of Law at the National and International Levels: Contestations* *and Deference*, 2016, pp. 99–113; "L'État situé dans le droit international de l'investissement", *in* Shotaro Hamamoto, Akiho Shibata and Hironobu Sakai eds., *"L'être situé", Effectiveness and Purposes of International Law: Essays in Honour of Professor Ryuichi Ida*, 2015, pp. 3–22; "Parties to the 'Obligations' in the Obligations Observance ('Umbrella') Clause", (2015) 30 *ICSID Review-Foreign Investment Law Journal* 449–64. He holds LLB (Kyoto), LLM (Kyoto) and Docteur en droit (Paris II).

Francesc Hernandez-Sancho has a PhD in Environmental Economics and is an Associate Professor at the University of Valencia (UV). He is Head of the Water Economics Research Group in UV and Director of the Master on Water Management programme. He is also a Member of the Management Committee of the International Water Association (IWA) Specialist Group on Statistics and Economics and Leader of the IWA Water Economics Working Group. His research focuses on water tariffs and the economic efficiency of water utilities. He has also worked on economic feasibility studies of wastewater treatment and water reuse projects including the economic valuation of externalities. He has been a researcher in more than 25 Projects R&D related to water management and water economics and financed by Spanish Government and European Commission. He is Associate Editor of the reviews *Water Science and Technology: Water Supply* and *Water Economics and Policy Journal*.

Miharu Hirano is Research Fellow of Japan Society for the Promotion of Science and is currently enrolled in the PhD program at the Graduate School of Advanced Integrated Studies in Human Survivability, Kyoto University, Japan. He is also a Specially Appointed Researcher at the International Water Association. His research relates to the human right to water with the focus on the roles of various global regulations that have effects on domestic water governance.

Allen Yu-Hung Lai is currently the Principal, Real-World Evidence Solutions. In this role, he is responsible for the Health Economics & Outcomes Research (HEOR) practice, and is spearheading the IMS Institute for Healthcare Informatics. Allen has extensive experience across the full spectrum of health outcomes research. He holds concurrent appointments as President of the International Society of Pharmacoeconomics and Outcomes Research (ISPOR) Singapore Chapter and Advisor of Department of Health Promotion, Ministry of Health and Welfare (MOHW) in Taiwan. Allen is also a thought leader in healthcare, publishing more than 40 peer-reviewed articles, op-eds and book chapters over the course of his career. Allen holds a PhD and Masters in Public Administration from the Lee Kuan Yew School of Public Policy, National

University of Singapore. In addition, he holds a Master of Science in Public Health from National Taiwan University and is a certified medical doctor and surgeon by training.

Jonatan A. Lassa is a Senior Lecturer in Disaster and Emergency Management at the Charles Darwin University. He was based at the Center for Non-Traditional Security Studies (NTS) S. Rajaratnam School of International Studies Nanyang Technological University, Singapore. A former fellow of the Harvard Kennedy School, Jonatan Lassa is an interdisciplinary scientist focused on institutional and human dimensions of disaster risk reduction and climate adaptation. He has qualifications in civil engineering, development studies, environmental policy, and social science. He holds a master's degree in environment and international development from the University of East Anglia, UK, received his Ph.D. in geoinformation science from the University of Bonn, Germany, and was a PhD researcher based at the United Nations University, Institute for Environment and Human Security in Bonn, Germany. He has more than ten years of professional experience in civil society organisations (including UN agencies) and in the disaster governance sector. As an Indonesian Research Fellow at Harvard Kennedy School for the Spring 2011 semester, Lassa's research focused on the role of institutions and governance in the (dis)integration of climate adaptation and disaster risk reduction.

Preetha Mahadevan holds a LLM from New York University (NYU) School of Law. Preetha is now a Senior Associate at Lakshmikumaran & Sridharan Attorneys in Bangalore, India. She advises multinational corporate clients and the board of various universities and large private sector clients on risk management, sponsor compliance matters and the immigration aspects of business restructuring.

Thomas McDonnell is the in-house macroeconomist at the Nevin Economic Research Institute (NERI), an economic think-tank operating in both the Republic of Ireland and Northern Ireland. Dr McDonnell is the lead author of the NERI's Quarterly Economic Observer and a regular commentator on economic issues in the Irish media. He specialises in economic growth theory, the economics of innovation, the Irish and European economies, and fiscal policy. He previously worked as an economist at the equality think-tank TASC and before that he was a lecturer in the economics department at National University of Ireland Galway and at Dublin City University. He has taught political economy at the National University of Ireland Maynooth. His Master's Degree is in economic policy evaluation and planning and he obtained his PhD in economics from NUI Galway specialising in economic growth and the economics of broadband. He is a native of Limerick city in Ireland.

Sacchidananda Mukherjee is an Associate Professor at National Institute of Public Finance and Policy (NIPFP), New Delhi, which is the largest think-tank on public economics and policies in India and has made significant research contributions in the area of revenue and taxation, fiscal management, public expenditure, macro-economic policies, fiscal federalism and other public finance and policy issues both at Central and State level. Prior to joining NIPFP, he was a Senior Scientific Officer at International Water Management Institute (IWMI), Hyderabad and Senior Manager – Water Resources and Policy at World Wide Fund for Nature (WWF)-India, New Delhi. His area of specialization includes public finance and fiscal policy, water resources management and environmental economics. He received his Doctoral Degree in Economics from University of Madras (Madras School of Economics), Chennai and Post Graduate Degree in Economics from Jawaharlal Nehru University (JNU), New Delhi. Dr Mukherjee has carried out several research projects on public finance, water resources management and environmental economics for Central as well as State Governments in India and also for multilateral development agencies. He has coordinated many training programmes on taxation and tax administration for government officials. He has published his research papers in journals of international repute (including *Journal of Development Studies*) and co-edited two books *Environmental Scenario in India: Successes and Predicaments* and *Environmental Challenges and Governance: Diverse Perspectives from Asia* published in the UK and the USA. He reviews papers for many journals including *Water Resources Management*, *Water Policy*, and *Urban Water Journal*.

Sikander Ahmed Shah is an Associate Professor of Public International Law at the Lahore University of Management Sciences in Lahore, Pakistan. Between 2012 and 2013, he served as the Legal Adviser to the Ministry of Foreign Affairs of Pakistan. He recently authored the book *International Law and Drone Strikes in Pakistan: The Legal and Socio-political Aspects* (2015), which deals with the treatment of drone strikes under International Human Rights Law and International Humanitarian Law.

Virginie J.M. Tassin is an award-winning specialist in Law of the Sea (INDEMER Prize 2011). She holds a double-badged PhD from The University Paris I Panthéon-Sorbonne and Melbourne University. Virginie also holds a Research Master in International Economic Law from The Sorbonne University, a Graduate Diploma of Commerce from

Sydney University, and a Graduate Certificate in Advanced Learning & Leadership from Melbourne University. She also attended The Hague Academy of Public International Law and a pluridisciplinary postgraduate programme of the European Summer University. Virginie has worked for the French government, law firms (Blake Waldron Dawson, BHM-Penlaw, Eversheds), as well as for regional and international organisations (Secretariat of the Pacific Community, World Health Organization, World Bank and the International Tribunal for the Law of the Sea). Her multicultural and pluridisciplinary mindset led her to create in 2009 a sustainable development project dedicated to food security, climate change and urban development that was supported by the Melbourne Sustainability Society Institute. Since then, Virginie has worked in various pluridisciplinary think tanks in China, Australia and France on topics related to Future Knowledge Economy, Climate Change, Environment and Maritime Security. Virginie is currently working at the Research Group on International Resources Law at the Escola Superior Dom Helder Câmara (Brazil, DHC Research Prize 2015) and as an Associate Professor at the Arch Institute (Geneva). She has published extensively in French and English and speaks Spanish as well.

Catharine Titi is a Research Scientist at the French National Centre for Scientific Research (CNRS) and Faculty Member of the CREDIMI, Law School of the University of Burgundy. She holds a PhD in Law (Dr iur) from the University of Siegen, Germany. Besides her formal legal education, she holds a PhD from the Courtauld Institute of Art, London. Catharine has previously worked as a consultant at the United Nations Conference on Trade and Development (UNCTAD) and at the University Panthéon-Assas Paris II (CRED). She has published extensively in international law journals, such as *Arbitration International, European Journal of International Law, Journal of World Investment & Trade*, and contributed to edited volumes, such as the *Yearbook on International Investment Law & Policy* (Oxford University Press, 2015). Her monograph 'The Right to Regulate in International Investment Law' was published in 2014 and she is co-editor of a special issue of the *Journal of World Investment & Trade* on Latin America (with Katia Fach Gomez, forthcoming in 2016). In 2016, Catharine received the prestigious Smit-Lowenfeld Prize for the best article published on international arbitration.

Paolo Turrini, since 2013, has been Post-doc Research Scholar at the School of International Studies of the University of Trento, where he has been working on projects on the international law regime governing transnational virtual water flows, and the legal tools deployed by the European Union to counter the impact of climate change on water resources. He obtained both his BA and MA in International Relations from the University of Bologna, School of Political Sciences, and his PhD in International and European Law from the University of Florence. From 2009 to 2015 he has assisted the Chair of International Law at the School of Law of the University of Bologna, a task that he is now carrying on at the School of Law of Trento. He is editorial secretary of the book series *The Search for Law in the International Community*, and is Advisory Editor of the University of Bologna Law Review. He is also a member of the team of italyspractice.info, a website devoted to translating into English the Italian practice on international law. His areas of academic interest include the history and theory of international law, international economic law with a particular focus on natural resources, and international environmental law.

Foreword

As one of the fundamental resources for the maintenance of life and its diversity, the physical management of water has been a preoccupation of men and women around the world for centuries. With the development of agriculture, towns and cities, together with industrial societies, the legal and economic aspects of water in terms of conservation, allocation and sustainable use have become essential characteristics of water management. Because of its fundamental importance, and in many places, limited availability, water resources have also been a cause of international, national and local community conflicts.

For these reasons, comprehensive regulatory regimes are found in almost all national and sub-national jurisdictions. With respect to the non-navigational use of internationally shared watercourses and lakes, these regimes comprise water agreements between national governments. Within nations, agreements exist between sub-national provincial/state governments as well as public-private partnership agreements, especially in relation to large-scale developments such as hydropower dams.

In addition, with increasing growth in foreign investment in general, multinational corporations increasingly invest and function in the field of water, and water resources are the subject of international trade disputes and arbitration processes. These developments underline the globalization of water allocation and water services.

One of the particular issues that has arisen in the past few decades as a result of major interventions by governments and the private sector is the rights of people and communities to water in general and safe drinking water in particular. The latter was underlined by the declaration of the United Nations (UN) Decade for Water 2005–2015, which culminated in UN Resolution 64/292 in July 2010, which explicitly recognized the human right to water and sanitation and acknowledged that clean drinking water and sanitation are essential to the realisation of all other human rights.

In acknowledgment of the growing recognition of water resources as a part of environmental protection regimes and global sustainable development concerns, we have seen United Nations fora and their

Declarations focusing substantial attention on water issues from the 1972 Stockholm Conference on the Human Environment through to the 1992 Rio Conference on Environment and Development, the 2002 Johannesburg World Summit on Sustainable Development and the 2012 Rio+20 Conference on Sustainable Development. In 2015, the UN generated the most comprehensive document to address sustainability issues, entitled *Transforming our World: the 2030 Agenda for Sustainable Development*.

This Agenda contains a set of Sustainable Development Goals (SDGs) which are much more ambitious and detailed than the UN Millennium Development Goals of the early 2000s. In the context of water, the Agenda's vision includes a commitment to the human right to safe drinking water and sanitation. Goal 6 of the SDGs is specifically focused on this objective, providing a challenging set of targets that all nations are expected to address.

Most of the targets are intended to be achieved by 2030. They include (in summary) achieving universal and equitable access to safe and affordable drinking water, adequate and equitable sanitation and hygiene, improving water quality by reducing pollution, eliminating dumping and minimizing release of hazardous chemicals and materials, increasing recycling and safe reuse, increasing water use efficiency, reducing the number of people suffering from water scarcity, and implementing integrated water resources management at all levels, including through transboundary cooperation.

Importantly the targets also include expansion and international cooperation and capacity-building support to developing countries in water and sanitation-related activities and programmes, including water harvesting, desalination, water efficiency, wastewater treatment, recycling and reuse technologies. The targets also have a shorter-term goal (by 2020) of protecting and restoring water-related ecosystems, including mountains, forests, wetlands, rivers, aquifers and lakes. Finally, they also aim to enhance the participation of local communities in improving water and sanitation management.

This book provides detailed analyses of many of the issues raised in the 2030 Agenda and the SDGs. Each chapter has its own particular value, many bringing forward original research and making insightful observations, giving a snapshot of water regimes, the varieties of water services, water use conflicts and their resolution from a wide range of jurisdictions. As a collection of learned papers from a number of disciplinary perspectives, the work represents a valuable contribution to this vital area of research, and the editors are to be commended for bringing this project to fruition.

Ben Boer
Distinguished Professor, Research Institute of Environmental Law, Wuhan University, People's Republic of China, Emeritus Professor, University of Sydney and Deputy Chair, International Union for Conservation of Nature and Natural Resources (IUCN) Commission on Environmental Law, 2012–16.

Table of cases

INTERNATIONAL

African Court of Human Rights
Social and Economic Rights Action Centre (SERAC) v Nigeria (2001) AHRLR 60 (ACHPR 2001) 209

GATT
Belgian Family Allowances (Allocations Familiales), Report of the Panel, adopted on 7 November 1952, BISD 1S/59 (1955) 412 32

ICJ
Gabčíkovo-Nagymaros Project (Hungary v Slovakia) (Judgment, ICJ Reports 1997) 72, 201
Legality of the Threat or Use of Nuclear Weapons (Advisory Opinion, ICJ Reports 1996) 119, 120
Pulp Mills on the River Uruguay (Argentina v Uruguay) (Judgment, ICJ Reports 2010) 72, 120

International Centre for Investment Disputes (ICSID)
Aguas del Tunari SA v Bolivia (Cochabamba), ICSID Case No ARB/02/3 (2003) 10, 52, 92, 101, 146, 156, 168, 198, 217, 218, 229
Aguas Argentinas/Suez: Aguas Argentinas, S.A., Suez, Sociedad General de Aguas de Barcelona, S.A. and Vivendi Universal, S.A. v. Argentina, Case No ARB/03/19, Decision on Liability (2010) 12, 92, 102, 115, 146, 153, 156, 157, 160, 161, 162, 163, 164, 230
Aguas Argentinas, S.A., Suez, Sociedad General de Aguas de Barcelona, S.A. and Vivendi Universal, S.A. v Argentina, Case No. ARB/03/19, Order in Response to a Petition for Transparency and Participation as Amicus Curiae (2005) 151, 152
Aguas Cordobesas SA, Suez, and Sociedad General de Aguas de Barcelona SA v Argentina, Case No ARB/03/18 146
Aguas Provinciales de Santa Fe/Suez: Aguas Provinciales de Santa Fe S.A, Suez, Sociedad General de Aguas de Barcelona, S.A., and InterAguas Servicios Integrales del Agua, S.A. v Argentina, Case No ARB/03/17, Decision on Liability (2010)................................. 12, 146, 153, 157, 160, 163, 164
ATA Construction, Indus. and Trading Co v Jordan (2010) 92
Azurix Corp v Argentina, Case No ARB/01/12, Award (2006) 12, 92, 146, 153, 157, 162, 164, 229
Azurix Corp v Argentina, Case No ARB/03/30, Decision on Preliminary Questions on Jurisdiction [2003] 146

Biwater Gauff (Tanzania) Ltd v Tanzania, Case No ARB/05/22, Award (2008) 12, 92, 146, 156, 161, 229
Compañiá de Aguas del Aconquija SA and Vivendi Universal SA v Argentina Case No ARB/97/3, Award (2007) 92, 146
EDF v Romania, Case No ARB/05/13, Award (2009) 160
El Paso v Argentina, Case No ARB/03/15, Award (2011) 160
Enron v Argentina, Case No ARB/01/3, Award (2007) 160
Gelsenwasser AG v Algeria, Case No ARB/12/32, Notice of Arbitration (2012) 146
Impregilo SpA v Argentina, Case No ARB/07/17, Award (2011) 12, 92, 146, 160, 161, 162, 163, 164
Marvin Feldman v Mexico, Award (2002) 94
Metalclad Corporation v Mexico, Case No ARB(AF)/97/1, Award (2000) 151, 152, 159
Pac Rim Cayman LLC v El Salvador, Decision on the Respondent's Jurisdictional Objections (2012) 103
Parkerings-Compagniet AS v Lithuania, Case No ARB/05/8, Award (2007) 12, 160
Piero Foresti, Laura de Carli & Others v South Africa, Case No ARB(AF)/07/01, Award (2010) 150, 152
Salini Costruttori v Jordan, Award (2006) 92
SAUR International SA v Argentina, Case No ARB/04/4, Décision sur la competence et sur la responsabilité (2012) 99, 100, 101, 146, 153, 157, 160, 162
SAUR International v Argentina, Award (2014) 12, 92
Total S.A v Argentina, Decision on Liability (2010) 95
United Utilities (Tallinn) BV and Aktsiaselts Tallinna Vesi v Estonia, Case No ARB/14/24, Procedural Order No. 1 (2015) 146
Urbaser SA and Consorcio de Aguas Bilbao Bizkaia, Bilbao Biskaia Ur Partzuergoa v Argentina, Case No ARB/07/26, Decision on Jurisdiction (2012) 146

International Tribunal for the Law of the Sea (ITLOS)
Responsibilities and Obligations of States with Respect to Activities in the Area, Advisory Opinion (ITLOS Reports 2011 1 February 2011) 121, 133
Southern Bluefin Tuna Cases (New Zealand v Japan; Australia v Japan) 121

UNCITRAL
Anglian Water Group (AWG) v Argentina, Decision on Liability (2010) 98, 102, 103
Chemtura v Canada, UNCITRAL (1976), Award (2010) 12, 159
Ethyl Corporation v Canada, UNCITRAL (1976), Award on Jurisdiction (1998) 12, 151, 152
Methanex Corporation v USA, UNCITRAL (1976), Decision of the Tribunal on Petitions from Third Persons to Intervene as 'amici curiae' (2001) 12, 157, 159, 160
Philip Morris Asia Ltd v Australia, UNCITRAL (1976), PCA Case No. 2012–12 .... 150, 152
Saluka Investments BV v Czech Republic, UNCITRAL (1976), Partial Award (2006) 159, 160
Ulysseas v Ecuador, UNCITRAL (1976), Final Award, 12 June 2012 160

WTO

Brazil: Measures Affecting Imports of Retread Tyres, WT/DS332/AB/R, 3 December 2007 44

Canada – Certain Measures Affecting the Automotive Industry: WTO Doc WT/DS139/R, WT/DS142/R (2000) (Report of the Panel); WTO Doc WT/DS139/ AB/R, WT/DS142/AB/R, AB-2000-2 (2000) (Report of the Appellate Body) §§10.247–8 42

China — Measures Affecting Trading Rights and Distribution Services for Certain Publications and Audiovisual Entertainment Products -AB-2009-3 (2009) 44, 47

EC — Bananas III: EC — Regime for the Importation, Sale and Distribution of Bananas, WTO Doc WT/DS27/R/USA (1997) (Report of the Panel); WTO DocWT/DS27/AB/R (1997) (Report of the Appellate Body) 43

EC – Seal Products: EC – Measures Prohibiting the Importation and Marketing of Seal Products, WT/DS400/AB/R and WT/DS401/AB/R (May 22, 2014) 44, 50

EU and its Member States – Certain Measures Relating to the Energy Sector – Dispute between EU and Russian Federation 47

US – Gambling: Appellate Body Report, United States – Measures Affecting the Cross-Boundary Supply of Gambling and Betting Services WT/DS285/AB/R, adopted 20 April 2005; Panel Report WT/DS285/R, adopted 20 April 2005 .... 8, 37, 44, 47–52, 54

United States – Import Prohibition of Certain Shrimp and Shrimp Products, 12 October 1998, 38 ILM 118 (1999) 44

NATIONAL

Brazil

Supreme Court, Adin 2340, Adin 2077 and Adin 1842 240

Ministério Público do Estado de São Paulo v Município de Sorocaba (Supreme Court, RE 254764/SP, August 24, 2010) 242

Companhia Estadual de Águas e Esgotos – CEDAE v Ministério Público Federal (Supreme Court, RE 417408 AgR / RJ, March 20, 2012) 242

Ministério Público do Estado de Rondonia v Companhia de Águas e Esgotos de Rondônia – CAERD (TJRO, Appeal n. 1100324-60.2008.8.22.0018, December 9, 2010) 243

Ministério Público Federal v Municipio de Barra do Sul, Cia Catarinense de Águas e Saneamento – CASAN, IBAMA and others (TRF4, Appeal n. 0002755-71.2003.404.7201/ SC, June 01, 2011) 243

SETEP construções SA v Companhia Catarinense de Águas e Saneamento – CASAN (Superior Court of Justice, AgReg na SS 2418, March 16, 2011) ... 244

Canada

British Columbia Supreme Court, partially setting aside the Metalclad award .. 159

India

Attakoya Thangal v Union of India (1990) 1 KLT 583 176
AP Pollution Control Bd II v Prof MV Nayudu (2001) 2 SCC 62 176

Kumar v Bihar, (1991) 1 SCC 598 176
MC Mehta v Kamal Nath, (1997) 1 SCC 388 177
MC Mehta v State of Orissa, AIR 1992 Ori 225 176
MC Mehta v Union of India (1988) 1 SCC 471 176
MC Mehta v Union of India (2004) 3 SCR 128, 48 176
Minerva Mills Ltd v Union of India AIR 1980 SC 1789 209
Narmada Bachao Andolan v. Union of India A.I.R. 2000 S.C. 375 217
Vellore Citizens' Welfare Forum v Union of India (1996) 5 SCC 647 176

Indonesia
Central Jakarta District Court, annulled concession contracts for water services,
    24 March 2015 146

New Zealand
Ioane Teitiota v The Chief Executive of Ministry of Business, Innovation and
    Employment (CA 50/2014 [2014] NZCA, 8 May 2014 106

Nigeria
Abacha v Fawehinmi (2000) 6NWLR (Pt 660) 228 .... 205, 206

Pakistan
Abdul Latif v Additional Sessions Judge, Sahiwal (2001 CLC 1139) 188
Benazir Bhutto v President of Pakistan (PLD 1998 Supreme Court 388) 186
Ch, Riaz Ahmad Yazdani v The Federation of Pakistan and 8 others (1990 CLC
    1406) 188
General Secretary, West Pakistan Salt Miners Labour Union (CBA) Khewra,
    Jhelum v The Director, Industries and Mineral Development, Punjab, Lahore
    (1994 SCMR 2061) 187
Mrs Anjum Irfan v Lahore Development Authority through Director-General
    and Others (PLD 2002 Lahore 555) 186–187
Shehla Zia and Others v WAPDA (PLD 1994 SC 693) 185, 186, 187

South Africa
Residents of Bon Vista Mansions v Southern Metropolitan Local Council 2002
    (6) BCLR 625 (W) 177
Mazibuko (Lindiwe) and Others v City of Johannesburg and Others 2008
    High Court of South Africa (Witwatersrand Local Division) Case No
    06/13865 3 177
Mazibuko and Others v City of Johannesburg and Others (CCT 39/09) [2009]
    ZACC 28; 2010 (3) BCLR 239 (CC); 2010 (4) SA 1 (CC) (8 October
    2009) 208
South Africa v Grootboom, 2000 (11) BCLR 1169 (CC) 178

United States of America
Griswold v Connecticut 381 US 479 (1965) 185
Munn v Illinois, (94 US 113 (1877) 185

Table of legislation

Treaties, Conventions and other Instruments

1994 Agreement: Agreement relating to the Implementation of Part XI of the UN Convention on the Law of the Sea 1982 132
   Preamble 132
   s 1 para 5k) 132
   s 1 para 5g) 132
Abuja Declaration by African Ministerial Conference on Water (AMCOW), April 29–30, 2002 201
African Charter on Human and Peoples' Right (adopted 27 June 1981, entered into force 21 October, 1986 204
   Art 16 .... 205, 209
   Art 24 .... 205, 209
African Charter on Human and Peoples' Rights on the Rights of Women in Africa (13 September 2000) CAB/LEG/66.6 Protocol Art 15 217
African Charter on the Rights and Welfare of the Child (July 1990) CAB/LEG/24.9/49
   Art 14 217
Agenda 21, adopted at the United Nations Conference on Environment and Development, Rio de Janeiro ... 134
   Art 18, para 2 110
Agreement on Agriculture (AoA), WTO 24, 25–27, 69
   Art 6.2 29
   Arts 8–10 67
   Annex 2 .... 60, 67
Agreement on Technical Barriers to Trade (TBT), WTO 31, 32–34

Agreement on the Encouragement and Protection of French Investments in Indonesia (1973) 985 UNTS 258
   Art 10 147
American Convention on Human Rights, art 44, Nov 22, 1969, 1144 UNTS 144 (AMCHR) 169
Code of Conduct on Accidental Pollution of Transboundary Inland Waters 1990
   Section II, para 1 120
Convention against Torture, Cruel, Inhuman and Degrading Treatment or Punishment (CAT) 178
Convention for the Elimination of All Forms of Discrimination against Women (CEDAW), UN GAOR, UN Doc A/34/46, (adopted 18 December 1979 entered into force 3 September, 1981) 109, 171, 178, 204, 216
   Art 14 216
   Art 14(2)(h) 171
Convention on Biological Diversity, UN, 1992
   Art 3 119
Convention on the High Seas, 1958 .... 126, 127, 128, 129
   Art 2 .... 127, 128
Convention on the Law of the Non-navigational Uses of International Watercourses 1997
   Art 6(b) 72
   Art 7 120
   Art 28 120
Convention on the Protection of Use of Transboundary Watercourses and International Lakes, 1992 109

Art 14 para 2 109
Annex III, c 109
Convention on the Regulation of
  Antarctic Mineral Resource
  Activities (not in force)
  Art 4, para 2 120
Convention on the Rights of
  Persons with Disabilities (CRPD)
  Art 28 216
Convention on the Rights of the
  Child (CRC), 1992 .... 109, 171,
                                     178, 238
  Art 24 216
Convention on the Settlement of
  Investment Disputes between
  States and Nationals of Other
  States (ICSID Convention)
  (1965) 575 United Nations
  Treaty Series 159 146
  Art 13(1) 148
  Art 54(1) 149
Convention on the Territorial Sea
  and the Contiguous Zone,
  1958 126
  Art 7, para 3 127
  Art 10 127
Convention on the Prohibition of
  Military or Any Other Hostile
  Use of Environmental
  Modification Techniques,
  1977, 31 UST 333, TIAS No
  9614 (ENMOD) 174
Declaration Granting Independence
  to Colonial Countries and
  Peoples and the Permanent
  Sovereignty Declaration on
  Natural Resources, 14
  December 1960 129
Declaration for the Establishment
  of a New International
  Economic Order A/RES/S-6/
  3201, United Nations General
  Assembly 1974 .... 129, 130
Declaration on the Right of
  Development, GA Res 44/128
  (1986)
  Art 8 172
Dublin Statement on Water and
  Sustainable Development,
  1992 172

Eritrea-Ethiopia Claims
  Commission, 2005 116
EU-Canada Comprehensive
  Economic and Trade
  Agreement's (CETA, version
  of 26 September 2014)
  Art X.08 99
European Charter on Water
  Resources 2001(17 October
  2001) CO-DBP (2001) 8 217
GATT 1947
  Art XVI 25
  Art XVI:3 25
  Art XX 31, 32
GATT 1994 .... 39, 40, 41, 47, 69
  Art III 32
  Art XI 32, 65
  Art XX 32, 34, 43, 44, 60,
                                           95
  Art XX(a)–(j) 34
  Art XX(g) 44, 63
General Agreement on Trade in
  Services (GATS) (Marrakesh
  Agreement Establishing the
  World Trade Organization,
  annex 1B, General Agreement
  on Trade in Services), 1869
  UNTS 183, 1995 .... 8, 36, 37, 38,
                                   39–47, 53, 54
  Art I 40
  Art I(i) 44
  Art I:3(b) 40, 46
  Art I:3(c) 40, 52
  Art II 41, 42
  Art II:1 41
  Art II:2 42
  Art III 41
  Art III(iv) 44
  Art VI 43, 50, 51
  Art VI:1 48
  Art VI:3 48
  Art XIV 43, 44, 49, 50
  Art XIV(a) 49, 50
  Art XIV(b) 44, 50
  Art XVI 36, 41, 42, 43
  Art XVI(a) 44
  Art XVI:1 49
  Art XVI:2 .... 42, 49
  Art XVI:2(a) 49
  Art XVI:2(c) 49

Art XVII 36, 41, 42, 43, 47
Art XVIII 42
General Comment 15: Economic and Social Council 2003, 'General Comment No 15 (2002) The right to water (Arts 11 and 12 of the ICESCR)', E/C.12/2002/11 .... 111, 112, 171, 172, 174, 179, 191
Geneva Conventions: Convention on the Territorial Sea and the Contiguous Zone; the Convention on the High Seas; the Convention on Fishing and Conservation of the Living Resources of the High Seas, 1958 .... 126, 127, 128, 129
Geneva Convention (I) for the Amelioration of the Condition of the Wounded and Sick in Armed Forces in the Field, 12 August 1949 179
Geneva Convention (II) for the Amelioration of the Condition of Wounded, Sick and Shipwrecked Members of Armed Forces at Sea, 12 August 1949 179
Geneva Convention (III) relative to the Treatment of Prisoners of War, 12 August 1949, 6 UST 3316, 74 UNTS 135 179
  Art 20 173
  Art 26 173
  Art 29 173
  Art 46 173
Geneva Convention (IV) Relative to the Protection of Civilian Persons in Time of War 1949, 75 UNTS 287, 6 UST 3516 (Fourth Geneva Convention)
  Art 89 173
Geneva Conventions of 12 August 1949, Protocol I, and relating to the Protection of Victims of International Armed Conflicts, 8 June 1977 179
  Art 54 116
  Art 55 117

Geneva Conventions of 12 August 1949, Protocol II relating to the Protection of Victims of Non-International Armed Conflicts, 8 June 1977 179
Hague Resolutions: Convention Respecting the Laws and Customs of War on Land art 23 (a), annexed to Convention [No IV] Respecting the Laws and Customs of War on Land, Oct 18, 1907, 37 Stat 2277 173
ILC (International Law Commission) Articles on State Responsibility
  Art 1 119
  Arts 20–27 96
  Art 20 96
  Art 21 96
  Art 22 96
  Art 23 96
  Art 24 96
  Art 25 .... 96, 97
ICSID Convention 229
ICSID Arbitration Rules 230
  r 32(2) 229
  r 37(2) 229
International Convention on Elimination of all forms of Racial Discrimination (adopted 21 December 1965 entered into force 4 January 1969) 204
International Covenant on Civil and Political Rights (ICCPR), (adopted 16 December 1966, entered into force 3 January, 1976) 112, 178, 204
  Preamble, para 1 204
  Art 4 109
  Art 6, para 1 109
International Covenant on Economic, Social and Cultural Rights (ICESCR), 1966 .... 112, 178, 204
  Art 2(1) 154, 175
  Art 11 .... 110, 112, 113, 152, 171, 174, 200, 205
  Art 11(1) 109, 152, 216

Art 11(2) 191
Art 12......110, 112, 171, 174, 200, 205
Art 12(2) 109
First Optional Protocol, Dec 16, 1966, 999 UNTS 302,
  Art 1 169
International Seabed Authority (ISA) 2013, Recommendations for the guidance of contractors for the assessment of the possible environmental impacts arising from exploration for marine minerals in the Area, ISBA/19/LTC/8 (Recommendation of 2013)............................... 135, 136
International Seabed Authority (ISA), Regulations on Prospecting and Exploration for Cobalt-Rich Manganese Crusts and Polymetallic Sulphides (ISBA/18/A/11 and ISBA/16/1/12/Rev1).......... 134, 135
  Reg 1 134
Law of the Sea Convention (LOS Convention): UN Convention on the Law of the Sea (UNCLOS), 1982 .... 125, 127, 128, 129, 130, 131
  Pt V..126
  Art 1.1(1) 130
  Art 1(4) 128, 129
  Art 49...128
  Art 56.1(a).....................................128
  Art 66...128
  Art 76...126
  Art 78   128
  Art 85...128
  Art 135..128
  Art 140(1)......................................130
  Art 140(2)......................................130
  Art 144(1)......................................131
  Art 144(2)......................................132
  Art 145..131
  Art 150..132
  Art 192................................. 128, 138
  Art 194..128
  Art 194(1)......................................120
Art 212..139
Art 257..128
Lisbon Charter: Guiding the Public Policy and Regulation of Drinking Water Supply, Sanitation and Wastewater Management Services (2015), International Water Association...............................164
Marrakesh Agreement
  Art 23(3)(f)(ii) annex 2 (Understanding on Rules and Procedures Governing the Settlement of Disputes) (DSU)....................................... 49
NAFTA
  Preamble 96
OECD Guidelines of Multinational Enterprises 10
OECD Principles on Water Governance under the Water Governance Initiative (C/MIN(2015)12) 164
Paris Agreement, FCCC/CP/2015/L.9/Rev.1, 12 December 2015................. 140, 142
  Art 2, para 1 142
Protocol on Water and Health, 1999..109
Rio Declaration 1992.............. 119, 200
  Principle 3 118
  Principle 15 .... 118, 133
  Principle 19 118
Statute of the International Court of Justice (ICJ)
  Art 38... 5
  Art 38 para 1 112
  Art 38(1)(b).................................120
Stockholm Declaration: UN Conference on the Human Environment, Stockholm, 1972
  para 3 .... 115, 140
  Principle 1 118
Subsidies and Countervailing Measures (SCM) Agreement, WTO 24, 27–28
Subsidies Code, 1979........................ 25
UN Charter 110
UN Convention on Biological Diversity, 1992................. 133, 134

UN Convention on Non-Navigational Uses of International Watercourses, 1997 174
UN Framework Convention on Climate Change, 1992..............139
  Preamble 118
UN General Assembly Resolution 58/217, adopted 23 December 2003.................................. 200, 219
UN General Assembly Resolution 64/292.......... 13, 98, 113, 153, 200, 216, 238, 298
UN General Assembly Resolution 68/157 (2013) 152, 153
UN Human Rights Council, Resolution A/HRC/RES/15/9................................. 298, 311
UN Human Rights Council, Resolution 15/L.14 113
UN Millennium Declaration, UN G A/RES/55/2 ....110, 111, 123, 125, 219
UN Millennium Development Goals 7, 114, 188, 220, 233, 257, 321
  MDG 7C............................. 257, 258
UN Stockholm Declaration 1972
  Principle 21 117
  Principle 22 117
UN Sustainable Development Goals (SDGs), Resolution A/RES/70/1 (2015) 30, 114
  SDG 6 ....23, 114, 115, 123, 257
  SDG 14 115
Universal Declaration of Human Rights (UDHR) (adopted 10 December 1948).......................204
  Art 25..170
Victims of International Armed Conflicts: Protocol Additional to the Geneva Conventions of 12 August 1949, and Relating to the Protection of Victims of International Armed Conflicts, opened for signature Dec 12, 1977, 1125 UNTS 3
  art 54 173
Vienna Convention on the Law of Treaties, 23 May 1969, 1155 UNTS 331; 8 ILM 679
  Art 31....................................96, 112
  Art 31(3)(c) 148
  Art 32... 49
WHO Guidelines for the Safe Use of Wastewater in Agriculture 2006..124

Free Trade Agreements (FTAs) and Bi-lateral Investment Treaties (BITs)
Argentina –France, 1991 .... 101, 163
Argentina –Italy BIT.......................163
Argentina –Spain BIT.....................163
Argentina –United Kingdom, 11 December 1990.......................... 80
Argentina-US 157
Canada-Colombia FTA (2008)
  Preamble 96
Canada-Costa Rica FTA (2001)
  Preamble 96
Canada-Peru FTA (2008)
  Preamble 96
Canadian Model BIT (2012)
  Art 15.. 95
France-Indonesia BIT.....................146
German Model BIT 2009
  Art 3(4) 94
Japan-Papua New Guinea BIT (2011)
  Preamble 96
Netherlands–Bolivia BIT (1992) 10, 229
United Kingdom –Tanzania, 7 January 1994.......................... 80
US-Chile FTA (2003)
  Preamble 96
US-Croatia BIT (1996)
  Preamble 96
US Model BIT of 2012
  Art 18.. 95
  Annex B(4)(b) 98
US-Peru TPA (2006)
  Preamble 96

NATIONAL LEGISLATION

Bolivia
Ley No 1544 – Ley Marco de Capitalización of 21 March 1994...101

Brazil
Constitution 1988
    Art 6 .... 238, 240
    Art 21, XX 240
    Art 23, IX 239
    Art 25, para 3 240
    Art 30, V 240
    Art 196 .... 238, 240
    Art 220 238
    Art 225 240
Law 6528/78 (Public sanitation) 239
Law 8080/90 238
2007 Federal Law 11445 240
Federal Decree 7217/2010 240
Federal Decree 8211/2014 240

Gambia
Constitution
    Art 216(4) 177

India
Constitution
    Art 37 176
    Art 38 175

Ireland
Water Services Act 2014
    s 3 318

Kenya
Constitution, 2010 217

Nigeria
Constitution 1999 .... 204, 211
    Ch 2 .... 206, 209
    s 1 .... 196, 206
    s 2 196
    s 3 196
    s 6(6) 209
    ss 13–24 206
    s 14(1) 206
National Environmental Standards and Regulations Enforcement Agency (Establishment) Act, 2007 204
River Basins Development Act 1987 204
Water Resources Act 1993 204

Pakistan
Constitution
    Art 9 .... 179, 185, 186, 187
    Art 14 .... 179, 185, 187
    Art 25 179
    Art 38 191
    Art 155 180
    Art 184 .... 180, 185, 186
Cantonment Pure Food Act 1966,
    Explanation 1 192
Environmental Protection Act 1997
    180, 187
    Art 14 180
    Art 16 190
    Art 20 180
Factories Act (Amended) 1934 180
Penal Code (PPC), (Act XLV of 1860)
    Ch XIV 180
    s 11 187
    s 277 187
Pure Food Ordinance 1960
    s 2 192
Safe Drinking Water Act 182
Water Resources Act 2007 181
Baluchistan Ground Water Rights Administration Ordinance, 1978 Ordinance IX of 1978 182
Punjab Local Government Ordinance 2001 183
    Art 52 184
    Art 53 185
    Art 76 185
    Art 94 185
    Art 95 185
Punjab Food Safety and Standards Act 2011
    s 2(f) 192

South Africa
Constitution 1996 .... 170, 217
    §27(1) 177
    §27(2) 177, 178

Uganda
Constitution, National Objectives and Directive Principles of State Policy XIV 177

United Kingdom
Water Industry Act 1991
  s 61(A)...218

United States of America
United States Code (Wire Act), s
  1084 of Title 18 48
United States Code (Travel Act), s
  1952 Title 18 48
United States Code (Illegal
  Gambling Business Act), s
  1955 of Title 18 48

Colorado Revised Statutes ss 18-10-
  103 48
Louisiana Revised Statutes
  (Annotated) s 14:90.3 48
Annotated Laws of Massachusetts s
  17A ch 271 48
Minnesota Statutes (Annotated) s
  609.755(1) and subdiv. 2–3 of s
  609.75 48

New Jersey Constitution, para 2 of
  s VII of art 4.............................. 48
New Jersey Code s 2A: 40–1 48
New York Constitution, art I,
  s 9 48
New York General Obligations Law
  s 5-401 48
South Dakota Codified Laws ss
  22-25A-1–22-25A-15 48
Utah Code (Annotated) s 76-10-
  1102 48

Zambia
Constitution (Constitution Act
  1991)
  Art 112(d). 177

EUROPEAN

Water Framework Directive –
  2000/60/EC 39

1. Introduction
Julien Chaisse

This book is about the issues, challenges and directions concerning water as an essential resource for mankind. The book aims to provide a finer understanding of the future regulation of water. It does so by inviting a number of leading experts to comment on the main problems confronting the actors of the water world such as governments, companies, international organizations, and citizens.

The world's water resources are sufficient to meet present requirements, from the simple mathematical point of view. However, this finding is not as reassuring as it may seem.[1] Most countries will encounter freshwater management issues because their water quality has been (or will be) degraded by pollution[2] and its per capita amount will decrease under pressure from population growth.[3] Also, changes in the world's climate and the new regional imbalances they cause will further aggravate this situation. Water remains a vital resource and the way it is used evolves over time; it is also an indispensable engine for economic activities, and moreover it is one which some countries do not hesitate to turn into a weapon of war. It is an economic, strategic and geopolitical issue. While water consumption is stabilizing or even decreasing in Western European countries, demand is increasing both in developing and emerging countries, thereby

---

[1] See Elliot Curry, Water Scarcity and the Recognition of the Human Right to Safe Freshwater, *Northwestern University Journal of International Human Rights*, Vol. 9, Issue 1 (Fall 2010), pp. 103–122. See also, *e.g.*, Raewyn Peart, Innovative Approaches to Water Resource Management: A Comparison of the New Zealand and South African Approaches (2001) 5 *NZJEL* 127.

[2] See Robin Kundis Craig, and Anna M. Roberts, When Will Governments Regulate Nonpoint Source Pollution: A Comparative Perspective, *Boston College Environmental Affairs Law Review*, Vol. 42, Issue 1 (2015).

[3] Over the past 50 years, the earth's population has grown from 2.5 billion to over 6.5 billion. 9.2 billion of people are expected by 2050. See United Nations, Department Of Economic and Social Affairs, Population division, *World Population Prospect: The 2006 Revision, Highlights*, Working Papers ESA//P/WP 202 (2007) (for more estimation in numbers, statistics and more on the worldwide population).

demonstrating a positive progress in health. However, too many people in the world do not have access to running water.[4]

Water is becoming an economic, geostrategic and even a political weapon in many parts of the world, such as the Middle East, North Africa, Sub-Saharan Africa, North America, Central America, Southeast Asia, China, etc.[5] The Aswan Dam gave flood management to the Nile thereby helping to ensure the water supply to a growing population. The dam was built by the Soviets, and it was also intended to affirm the identity of newly independent Egypt. However, several decades after its construction, the questions raised about the dam, intended to sustain the fertility of Egypt's Nile valley and Nile delta, remain.[6] China has recently recognized the design errors of the Three Gorges Dam, the largest in the world. Beijing seems to be already anticipating the management of a dramatic national water shortage over the next decade. In these examples, national political ambitions primarily reflect the geostrategic positions worldwide.

CHANGING PATTERNS OF THE SUPPLY AND DEMAND

The global demand for water has increased considerably during the last three decades, while, at the same time, the world population has also significantly increased.[7] This growth also reflects considerable progress

---

[4] Between 1990 and 2010, over 2 billion people gained access to improved drinking water sources, such as piped supplies and protected wells. See UNICEF/WHO (2012) Millennium Development Goal drinking water target met – Sanitation target still lagging far behind, Joint news release: UNICEF/WHO 6 MARCH 2012 http://www.who.int/mediacentre/news/releases/2012/drinking_water_20120306/en/ See also Stephen Diamond, Water Ethics and Commodification of Freshwater Resources, 6(1) *Santa Clara J. Int'l L.* 15 (2008) pp. 15–32.

[5] See, *e.g.*, Scott C. Armstrong, Water Is for Fighting: Transnational Legal Disputes in the Mekong River Basin, *Vermont Journal of Environmental Law*, Vol. 17, Issue 1 (Fall 2015), pp. 1–26. See generally Christina Leb, and Mara Tignino, Freshwater and International Law, *Environmental Policy and Law*, Vol. 41, Issue 4–5 (October 2011), pp. 218–21.

[6] See, *e.g.*, Adel Darwish, Analysis: Middle East Water Wars, BBC News, Friday 30 May, 2003 at: http: /news.bbc.co.uk/ 1 /hi/world/middle east/2949768.stm and Patrick McLoughlin, Scientists Say Risk of Water Wars Rising, *Planet Ark*, 23 August, 2004 at: http://www.planetark.com/dailynewsstory.cfm/newsid/26728/story.htm (all websites last accessed 31 January, 2016).

[7] See Francois Bouguignon, Inequality and Globalization: How the Rich Get Richer as the Poor Catch Up, *Foreign Affairs*, Vol. 95, Issue 1 (January/February 2016), pp. 11–15.

for humanity, particularly in the fields of hygiene and economic activity. Ample population movements are compounding the consequences of the unequal distribution of water resources.

The growth of the urban population is a major factor. The rural exodus, a constant in Europe from the beginning of the Industrial Revolution to the end of the 1960s, has had a geometric growth in developing countries over the last 40 years.[8] Capital cities and regional cities have seen their populations hypertrophy, often settling in highly precarious conditions in unserved slums, the incomers being, for example, the victims of landslides or floods. Cities are becoming larger and the need for water will begin to exceed the available supply so that these cities will have to invest not only in maintaining their water infrastructures but also in catering for increasing demands.

Many industries are also becoming less dependent on the use of water, but as they produce more, the total amounts of water needed are increasing. Companies which offer analysis and means of various measures should also benefit from the increase in the market valuation of water.[9] Existing water distribution networks (often undersized), the depletion of groundwater, the introduction of precarious water supply systems (terminals, fountains), and the use of water carriers – some aspects of water supply are becoming unmanageable both for government and local authorities.

During the same period, the democratization of transport followed by the explosion of mass tourism has led to the construction of large resorts and recreation facilities incommensurate with the water resources in countries where water is a scarce commodity. Establishments with hundreds or thousands of rooms, and with swimming pools and golf courses, compete with local and more traditional consumption. Consequently, local people are deprived of significant amounts of water or are forced to pay for infrastructure investments from which they get little or no benefit. For instance, each year millions of tourists stay on the coast of Thailand, Indonesia, the Philippines or Malaysia and the resorts, swimming pools, watering recreational places (gardens, lawns) together with golf courses, consume huge volumes of water, and this is detrimental to the local native population. Although these centres generate jobs, it remains to precisely

---

[8] See David R. Hansen, *et al.* Solving the Orphan Works Problem for the United States, *Columbia Journal of Law & the Arts*, Vol. 37, Issue 1 (2013), pp. 1–56 at 21.

[9] Thus, in 20 years, Sao Paolo has doubled the number of its inhabitants and quadrupled its size. The city grew from 9.3 million in 1973 to over 17 million in 2003.

ascertain the social profit and loss on water quantity and quality and the number of real local jobs.

NEW PARAMETERS: FOOD AND ENERGY CRISIS

Recently, it has become necessary to consider two additional factors, namely, the food crisis and the energy sector.[10] The expected growth of the world population confronts us with new challenges. This continuous and steady increase raises the question of people's livelihoods, especially when the extent of arable land is considered. The rapid expansion of cities and the multiplication of infrastructure correlation with economic development come at the expense of agricultural land, which is already known to be limited.

The availability of water essentially also influences world security, and this will be exacerbated by the expected climate changes.[11] By 2050, it is estimated that agricultural production will double.[12] The globalization of trade and production is not enough to ensure food security (given the cost of transport, climatic uncertainty, and speculation on food). The return to self-sufficiency for many countries, particularly in Africa, does not seem to be as easy as one would hope; this is because local agricultural produce, intended for local markets, has often been diverted towards lucrative exports.

THE ROLE OF RULES AND INSTITUTIONS

The recognition that when water supplies run out socio-economic disorders will develop, appeared for the first time at the conference of Mar del Plata in 1977.[13] On that occasion, the states proclaimed water to be a "global resource".[14] At the centre of the international debate is the major question

---

[10] See Joanne Hawkins, Fracking: Minding the Gaps, *Environmental Law Review*, Vol. 17, Issue 1 (March 2015), pp. 8–21.
[11] See Noah D. Hall, Bret B. Stuntz, and Robert H. Abrams, Climate Change and Freshwater Resources, *Natural Resources & Environment*, Vol. 22, Issue 3 (Winter 2008), pp. 30–35.
[12] See Susan A. Schneider, Predicting the Future: Our Food System in 2025, *Journal of Food Law and Policy*, Vol. 11, Issue 1 (Spring 2015), pp. 21–30.
[13] See, generally, Lilian Del Castillo LaBorde, Legal Regime of the Rio de la Plata, *Natural Resources Journal*, Vol. 36, Issue 2 (Spring 1996), pp. 251–96.
[14] See Alexandre Kiss and Dinah Shelton, *International Environmental Law*, 2nd ed. (Transnational Publishers Inc, USA, 2000), 395.

of ensuring that people adapt to each river basin and participate in the resolution of water disputes between residents. The most significant result of these debates was the establishment of legal processes relating to watercourses and the management of potential conflicts, and also of the wider problems that may affect them. By legal processes regulating the water management we mean the set of rules established at the international level to overcome the difficulties related to the management of scarce water resources.

The concept expressed by water resources "international", "shared" or "border" is synonymous. The term may refer to shared water, air, surface or ground between two or more states.[15] Worldwide, there are more than 240 international river basins and an undetermined number of rivers which are shared between two or more sovereign states. Any significant interference in the rivers or lakes could have beneficial or harmful effects on the territory of another state upstream or downstream. The international law of water resources, as part of international law, regulates relations between states with regard to the use of water resources "shared", "common" or "border".[16] Any of the world's rivers can be considered from the geographical and the international legal perspective.

Sources of international river law, they being no exception to the conventional sources of international law in general, are determined by the charter establishing the International Court of Justice. Indeed, according to Article 38 of the Statute of the International Court of Justice, which settles disputes between sovereign states, sources of international law are: (1) the international treaty law or the law of treaties; (2) the international customary law or state practice; (3) the general principles of law recognized by civilized nations; and (4) judicial decisions or international law and the teachings of the most highly qualified publicists, as a subsidiary source.

These sources are in a certain reading of the charter, listed in order of precedence, and they are also the basic sources of international environmental law in general, and therefore of the specific duties which relate to watercourses. It must still be noted that the technical nature relating to the wider environment and to streams in particular, inevitably makes it very

---

[15] See Jesse H. Hamner, Patterns in International Water Resource Treaties: The Transboundary Freshwater Dispute Database, *Colorado Journal of International Environmental Law and Policy*, Vol. 9, 1997 Yearbook (1997), pp. 157–77.
[16] See Barry Sadler, Shared Resources, Common Future: Sustainable Management of Canada-United States Border Waters, *Natural Resources Journal*, Vol. 33, Issue 2 (Spring 1993), pp. 375–396. See also Kerstin Mechlem, Shared Resources: Transboundary Groundwaters, Environmental Policy and Law, Vol. 34, Issue 4–5 (July 2004), pp. 162–75.

technical, and therefore international law needs to continuously evolve in line with contemporary conditions, and this leads us to consider the history of international river law.[17]

More precisely, many different treaties may apply to water. Countries have begun to increasingly rely on private sector participation in the water supply sector and the provision of sanitation services. This is due to budget pressures, a drive for greater efficiency in service delivery, and because of the promotion by agency donors of greater private sector participation.[18] A range of options for private sector participation in water supply and sanitation exists, ranging from the service contracts for functions such as billing and the concessions for complete operations for maintenance and network expansion.[19] Investing in water services can be a very delicate, laborious and unpredictable task. While the definition of investment inevitably involves some risks, water services appear to be a singular type of investment. Indeed, they simultaneously involve technological methods and knowledge, financial funding, and a panoply of laws including investment law, international law, human rights standards, contractual rights and obligations, national laws, and others.[20]

CHARTING THE WATER REGULATORY FUTURE

In this context, the access and management of water become a lever for sustainable development which are even more powerful. For these less developed countries (LDCs), to a lesser extent perhaps than for developed

---

[17] Eyal Benvenisti, Collective Action in the Utilization of Shared Freshwater: The Challenges of International Water Resources Law, *American Journal of International Law*, Vol. 90, Issue 3 (July 1995), pp. 384–415.

[18] See Julien Chaisse, Debashis Chakraborty and Jaydeep Mukherjee. 2013. Deconstructing Service and Investment Negotiating Stance. *Journal of World Investment & Trade*, 14, pp. 44–78.

[19] Amy K. Miller, Blue Rush: is an International Privatization Agreement a Viable Solution for Developing Countries in the Face of an Impending World Water Crisis, 16 *Indiana International & Comparative Law Review* 227 (2005) at note 121.

[20] See Markus A. Goll, Desalination in Texas: Struggling to Cope, *Texas Environmental Law Journal*, Vol. 45, Issue 1 (February 2015), pp. 51–86. See also Rebecca Bates, The Trade in Water Services: How Does GATS Apply to the Water and Sanitation Service Sector, 31 *Sydney Law Rev*iew 121 (2009) (exploring the stakes and impact of general investment agreement on water and sanitation services sector). See, generally, IWA Specialist Group on Statistics and Economics, *International Statistics for Water Services, Information every Water Management should know about*, International Water Association, 1 (2012).

countries, relocation of part of the national economies in which food production is essential. All these factors make it necessary to question the management of water and the ability of all actors to respond to new challenges.

We must re-examine the water management models. Can public or private management concepts even justify their specificity before such a huge issue? We must also ask ourselves about the problems of scales and jurisdictions between the public and the private, and between meeting a social need or a humanitarian one, and the profitability requirements inherent in a market economy; the choices may become conflictual.[21] As for the management of energy, it is to initiate a change in global behaviour at all stages of the economic chain involving consumers, distributors and directors. This is part of the dynamics of sustainable development, initiated by the Millennium Development Goals (MDG), and we cannot afford the luxury of reflecting on new devices, new structures, and new businesses that should help to consolidate the necessary link between the small and the large water cycle, that is to say between the water management and the overall consideration of the resource that is essential to all players as the new paradigm for water management in the coming years.

The contributions in this book are grouped around specific themes. In the Part I, the contributions address the challenges which water poses to public international law. In the Part II, the authors explore the most pressing ethical, legal, and social issues. In the Part III, the discussion covers the economic drivers shaping the future of water.

PART I THE WATER CHALLENGE TO PUBLIC INTERNATIONAL LAW

Part I contains contributions that analyse the regulatory foundations of the global water and sanitation services. *Manzoor Ahmad* provides a comprehensive overview of the international trade rules that address global water services with a particular focus on the WTO. He critically reviews the relevant rules and disciplines and provides a number of recommendations concerning the way the regulatory framework thereof should be reformed with the goal of promoting global water-use efficiency. He further observes that there are provisions in the WTO rules for the

---

[21] See Noah D. Hall, Protecting Freshwater Resources in the Era of Global Water Markets: Lessons Learned from Bottled Water, *University of Denver Water Law Review*, Vol. 13, Issue 1 (Fall 2009), pp. 1–54.

better management of water. However, they are not invoked and currently there is no transparency for peer-reviewing policies of member states in this important area. The WTO can play a more significant role in preventing wastages of water and in closing the gap between the quantities available and the quantities required. The WTO rules could provide incentives to member countries to adopt more environmental friendly ways of using water. Therefore, there may be a need to explore what further changes to the WTO rules and practices should be made so that trade rules can play their due role in overcoming the future water shortages.

Rebecca Bates further refines the legal analysis by focusing on the trade in water services. She answers the question of How Does GATS Apply to the Water and Sanitation Services Sector? The chapter explores the potential impact of the General Agreement on Trade in Services (GATS) on the water and sanitation services sector. She argues that water and sanitation require special consideration in the liberalization debate given their essential role in promoting human health and survival and their position as a human right. GATS has the potential to benefit the sector through creating increased efficiencies and encouraging additional funds to expand dilapidated infrastructures. Conversely, the sometimes punitive nature of trade laws risks undermining individual human rights and national legislation. At present there is some uncertainty as to how the GATs will apply to this sector because no WTO members have nominated their water sectors for liberalization. The recent *US — Gambling* decision demonstrates the power of the WTO to define and to potentially extend a member state's original commitment. Similarly, it has been argued that certain provisions have the scope to trigger a commitment without the consent of the member state. This chapter argues that given the essential role of water and sanitation, greater certainty must be provided to ensure the effective operation of trade laws, the validity of national legislation and the protection of water consumers.

Paolo Turrini extends the analysis by looking at Virtual Water and International Law. Water is a scarce resource and, at the same time, one of the most valuable and most necessary. Thus, according to the laws of economics, it should be a costly commodity. On the contrary, the price of water seems not to reflect adequately its limited availability. Moreover – and as a consequence – human behaviour quite often appears to be irrational, given that people who can count on a small amount of water sometimes use it to produce goods which end up being sold to people who are much richer in water. From the domestic standpoint of the former, this does not make sense, at least apparently. This is the reason why the notion of virtual water has been conceived. Its aim is to raise awareness of the scarcity of water resources globally, and of the need to exploit them rationally. Virtual water is the water used in the process of growing or manufacturing a product. If this product is then sold abroad, we witness an international flow of virtual water from the seller to the buyer. And if the product is water-intensive (*i.e.*, its production required a significant amount of water), as many agricultural goods are, then the flow of water from one country to another can become politically and economically relevant, especially if the water assets of the state transferring a part of them are poorer than those of the state acquiring them (even if only in virtual form). As a consequence, the need might emerge to reshape the rules of the game in order to discourage this kind of transaction. Rules – international rules – are what will be addressed here.

Julien Chaisse engages the debate with regard to investment treaties and their application to cross-border water services. The world of water services has changed significantly over the last two decades, opening it up to new business possibilities, as promoted by different international financial institutions. Such prospects have arisen in the face of extraordinary population growth and intensified water expansion needs. Accordingly, a vast increase of water-services privatization contracts between foreign investors and states has ensued. Today, 10 per cent of global consumers receive water from private companies. Inevitably, disputes have emerged regarding these privatization contracts, with there being little indication of their subsiding in the near future. In the absence of a specialized international regime to regulate these fast-growing activities, both investors and host states have filed 21 investment claims to investment tribunals in less than two decades. These filings have invited tribunals to interpret foreign investments in the water industry. The tribunal interpretations have generated the embryonic international regulatory and jurisprudential regime on water services analysed in this chapter. Governments must design water-related policies that comply with investment treaties because failure to do so results in higher water costs and deters foreign investors from providing much-needed high-quality services to local populations and industries. Although the investment jurisprudence may be seen as progress towards the regulation of an important service, it also emphasizes the lack of a true global holistic approach to regulate water services.

Catharine Titi looks at the Right of the Host State to Regulate Water Services. Since the inception of the international investment law system, investment promotion and protection have been the international investment agreements' (IIAs) principal *raison d'être*. By concluding these agreements, states have offered investors safeguards such as fair and equitable treatment, full protection and security, protection in case of expropriation, most-favoured-nation treatment, and guarantees of free capital transfers, in tandem with the possibility of recourse to investor–state dispute settlement, and in so doing they have confined their policy space and their capacity to adopt measures for the protection of the public interest. Against this background, the preoccupation with reserving the host state's right to regulate has gradually moved centre-stage with states starting to look at ways in which to safeguard their regulatory powers and to guide – and delimit – the interpretative freedom of arbitral tribunals by addressing their right to pursue specific public policy goals. A new generation of investment treaties, first launched with the US and Canadian model bilateral investment treaties (BITs) of 2004, have started to provide contracting parties with a modicum of flexibility, and the question is being asked for the first time of the pertinence of this right to regulate the state measures relating to water services. State regulation concerning water may adversely affect an investor not only where investment is made in water utilities, such as the supply of drinking water and sanitation services, but also where the investor is engaged in water-intensive activities, such as in the agricultural, industrial, energy, and in the mining and oil sectors. Water regulation may also become relevant where the investment pollutes or damages the water in the local environment. Environmental legislation and state measures for the protection of human, animal or plant life or health, are two fields of public policy-making that may have a direct impact on a foreign investment and they may overlay the investment protections afforded by the regulating state in an investment agreement. Given this potential for overlap between water, investment and state regulation, it is not surprising that several water-related claims have been initiated, against both developing and industrialized countries. Among the numerous examples, one may cite the famous *Aguas del Tunari, S.A.* dispute against Bolivia and related events that became known as Bolivia's Water War as well as the first *Vattenfall* dispute against Germany, involving local authorities' measures relating to compliance with the cleanness of river water targets of EU legislation, earning Germany its first-ever investment arbitration. The present chapter will query the extent to which a host state is able to adopt measures relating to water services affecting, *inter alia*, the economic value of an investment without violating an IIA to which it is a party and, significantly, without incurring financial liability vis-à-vis the foreign investor. To do so, it will focus on the presence of the right of the host state to regulate water in conventional law and customary international law, and it will look further into the potential impact of soft law instruments, such as the OECD Guidelines of Multinational Enterprises.

Finally, in this Part I devoted to the challenge of water to public international law, *Virginie Tassin* completes the analysis of the regulatory landscape and analyses the Regulation and Protection of Water as a

Resource: Terrestrial and Maritime Perspectives. The present chapter will examine in detail the fragmentation of the regulation and protection of water as a resource and highlight the large ambivalence between the terrestrial and maritime regimes. This study will provide necessary tools to understand the benefits and challenges of an international fragmented water regime. It will also analyse the limits attached to the human rights to clean water and confront it to the emerging human rights to a safe environment. Recommendations will be given on the water status in international law.

PART II ETHICAL, LEGAL AND SOCIAL ISSUES

Part II goes beyond the law and the current regulatory framework to identify issues at the social, ethical and legal levels. This group of chapters seeks to present the Emerging Issues and Challenges for the Future of Water Governance.

Miharu Hirano and *Shotaro Hamamoto* explore the geography of investment disputes in water services. Recently a group of UN experts, including the Special Rapporteur on the human right to safe drinking water and sanitation, voiced concerns about the adverse impact of investment agreements on human rights.[22] The issued statement draws attention to "the potential detrimental impact these treaties and agreements may have on the enjoyment of human rights", including the human right to water. This statement goes one step further than the 2007 Report of the UNHCHR on equitable access to safe drinking water, which had simply noted that further analysis was needed in this field.[23] In fact, urban water utility has recently become rather a common sector in the list of publicly available investment arbitration proceedings. This chapter examines these awards to identify the standard which the tribunals have employed to strike an appropriate balance between the protection of investors/investments and domestic public interests. This study also compares the water-related awards with those dealing with other public interest issues, such as environmental protection, so as to identify the water-specific considerations paid by the tribunals. This chapter begins with the big

---

[22] "UN experts voice concern over adverse impact of free trade and investment agreements on human rights" (2 June, 2015), http://www.ohchr.org/EN/NewsEvents/Pages/DisplayNews.aspx?NewsID=16031&LangID=E.

[23] "Report of the United Nations High Commissioner for Human Rights on the Scope and Content of the Relevant Human Rights Obligations related to Equitable Access to Safe Drinking Water and Sanitation under International Human Rights Instruments", A/HRC/6/3 (2007), paras 63–64.

picture of arbitral cases on water service. It is now a widely shared view that treaty-based arbitration may have certain impacts on the domestic regulatory authority. The structure of the treaty-based arbitration allows a foreign private entity to bring a case on the conformity of governmental measures with the applicable investment treaty, which will be examined by arbitrators appointed by the disputing parties. Arbitral decisions are enforceable, including in third states. In case of water services, conflicts between the investor and the host state often centre on the unwillingness of the local authorities to increase tariffs despite inflation or on the validity of modifications to the original concession unilaterally introduced by the host state. In such cases, the tribunals have consistently rejected the arguments of the respondent state that its measures in question should be justified in the light of the human right to water. At the same time, they made it clear that obligations under the applicable investment treaty and the human rights norms were not inconsistent, contradictory, or mutually exclusive. They also accepted *amicus curiae* briefs. It is thus an overstatement to say that arbitral tribunals ignore the public interests of the host state. The tribunals find violations of the applicable treaty, when they detect disguised illegitimate motives in the host state's measures allegedly taken to protect public interests (water-related cases: *Azurix v. Argentina*, *Vivendi v. Argentina*, and *SAUR v. Argentina*; non-water cases: *S. D. Myers v. Canada* and *Tecmed v. Mexico*). The tribunals consider that the host state has a legitimate power to regulate water tariffs (*Biwater v. Tanzania*, *Aguas Argentinas/Suez v. Argentina*, *Aguas de Santa Fe/Suez v. Argentina*, and *Impregilo v. Argentina*), as well as environmental and health protection or archaeological preservation (*Chemtura v. Canada*, *Methanex v. USA*, and *Parkerings v. Lithuania*), and examine whether the host state exercised such powers in an appropriate or proportionate manner in the light of the legitimate policy goal. In this respect, the eventual question becomes the standards set by the tribunals to review domestic regulatory measures. Finally, this chapter gives some accounts of the water-specific reasoning of the tribunals. *Aguas Argentinas/Suez* and *Aguas de Santa Fe/Suez v. Argentina* explicitly stated and recognized the difficulties in striking the right balance between the financial viability and the consideration to the most vulnerable parts of society. The tribunal went on to point out alternative measures that Argentina could have taken during the economic crisis to safeguard the interests of investors and citizens. This argument by the tribunal may be considered as an incursion into the regulatory autonomy of the host state. In recent years, intensive exchanges of good practices have been witnessed, failed projects have been analysed in detail, and various policy guidelines have been proposed at a variety of fora with respect to water governance. As such, the research continues; future tribunals will have access to materials on which they can rely when assessing the conformity of the measures in question with the applicable treaty. It is then not likely that the deference accorded to states would increase.

Sikander Ahmed Shah takes a human-rights-based approach to discuss the Provision and Violation of Water Rights in Pakistan. Water is not only a vital resource for development, but access to clean drinking water and sanitation has been recognized as essential to the realization of all human rights. This declaration accompanied the explicit recognition of water as a human right in the UN General Assembly Resolution 64/292. The scarcity of water, the exploitation of water resources, and the consequent lack of quality water supply in developing countries are barriers to the realization of this human right. With governments lacking the capacity to provide water services of a high quality, private players (mostly foreign investors) are playing an increasingly large role in the provision of such services. A concurrent narrative is of the increasing number of investment disputes filed against developing countries over the last decade. This study will document a subset of these disputes based on water services agreements between developing countries and foreign investors.

Cosmas Emeziem extends the analysis of the human right to water by looking at its implementation in Nigeria. This is an important contribution because there is still a gap between this perception and the reality of access to clean water and sanitation as a human right. The norms postulated do not yet translate into sustainable clean water access for many. Undoubtedly, the actualization of this premium right lies essentially at the local level, hence the need to act locally while thinking globally. This work – which is in four parts – explores the human right to clean water and sanitation by using Nigeria as the prism. It is aimed at ascertaining the extent of availability and enjoyment of this essential human right in Nigeria. Query: how do we move from slogans to real programmes on the human right to clean water and sanitation? Part I introduces the chapter. Part II looks at the legal regime on water and sanitation in Nigeria. Part III considers judicial enforcement and the challenges facing the right. Part IV argues and recommends that a social justice approach should be applied as a key tool towards realizing this human right. It is hoped that a study of the Nigerian situation will help towards formulating a sustainable human right to clean water and sanitation policy and also facilitating its full realization.

Preetha Mahadevan focuses her analysis on the impact of the private sector in implementing the right to water. Her chapter examines methods such as privatization, public–private partnerships, concession agreements, and international investment agreements with the support of international organizations such as the World Bank, and the outcome of such efforts. Finally, the chapter explores how corporate entities – the private sector involved in the construction of facilities and providing water utilities to the public on behalf of the government and the entities involved in packaging/ bottling of water for commercial sales – embrace the "right to water", and the efforts in developing clear standards and metrics for what companies should or must do; ways to evaluate their compliance with these standards, and the consequences of failure to comply.

Ana Paula de Barcellos focuses on the interaction between Sanitation Rights, Public Law Litigation, and Inequality. She does so in A Case Study from Brazil. Public law litigation has been used to advance human rights for decades; in Brazil, it has also been employed since the 1990s in the advancement of rights related to health. Such bids are usually accomplished through requests that the government pay for pharmaceuticals and medical procedures to individuals. This kind of litigation ends up concentrating resources on pharmaceuticals, instead of on other public health needs. The literature has been pointing out the inequitable effect which litigation may produce by particularly benefiting successful plaintiffs, who will receive goods usually not available to the rest of the population. But could litigation play a role in shaping public health policies? To explore this question, the chapter focuses on lawsuits involving the determinants of health, namely, water and sanitation public policies. The chapter presents and analyses the results of an empirical study of 258 Brazilian court orders, issued between January 2003 and March 2013, that address requests for sewage collection and treatment. The data show that, on the one hand, the Brazilian judiciary is willing to improve access to sanitation services, but that, on the other hand, litigation has addressed fewer than 177 out of the 2,495 Brazilian municipalities that lack both sewage collection and treatment systems, and lawsuits are concentrated in the richer cities, not in the poorest ones. This chapter suggests that public law litigation can be used to foster public health policies similar to the way in which structural reform litigation and the experimentalist approach between courts and defendants have influenced public policies and achieved institutional reform in schools and prisons. However, greater effort is needed to target initiatives that would reach the most disenfranchized communities.

PART III ECONOMIC DRIVERS SHAPING THE FUTURE OF WATER

Part III offers a comprehensive analysis of the economic drivers and determinants of a global market for water services.

Sacchidananda Mukherjee and *Debashis Chakraborty* consider the demand for Infrastructure Investment for Water Services and the Key Features and Assessment Methods. Assessing the present state of water services (access to water supply and sanitation) is important when estimating the demand for infrastructure investment. Different countries and regions are generally in different stages of providing access to safe drinking water and sanitation. Unlike other infrastructure investment, investment in water services infrastructure however has direct (public health benefits) and indirect benefits, as access to safe water and sanitation is the key to achieving economic growth and human development. Hitherto, public investment is the dominant source of financing water services infrastructure, at least in developing countries and LDCs. Limited fiscal space of the governments, in most of the developing countries and the LDCs, and growing demands from other developmental sectors (*e.g.*, education, health), often lead to low investment in water services infrastructure. Inadequate access to safe drinking water and sanitation leads to various public health hazards and costs to the economy. Unless there are specific rules and regulations to charge the beneficiaries for water services, the revenue stream from investment in water services infrastructure will be uncertain. Unlike other infrastructure services, private participation in water services infrastructure investment is low and that is also restricted to a few regions in the world. The challenges to the provision of water services shift as a country climbs the development ladder. Growing competition across the sectoral usage pattern of water – *i.e.*, among agriculture, industry, ecology – also puts additional challenges of securing sustainable sources for the long-term sustainability of water infrastructure projects. Growing urbanization, industrialization and the intensification of agriculture and not having adequate wastewater treatment facilities – at least in developing countries and LDCs – impose additional demands for investment in the water treatment infrastructure in order to meet the requisite quality standards. Sustaining safe sources of drinking water is a challenge not only for developing but also for their developed counterparts. Investment in the collection, treatment and disposal of domestic wastewater (including sewage and sanitation) is an integral part of investment in the water supply infrastructure. Challenges in meeting the demand for water services infrastructure are many and this chapter highlights the concern areas in a wider developmental context. The chapter explores various methods of estimating the demand for water services infrastructure and attempts to identify an appropriate one based on an extensive literature review. Investment in the maintenance of water services is as crucial as investment in greenfield projects, and therefore this chapter separately estimates the demand for investment in maintenance. Finally, the analysis also explores the possible sources of financing the infrastructure in water services with a special reference to developing countries and LDCs.

Thomas A. McDonnell analyses the residential Water Charges in Ireland: Policy Objectives and Funding Models. Water services provision is a natural monopoly. This drives a propensity to market failure and economic inefficiency in the absence of robust regulatory measures. In this context he discusses the appropriate role of the regulator and the advantages and disadvantages of pricing water usage. Beyond subsidies funded from general taxation there are potentially three main sources of revenue for a water utility. These are: connection fees; recurrent fixed charges; and volumetric charges based on usage. Water policy pursues multiple objectives, and a wide variety of pricing structures are employed within the OECD. These objectives can be structured around four sustainability dimensions: environmental sustainability; financial sustainability; economic efficiency; and social concerns, including affordability. There are trade-offs between each of these policy objectives. Full cost recovery through usage-based tariffs creates affordability and equity concerns, while the most efficient allocation of water may not be consistent with water saving and environmental concerns. This chapter considers the main trade-offs as a set of dilemmas and discusses the advantages and disadvantages of a variety of different water charging models. Water charges for domestic users were introduced in Ireland in 2014. In this context he considers various charging models. Water affordability and water poverty are important concerns. A universal free allowance was proposed as a means of protecting households from water poverty. However, a small free universal allowance will not address the affordability issue while a large free allowance undermines other policy objectives – notably, economic, financial, and ecological sustainability. Instead, a system of income-related water credits is described as a potential alternative. Combining a volume-based pricing structure with a system of income-related water credits may best reconcile the four main policy objectives, and it would also address the affordability issue at a much lower cost to the exchequer than a universal allowance; if properly designed, it would ensure that a combination of water charges and low income does not become a barrier to vulnerable households accessing water and wastewater services.

Tihomir Ancev, Samad Azad, and *Francesc Hernandez-Sancho* analyse the role of multinationals in providing water services. They answer the key questions of whether they are more efficient. Over the last two decades there has been a significant trend of multinational companies operating water provision services in many different countries, both developed and developing. One of the key arguments in favour of the participation of these foreign-owned corporations in the provision of water services in host countries is that they can lead the way in improving the efficiency and service quality of water utilities. It is commonly assumed that opening up the market for water services to foreign corporations will bring much-needed infrastructure investment as well as best international practice, and that this will ultimately result in an efficient water sector that will provide services to the public at acceptable cost. This chapter will empirically test the validity of this assumption. A key objective of the study is to measure the productivity and efficiency of water utilities in several countries, and to relate the observed differences in efficiency to the majority ownership of the analysed utilities, whether they be domestic or international. The standard productivity and efficiency measurement approach will be followed to measure the productivity and economic performance of water utilities. This will involve gathering data on output (*e.g.*, drinking water delivered, quantity of pollutants removed from wastewater) and input quantities (*e.g.*, labour, capital, energy) of the water utilities in several developed (*e.g.*, UK, Japan) and developing (*e.g.*, Kenya, Uganda) countries. The data will then be used to measure the efficiency of individual utilities in relation to an overall efficiency frontier for a particular country. The resulting efficiency scores will be related by applying econometric methods to the variable denoting the majority ownership of a given utility. Statistical tests will be run to determine whether foreign ownership is indeed related to increased efficiency scores, as hypothesized. These findings are subsequently related to the key motivations for the internationalization of water services, as well as to the issues around water pricing and water access, especially from the perspective of adaptation to climate change.

Allen Yu-Hung Lai and *Jonatan A. Lassa* complement the analysis by looking at the Microfinance in Water and Sanitation Services and they identify best practices. The existing literature on adaptation financing still lacks empirical evidence. This research contributes to the debate and literature of adaptation financing at the grassroots level in the urban context. This chapter presents good practice and lessons from a community-based sanitation micro-finance initiative that was recently adopted by Kemijen Village in the eastern part of Semarang City, Indonesia. The chapter argues that micro-finance can be used in the community-based adaptation planning in Semarang City if the conditions for sustainability can be controlled by local actors at the village level. An adaptation fund through a community-based micro-finance mechanism is a possible adaptation and risk management path. The participatory design of a "sanitation credit" mechanism suggests promising results because such a mechanism not only allows the local community to find a "best fit" approach in solving immediate problems such as water access and community sanitation but it also opens up the possibility for community climate and disaster financing.

The globalization of water services illustrates the clash of foreign investors' protection with human rights protection as being the state's responsibility. This tension is only now emerging but it will intensify rapidly because more investors will seek access to fresh water in new countries. There is therefore an impending risk of the emergence of global monopolies in this scarce commodity, which would be detrimental to many people, especially under climate change. These risks and challenges demand a proper regulatory answer, which should include an economic, legal, and human rights perspective on water services.

This is the context in which The Chinese University of Hong Kong (CUHK) Faculty of Law and Centre for Financial Regulation and Economic Development (CFRED) decided to organize a conference "Managing the Globalization of Water Services in a World Affected by Climate Change: Regulatory and Economic Challenges" in March 2015. This conference brought together more than 40 experts from the five continents comprising academics, government officials, NGOs, business representatives, and lawyers specializing in water services.

The conference examined the international law that governs the globalization of water services, to identify gaps and the need for various changes, and to relate the legal framework to the economic issues surrounding water provision including the economic rationale for the protection of foreign investments. The project contrasts the economic–legal view on international investment to the nascent notion of water as a human right. The two aspects of water – investment protection and human rights – are increasingly contradictory. While the scarcity of water has intensified the movement towards a human right to water, private control over water utilities is increasing. In particular, the conference participants examined how arbitral tribunals have dealt with the failure of contracts to manage privatized water supplies by focusing on economic cases of water privatization.

We decided to publish a selection of the papers presented at the conference as a way of contributing to the research and debate on the future of water. The CFRED would like to thank the team in the centre (Ms Susanna Leung, Ms Bonnie Leung and Mr Noel Chan) for organizing the conference and Ms Sammi Lee and Mr Keith Ji for their contributions in editing the volume.

PART I

The water challenge to public international law

2. Promoting global water-use efficiency – promises and shortcomings of international trade rules

Manzoor Ahmad

INTRODUCTION

Water has been identified as one of the top three concerns in terms of impact in the Global Risks 2016 report recently issued by the World Economic Forum. The US National Intelligence Council considers the nexus of food, water, energy and climate change as one of four overarching megatrends shaping the world by 2030. Furthermore, the 2030 Agenda for Sustainable Development includes a dedicated goal on water and sanitation (SDG 6) that sets out to 'ensure availability and sustainable management of water and sanitation for all'.[1]

Water use is growing at twice the pace of population growth and, as a result, water requirements are expected to outgrow sustainable water supplies by 40 per cent over the next 15 years.[2] Although there may not be global scarcity of water, in just ten years, two-thirds of the world population may be experiencing water 'stress'.[3] Not only are water shortages becoming acute in some countries but they are also wasting more water. Few countries price water appropriately and there is little or no transparency about subsidies given to the sector. Wrong incentives and policies are leading to unsustainable management of a scarce and precious resource.

In order to cope with this challenge, various national and international agencies are exploring options and devising strategies. This chapter aims

---

[1] See https://sustainabledevelopment.un.org/sdgs.
[2] J Kirsty, Washington, DC: World Resource Institute (2011).
[3] See generally J Chaisse and M Polo, 'Globalization of water privatization: Ramifications of investor-state disputes in the "blue gold" economy', (2015) 38 *Boston College International and Comparative Law Review* 1.

to explore if existing international trade rules can effectively promote efficient water use and, if not, could the World Trade Organization members negotiate new rules for this purpose. Some 70 per cent of the fresh water used is for agricultural purposes and evidence suggests that removing water subsidies could reduce its use by 20–30 per cent.

The primary focus of this chapter is on international trade rules, which may be applicable to disciplining subsidies given to water used in irrigation. Not discussed here is trade in water-related services as that is being covered in Chapter 3 by Rebecca Bates. Also trade in bottled water is not covered as the current WTO rules adequately deal with it, and which, in any case, is insignificant compared to the total use of water. This chapter consists of two parts, which cover WTO rules on the supply side and those applicable on the demand side. The first part focuses on irrigation subsidies which are responsible for most of the fresh water wastages while the second part discusses virtual water (discussed by Paolo Turini in Chapter 4) trade (trade in embedded or embodied water) and measuring water footprints and the role that an increased knowledge in this field could play in making the WTO rules more effective for promoting sustainable use of water.[4]

1. Supply Side Issues

There is no specific WTO or any other multilateral environmental agreement (MEA) relating to trade-related aspects of water, although there are some general rules which can also apply to water. However, the main issue on the supply side is subsidies given to irrigation water, which results in under-pricing and encouraging the overuse of fertilizers and inefficient use of water. In this respect, there are two WTO agreements which impose restrictions on agricultural subsidies: (1) the Subsidies and Countervailing Measures (SCM) Agreement and (2) the Agreement on Agriculture (AoA). While the SCM agreement applies to all subsidies and generally forbids both export subsidies and other subsidies that can be shown to have trade-distorting effects, the AoA is more specific for agricultural subsidies.

---

[4] For a more exhaustive discussion on the existing WTO rules, please see: L.A. Jackson, C. Pene, M.-B. Martinez-Hommel, C. Hofmann and L. Tamiotti, 'Water policy, agricultural trade and WTO rules' in Pedro Martinez-Santos, Maite M. Aldaya, M. Ramón Llamas, *Integrated Water Resources Management in the 21st Century: Revisiting the Paradigm* (London: CRC Press 2014) 322. This chapter builds on those ideas and is meant to suggest the way forward for making the existing rules more effective.

The Agreement on Agriculture and the regulation of subsidies

Trade in agriculture amounts not only to one of the most contentious but also to one of the most complex regulatory areas. To one extent, this is due to the historical legacy described supra. To the other, this is due to the fact that agriculture is critical for long-term global food security (biodiversity), sustainable development and the combat of hunger and poverty. Trade liberalization on its own, and without appropriate flanking policies, merely serves a small number of efficient producer countries but does little or nothing for the subsistence of the millions of farmers in a majority of developing countries.[5] The present tools and instruments available in the WTO will need further development much beyond the reduction of tariffs and subsidies to which the rules of the present AoA are essentially dedicated. The objective of the AoA, as given in its preamble, is 'to establish a fair and market-oriented agricultural trading system . . .' and to 'provide for substantial progressive reductions in agricultural support and protection sustained over an agreed period of time, resulting in correcting and preventing restrictions and distortions in world agricultural markets'.

The AoA distinguishes between two types of subsidies: (1) 'domestic support' and (2) 'export subsidies'.[6] Under the GATT 1947, all domestic support measures were allowed pursuant to Article XVI. The 1979 Subsidies Code, a plurilateral agreement, also did not contain any substantial obligations of the Contracting Parties to refrain from granting financial support to domestic agricultural producers.

Domestic support subsidies are categorized in three 'boxes' – the Amber

---

[5] See Julien Chaisse and Mitsuo Matsushita, Maintaining the WTO's Supremacy in the International Trade Order – A Proposal to Refine and Revise the Role of the Trade Policy Review Mechanism 18 *Journal of International Economic Law* 9 (2013).

[6] Under the GATT 1947, granting subsidies contingent on the export of primary products – i.e. most agricultural products – was allowed, as long as the Contracting Party concerned did not gain a 'more than equitable share of world export trade' by doing so (art. XVI:3). Even these rules remained largely ineffective after the US was granted, in 1955, a waiver from GATT obligations regarding their extensive export support policy. The 1979 Subsidies Code likewise exempted agricultural products from its ban on export subsidies. The widespread use of export subsidies, in particular on agricultural products, was at the root of the major 'trade wars' of the twentieth century. While tariffs and most non-tariff barriers to trade were continually being reduced during the GATT 1947 rounds, agricultural export subsidies kept increasing. Ultimately, the Contracting Parties were forced to address the problem of export subsidies in the context of the Uruguay Round. See Melaku Geboye Desta, *The Law of International Trade in Agricultural Products* (Kluwer Law International, The Hague, 2002), p. 240.

Box, the Blue Box, and the Green Box. The Amber box is used for all domestic support measures considered to distort production and trade.

These subsidies are subject to limits: 'de minimis' minimal supports must not exceed 5 per cent of agricultural production for developed countries and 10 per cent for developing countries. The Green Box covers subsidies that have no, or at most minimal, trade distorting effect on production. Therefore, there are no spending limits on these subsidies.

Although a majority of WTO members do not notify any irrigation subsidies, about a third who do so mostly list them under the Green Box. One of the reasons is that the governmental services programmes such as infrastructural services including water supply facilities are specially listed in the Green Box.

Strictly speaking, it is not correct to do so as the Green Box is meant for government service programmes, such as research, disease control, infrastructural services, including roads and other means of transport, market and port facilities, etc. Although the Green Box does include undefined water supply facilities, what is clearly meant by that is irrigation infrastructure and not subsidies for inputs or operating costs. Subsidies for input of irrigation water do not qualify as having 'no, or at most minimal, trade-distorting effects or effects on production'. Sheltering irrigation subsidies under Green Box is against the overall concept of this scheme.

There are many studies to show that most irrigation subsidies are trade distorting. For example, according to a report published in the *New York Times*:

> in the United States millions of dollars in farm subsidies for irrigation equipment aimed at water conservation have led to more water use, not less, threatening vulnerable aquifers and streams . . . water tables have fallen 150 feet in some areas – ranging from 15 per cent to 75 per cent – since the 1950s, scientists say, because the subsidies give farmers the incentive to irrigate more acres of land.[7]

Similarly the Global Subsidies Initiative (GSI) launched by the International Institute for Sustainable Development (IISD) has been examining issues relating to irrigation subsidies in different countries. Their estimates suggest that in Spain subsidies to irrigated agriculture may be about € 1 billion per year for the period 1998–2008. Even in low- and middle-income countries, there are huge government allocations

---

[7] R Nixon, 'Farm Subsidies Leading to More Water Use', *The New York Times* (2013) <www.nytimes.com/2013/06/07/us/irrigation-subsidies-leading-to-more-water-use.html?_r=0>.

for agricultural subsidies. In India, irrigation subsidies for 2010–11 were estimated at US$ 4.7 billion.[8]

In addition to major irrigation projects, many governments subsidize diesel or electricity for tube-wells to assist resource-poor farmers. In the AoA, such subsidies can be claimed under 'development box', and are permissible for low-income or resource-poor producers in developing countries as these producers are considered to have a relatively marginal impact on markets. However, there are some concerns about this at the WTO as well as at the national level. At the WTO, members are trying to reach some shared understanding of how Doha issues and other 'new' issues can best be addressed given the inability of the Nairobi ministerial to find a consensus on reaffirming previous negotiating mandates. At the national level, there is on-going debate as to how some reforms could be introduced for limiting such subsidies. Depletion of ground water and use of higher levels of fertilizer than would occur otherwise, are forcing the pace of reforms.

The Agreement on Subsidies and Countervailing Measures
Besides the AoA, the other major instrument in the WTO toolbox is the SCM Agreement. In order to understand the applicability of SCM vis-à-vis the AoA, it is important to understand its essential features and the relationship between the two agreements.

Subsidies consist of specific financial or equivalent benefits to public entities and economic operators. They are an important instrument in pursuing domestic policy goals and redistribution. At the same time, they exert distorting effects on competition, in particular to the detriment of foreign competitors who generally do not benefit from such measures. International trade regulation has been characterized by a progressive regulation of subsidies, tightening disciplines over time in order to avoid such distortions. These rules essentially seek to balance the need for redistribution and implementation of legitimate policy goals and to avoid protectionism and unnecessary distortions of conditions of competition on domestic markets. Trade-restrictive border measures apply to countervail unlawful subsidies but are not at the heart of legal rules relating to this important field of international trade law.

The SCM addresses two separate but closely related topics: multilateral

---

[8] A Hoda and A Gulati, 'India's Agricultural Trade Policy and Sustainable Development' http://www.ictsd.org/themes/agriculture/research/india%E2%80%99s-agricultural-trade-policy-and-sustainable-development Issue Paper No. 49. ICTSD (2013).

disciplines regulating the provision of subsidies, and the use of countervailing measures to offset injury caused by subsidized imports. Subsidy, as defined in this agreement, is a measure that meets three basic elements: it must be a financial contribution; it must be made by a government or any public body within the territory of a member; and it must confer a benefit. However, even if a measure is a subsidy under the definition of the SCM Agreement, it is not subject to the disciplines of the SCM Agreement unless the concerned subsidy is a specific subsidy.

The SCM Agreement creates two basic categories of subsidies: those that are prohibited, and those that are actionable (i.e., subject to challenge in the WTO or to countervailing measures). If the subsidy is contingent on export performance or upon the use of domestic over imported goods then such subsidies are prohibited.

The other category is actionable subsidies, which are not prohibited but are subject to challenge, either through multilateral dispute settlement or through countervailing action, in the event that they cause adverse effects to the interests of another member. If they cause injury to a domestic industry through subsidized imports, then they are countervailable. If they cause adverse effects (e.g., export displacement) in the market of the subsidizing member or in a third-country market, they can serve as the basis for a complaint related to harm to a member's export interests. Finally, there is nullification or impairment of benefits where the improved market access presumed to flow from a bound tariff reduction is undercut by subsidization.

Irrigation subsidies seem to meet the category of actionable subsidies and are thus challengeable either through multilateral dispute settlement or through countervailing action whether they conform to the AoA or not. However, it will require the complaining party to show the existence of a specific subsidy as well as 'adverse effects' to its interests. The fact that no one has so far challenged irrigation subsidies shows that most members find it difficult to establish causal link between the subsidy and the factor evidencing serious prejudice. They may also not do it for fear of tit-for-tat action.

Making WTO rules more effective
In order to make the current WTO disciplines more relevant and effective to deal with irrigation subsidies, the following steps are being proposed. The first phase would be to bring more transparency regarding notifying irrigation subsidies. At present, they are not accounted separately but included in the overall figures for agricultural subsidies, some of which can be genuinely regarded as Green Box. This mix makes it difficult to single out irrigation subsidies and apply WTO disciplines. Another step

would be to bring some uniformity in the way irrigation subsidies are calculated and notified to the WTO. Given that measuring subsidies to the irrigation sector is not an easy task, there has been no serious effort to tackle this issue. Recently the IISD has looked into this issue and made some recommendations for quantifying irrigation subsidies. Their methodology seems to be neutral and workable. If used globally, it could substantially improve reporting and making accounting of subsidies more uniform. Furthermore, there should be closer scrutiny of the notifications submitted to the WTO. The notifications should be up-to-date and harmonized so as to make them comparable.

While bringing more clarity regarding the scope of Green Box and ensuring better measurement and timely reporting of irrigation subsidies to the WTO would raise awareness on this issue, there may be a need for tougher disciplines and more cooperative efforts to meet the overall objective of reducing subsidies and thus improve sustainability. There should also be more clarity about the interpretation of 'low-income' or 'resource-poor' farmers. This would bring some disciplines for measuring the de minimus limit of 10 per cent exemption calculated under Article 6.2 of the AoA. For this purpose, two areas of negotiations can serve as models. One is the ongoing negotiations on fisheries subsidies and the other is the recently concluded Trade Facilitation Agreement (TFA). Fisheries subsidies have many parallels with irrigation subsidies. In both cases, there is very poor disclosure and notification of subsidies. Furthermore, it is difficult to find countries that could not be accused of granting subsidies in these two areas thus making it awkward for most governments to point fingers at others for granting such subsidies. It is, therefore, worth exploring if it is useful to follow the example of fisheries and have a separate agreement on water-related subsidies. Such an effort may not be acceptable to all the members as some may argue that this would not be within the WTO mandate which is concerned with trade-related matters and not sustainability. Furthermore, they could also argue that the WTO does not have the means to undertake the task.

Notwithstanding these concerns, and as a first step, it would be worthwhile for the WTO members to enter into an exploratory phase for possibly looking into clarification and improving WTO disciplines on irrigation subsidies. Such an agreement would enable members to put forward proposals that focus on improving sustainable use of water resources as is currently happening in case of fisheries subsidies. If a multilateral agreement is not considered achievable, a plurilateral agreement between reform-minded countries could be negotiated.

The negotiations on fisheries subsidies can also provide guidance as to the kind of issues that are likely to arise and what would be a more acceptable way to proceed. The TFA can provide guidance as to how a cooperative arrangement for achieving internal reforms can work through multilateral rules and by sharing best international practices. Other features of TFA, which can provide guidance, include having à la carte approach to determining the timing of implementation and joint action by donors to assist developing countries in implementing some of the provisions of the agreement.

No doubt having a water-specific agreement at the WTO would raise more awareness, bring transparency in policies and enable WTO members to peer-review one another's policies. This would be a major contribution towards achieving one of the strategic objectives of 2013, the International Year of Water Cooperation, which is to 'Strengthen international cooperation among institutions, users, social and economic sectors and others in order to reach a consensus on Sustainable Development Goals for the post-2015 era which will effectively address our future water needs.'[9] However, since water is not a commodity like most other items traded in international commerce, negotiating an agreement at the WTO would be very challenging and controversial. Some members of the civil society, who have strongly opposed including water-related services in the WTO Services negotiations, may oppose stricter disciplines on water use through WTO rules. At various WTO ministerial meetings, their slogan has been 'Don't let the WTO get hold of our water.' Their concern is that in the event of shortages in future, water 'will eventually be distributed and sold much like petroleum is today'. However, there is no unanimity and many groups and associations would welcome a more sustainable and equitable management of natural resources.

Another impediment may be that the WTO negotiations on new rule-making have almost come to a standstill. The only significant agreement on rule-making agreed so far relates to trade-facilitation and agricultural export subsidies, which was much less controversial. Handling difficult negotiations like that of water may be so daunting that most WTO members would be reluctant to make the effort needed for placing such an item on its agenda.

2. Demand Side and WTO Rules

For the demand side, an important factor for promoting sustainability of water is how it is used for producing goods for exports. For this purpose, understanding the concepts of virtual water and water-footprint are

---

[9] (UN General Assembly, 18 August 2014)

important. These concepts are relatively recent in their origin but have received considerable attention.

Virtual-water content of a product (a commodity, good or service) is 'the volume of freshwater used to produce the product, measured at the place where the product was actually produced'.[10] This refers to the idea that when goods and services are exchanged, so is virtual water. According to Professor John A Allan when a country imports one ton of wheat instead of producing it domestically, it is saving about 1,300 cubic metres of real indigenous water. If this country is water-scarce, the water that is 'saved' can be used towards other ends. If the exporting country is water-scarce, however, it has exported 1,300 cubic metres of virtual water since the real water used to grow the wheat will no longer be available for other purposes.

Another related notion is of water footprint, which is 'a way of assessing not only water use but potential environmental impacts related to water'. Now that a new standard (ISO 14046) has been developed which is designed to track how much water is consumed in producing goods and services, this concept perhaps will become as significant for estimating the impact of water-related activities as 'carbon footprint' is used for climate change.[11] Already businesses are using water foot-printing to improve water efficiency in their supply chains.

The scope of this chapter does not cover the pros and cons of using water footprint or virtual water but only discusses them as they would relate to the WTO rules. Despite the popular misperception in some circles, it is generally agreed that there is nothing in the WTO rules that restricts countries from pursuing legitimate environmental and sustainable development goals as long as these do not result in any arbitrary or unjustifiable discrimination, or create disguised restrictions on trade. The rules most often used by WTO members are the WTO Agreement on Technical Barriers to Trade (TBT) and GATT Article XX on General Exceptions. The relevance of TBT agreement is due to the fact that it provides a balance of rights and obligations for both labelling programmes and technical regulations.

However, it needs to be clarified that the WTO rules only apply if water footprints are taken up as part of government regulation, not if they are used as private standards. Also, recent WTO jurisprudence suggests that,

---

[10] Thomas Cottier, 'Renewable Energy and Process and Production Methods', E15 Think Piece August 2015 [available at: http://e15initiative.org/publications/renewable-energy-and-process-and-production-methods].

[11] ISO, *ISO 14046-2014 Environmental management – Water footprint – Principles, requirements and guideline* (Geneva 2014).

in principle, WTO members can apply import measures based on non-product related process and production methods (PPMs) under GATT Article XX, although no Member has successfully passed the hurdle of the chapeau yet.[12] The issue is still uncertain, and controversial.

The Agreement on Technical Barriers to Trade
The TBT Agreement was first developed in the Tokyo Round (then also called Standards Code) and was further improved in the Uruguay Round, with a view to strengthening the basic disciplines enshrined in Articles XI and XX of the GATT 1994. The TBT Agreement responds to two broad policy considerations. On the one hand, technical regulations and product standards, including packaging, marketing and labelling requirements, as well procedures for testing and certifying compliance with these regulations and standards, should not create unnecessary obstacles to international trade. On the other hand, members should, at the same time, be able to pursue appropriately legitimate policy objectives such as the protection of national security, the prevention of deceptive practices and the protection of human, animal and plant life and health. The agreement is designed according to these two underpinnings which are explicitly referred to in the preamble.[13]

Products are not only defined by their physical properties, their end uses in a given market and consumers' tastes and habits. They may also be defined by way they are produced, i.e., on the basis of PPMs. The distinction is of importance in particular in relation to environmental and human or labour rights' concerns. The question as to whether Article III of the GATT 1994 per se allows to differentiate between products that are physically similar but that are processed or produced differently is highly controversial. In the case of PPMs, the 'like product' test does not focus on the products themselves but rather on the way they have been processed or produced. Therefore, an isolated comparison of the end-products might not reveal any (physical) differences; it is the 'history' of the product that distinguishes it from other (physically similar) products. The PPM debate is equally relevant under the TBT Agreement and the Enabling Clause where tariff preferences are often granted to developing countries upon the fulfilment of certain conditionalities.[14]

---

[12] Cottier, (n 10).
[13] See generally Luan Xinjie and Julien Chaisse, 'The WTO Seals Products Dispute – Traditional Hunting, Public Morals and Technical Barriers to Trade' (2011) 22(1) *Colorado Journal of International Environmental Law and Policy* 79–121.
[14] A very early GATT 1947 panel report, Belgium Family Allowances, is often interpreted, and cited as an important precedence, to the effect that discrimination on the basis of how (physically similar) products are produced or processed is

While using the labelling schemes, the difficulty arises if labels are linked to PPMs. If the production method leaves a residue in the final product, it can be identified as to how it was produced. For example, if there is a pesticide residue in the imported rice, it is easy to test it and have a standard, which lays down the minimum residues. Thus import could be denied or a label could indicate this fact for a consumer to decide whether to buy such a product.

On the other hand, if rice is produced through flooding or another technique which saves substantial water, it is difficult to determine the production process at the import stage. For example, the International Rice Research Institute (IRRI) has developed an irrigation technique called alternate wetting and drying (AWD) that can cut down water use in producing rice by 25 per cent.[15] IRRI claims that by using this technique, it would take 1,500 litres of water per kilogram of rice as compared to normal usage of 2,000 kg. Similarly drip irrigation can save 40–50 per cent of water for many crops.

If a country introduces a labelling scheme that requires it to indicate whether the rice is produced through AWD or through the traditional technique, it is likely to run the risk of being non-compliant with WTO rules. Furthermore, such differentiation could also be used for protectionist purposes or for enforcing a country's own standards and production methods on others. This may also make it difficult for developing countries to compete in some cases. They may not be able to meet the standards developed and applied by more advanced countries. It would therefore be difficult for developing countries to accept such labelling schemes.

However, there are WTO compliant ways for making PPM more acceptable. WTO law, under the so-called 'Enabling Clause', allows for an exception to the WTO 'most-favoured nation' principle (i.e., equal treatment should be accorded to all WTO members). Currently many developed countries have schemes for allowing tariff preferences for products from developing countries, which meet certain specified criteria. For example, the EU's GSP Plus scheme allows duty free access for countries, which ratify and implement international conventions relating to human and labour rights, environment and good governance. Such

---

prohibited under the GATT: Belgian Family Allowances (Allocations Familiales), Report of the Panel, adopted on 7 November 1952, BISD 1S/59 (1955) 412.

[15] IRRI, From Far Eastern Agriculture Web site: <www.fareasternagriculture.com/crops/agriculture/irri-introduces-water-saving-technique-in-producing-rice> accessed 9 June 2014.

schemes could be used as incentives for making PPM distinctions more acceptable.

GATT Article XX on general exceptions
The other WTO rule, which is sometime invoked for keeping out environmentally unfriendly products, is GATT Article XX. In the practice of WTO law, Article XX of the GATT 1994 is one of the most important provisions. It justifies deviations from other rules, in particular, but not exclusively, from the principle of national treatment and from the prohibition of quantitative restrictions. Article XX is composed of two distinct parts: First, it contains an enumeration of specific motives and conditions for restricting trade, listed in paragraphs (a)–(j). Not all of them are of equal practical importance. The critical provisions which are frequently invoked in practice – as WTO members have become increasingly concerned with environmental and human health issues as well as with the protection of intellectual property rights – refer to measures necessary to protect human, animal or plant life and health (para. b), measures necessary to secure compliance with laws relating to the protection of patents, trademarks and copyrights, and the prevention of deceptive practices (para. d), and measures relating to the conservation of exhaustible natural resources (para. g). Moreover, protection of public morals is provided for (para. a). This latter paragraph may gain, along with banning imports from prison labour (para. e), increased importance in relation to the protection of human rights. Second, Article XX contains a general provision, the so-called chapeau, which applies in addition to the specific motives.[16]

Under the exceptions allowed in this article, WTO members can adopt measures for the conservation of exhaustible natural resources. However, such measures would be extreme for enforcing PPM standards and are not likely to pass the WTO test. Any such measures have to maintain a balance between market access obligations and the right of members to invoke the environmental justifications. Invoking GATT Article XX provisions for PPM will likely result in undoing market access commitments and thus disturbing the balance between the market access commitments and environmental obligations.

---

[16] See J Chaisse, 'Exploring the confines of international investment and domestic health protection – general exceptions clause as a forced perspective' (2013) 39(2/3) *American Journal of Law & Medicine* 332–61.

CONCLUSION

Trade regulation is often perceived, at the outset, as a set of rules simply bringing about progressive liberalization of trade and the dismantlement of trade barriers. Economic theory and empirical data show that open markets in principle bring about beneficial effects. However, as shown in this chapter, the matter is more complex as it entails and touches upon other policy areas and may also produce effects which are difficult to absorb in political terms.

There are provisions in the WTO rules for the better management of water. However, they are not invoked and currently there is no transparency for peer reviewing policies of member states in this important area. The WTO can play a more significant role in preventing wastages of water and closing the gap between its availability and requirements. WTO rules could provide incentives to member countries to adopt more environmentally friendly ways for using water. Therefore, there is a need to explore what further changes to the WTO rules and practices could be made so that trade rules can play their due role in overcoming the future water-shortages, especially in light of the increasing important role played by investment treaties.

3. The trade in water services – how does GATS apply to the water and sanitation services sector?

Rebecca Bates

INTRODUCTION

The General Agreement on Trade in Services (GATS)[1] is a complex and at times poorly understood agreement. These characteristics are a direct result of its negotiation history and the compromises made by the Member States of the World Trade Organization (WTO) to reach consensus regarding a services based agreement during the Uruguay round of negotiations.[2] As a result of this negotiation process the GATS was designed to be an 'opt in' agreement through which two of the main provisions, Article XVI (national treatment) and Article XVII (market access) only apply in circumstances where a Member State nominates the sector for liberalisation. This however requires the 'classification' of the service being nominated for liberalisation. The list of services sectors and their classification for liberalisation are broadly contained within two documents, the W/120 Scheduling Guidelines[3] and Central Product Classification (CPC).[4] The voluntary nature of the agreements and the non-exhaustive nature of the classification lists have done little to remove the uncertainty surrounding the document. The uncertainty is perhaps

---

[1] *Marrakesh Agreement Establishing the World Trade Organization*, opened for signature 15 April 1994, 1867 UNTS 3 (hereinafter 'Marrakesh Agreement'), annex 1B (*General Agreement on Trade in Services*) 1869 UNTS 183 (entered into force 1 January 1995) (hereinafter 'GATS').

[2] See generally Eric Leroux, 'Eleven Years of GATS Case Law: What Have we Learned?' (2007) 10 *Journal of International Economic Law* 749.

[3] See Services Sectorial Classification List, MTN.GNS/W/120, World Trade Organization (W/120).

[4] See Central Product Classification (CPC) Version 1.0, Statistical Papers Series, M, No. 77 Ver 1.0, United Nations 1998.

most pronounced in, but certainly not limited to, the area of water services where the very application of the agreement itself continues to be an issue.

The globalisation of water services is a multifaceted concept and process and is one inherently intertwined with the process of service liberalisation and privatisation. This chapter will explore the nature of water services and the content and application of the GATS. In particular, it will examine the key provisions of the agreement to the water services sector and assess how water services are classified under the W/120 and CPC. It will also explore the application of the limited GATS related case law, in particular the leading Appellate Body Decision, *US-Gambling*[5] and ask whether the decision and general uncertainty surrounding service classification raises the prospect of unintended liberalisation and whether changes to the service classification sectors may provide greater certainty.

THE GLOBALISATION, LIBERALISATION (AND PRIVATISATION) OF WATER SERVICES

Globalisation is an amorphous, multifaceted and multidimensional process that eludes simple definition. Broadly, globalisation may be taken to mean the total amount of economic, social, political and legal processes which transcend national boundaries and move freely between states.[6] The term was coined by Theodore Levitt, who used the expression in 1985 to describe the pervasive and rapid flow of investment, production and consumption of goods, services, technology and capital across the globe which he had observed occurring over the previous two decades. From this perspective, Mathias Finger and Jeremy Allouche note that the term globalisation was mainly employed by economic historians as a means of describing the changing global economy, a connotation that the expression maintained until recently.[7] Today, globalisation describes

---

[5] See Appellate Body Report, *United States – Measures Affecting the Cross-Boundary Supply of Gambling and Betting Services* WT/DS285/AB/R, adopted 20 April 2005; Panel Report WT/DS285/R, adopted 20 April 2005 (*US – Gambling*).

[6] See generally Frank Garcia, 'The Global Market and Human Rights: Trading Away the Human Rights Principle' (1999) 25 *Brooklyn Journal of International Law* 51, 56.

[7] See generally Theodore Levitt, 'The Globalisation of Markets' in AM Kantrow, *Sunrise... Sunset: Challenging the Myth of Industrial Obsolescence* (John Wiley and Sons 1985) 53; Matthias Finger and Jeremy Allouche, *Water Privatisation: Trans-National Corporations and the Re-Regulation of the Water Industry* (Spon Press 2002) 2.

the different types of changes occurring within not merely the economic dimension, but all aspects of human life.[8]

Globalisation challenges the concept of state boundaries, as national governments no longer possess total sovereignty in managing their economic affairs. The process also demonstrates the dominance of neo-liberal economic theory as it aims to remove global barriers to the free flow of commerce and trade.[9] Since the 1970s, it has not been a static process. Jurgen Habermas argues that, since the formation of the WTO, the rate of globalisation has rapidly augmented, as a result of the increased imposition of free trade imperatives on economic activity.[10] Consequently, national boundaries diminish in significance as the instruments of liberalisation take effect.

Globalisation and the tools of trade liberalisation, such as the GATS, have the potential to radically change the operation of water markets, in particular those that have traditionally operated under monopoly government control. The private provision of water services is not however a new phenomenon. The private sector was responsible for the first formal provision of water and sanitation services in Western Europe and North America in the nineteenth century and, from this time has expanded into a multi-billion dollar industry.[11] The responsibility for the provision of water services gradually shifted to the public sector over the nineteenth and early twentieth centuries as private firms failed to meet the needs of their consumers.[12] The government sector maintained its dominance in the water market from this time until the 1970s when Western political thought embraced neo-liberal economic theory and the concept of the 'free market'.[13] Private firms operate in over 120 countries around the world.[14] National governments generally rely upon regulation as the

---

[8] See e.g. Maude Barlow, *The Free Trade Area of the Americas: The Threat to Social Programs, Environmental Sustainability and Social Justice* (Council of Canadians 2001) 2.

[9] See Jurgen Habermas 'The European Nation-State and the Pressures of Globalisation' (1999) 235 *New Left Review* 46.

[10] See ibid., 52.

[11] See Jessica Budds and Gordon McGranahan, 'Are the Debates on Water Privatisation Missing the Point? Experience from Africa, Asia and Latin America' (2003) 7 *Environment & Urbanisation* 87.

[12] See generally James Salzman, 'Thirst: A Short History of Drinking Water' (2006) 31 Duke Law School Working Paper Series 1.

[13] See Budds and McGranahan, n. 11 above.

[14] See Jason Morrison and Peter Gleick, *Freshwater Resources: Managing the Risks Facing the Private Sector* (Pacific Institute, 2004) 5; Vandana Shiva, *Water Wars: Privatisation, Pollution and Profit* (South End Press 2002) 97.

primary means of ensuring a balance between consumer and corporate interests.[15]

Despite the high level of private sector participation in the water services sector, the industry operates generally outside the direct influence of trade liberalisation and the WTO. At present there are no specific commitments with respect to water services under the GATS. There is also a significant degree of uncertainty as to how the agreement classifies a water service if a specific commitment were to be made by a Member State.[16] It is therefore important to understand how GATS applies to the water services sector and whether there is any scope to provide greater clarity to their relationship.

OVERVIEW OF THE GENERAL AGREEMENT ON TRADE IN SERVICES (GATS)

The GATS is the sector specific agreement negotiated by WTO Member States during the Uruguay Round of negotiations. It formed part of the 'new' WTO replacing the previously standalone General Agreement on Tariffs and Trade (GATT). The GATS is responsible for establishing 'binding rules' on the international trade of services.[17] Eric Leroux argues that this Agreement is 'somewhat complex' as a result of the substantial challenges faced by the negotiators in achieving their goal of drafting a 'comprehensive set of disciplines governing the multilateral trade in services'.[18] As a result, the GATS is a mixture of mandatory and voluntary obligations, which at times create substantial interpretative difficulties.[19] Interestingly, the Agreement does not define the meaning of 'services'[20]

---

[15] See e.g., OFWAT, 'the economic regulator of water services in England and Wales' <http://www.ofwat.gov.uk/about-us/our-duties/> accessed 4/8/15 1. See also as an example of regional water regulation: Water Framework Directive – 2000/60/EC.

[16] See Rebecca Bates, 'The Trade in Water Services: How Does GATS Apply to the Water and Sanitation Services Sector?' (2009) 31 *Sydney Law Review* 121.

[17] World Trade Organization, 'The General Agreement on Trade in Services (GATS): objectives, coverage and disciplines' < https://www.wto.org/english/tratop_e/serv_e/gatsqa_e.htm> accessed 4/8/15, 1–2; David Hunter and others, *International Environmental Law and Policy* (4th ed, Foundation Press 2011) 1216.

[18] See Leroux, n. 2 above.

[19] See ibid.

[20] The exclusion of a definition was the intention of the drafters: Aly K Abu-Akeel, 'Definition of Trade in Services Under the GATS: Legal Implications' (1999–2000) 32 *George Washington Journal of International Law and Economics* 189.

within its text. However, it is clear that GATS applies to all forms of trade in services and ensures that the liberalisation commitments made by Member States apply to all services nominated by a Member for liberalisation.[21] The Agreement does, however, define the meaning of 'trade in services' as being the supply of a service:

(a) from the territory of one Member into the territory of any other Member;
(b) in the territory of one Member to the service consumer of any other Member;
(c) by a service supplier of one Member, through commercial presence in the territory of any other Member;
(d) by a service supplier of one Member, through presence of natural persons of a Member in the territory of any other Member.[22]

Article I:(3)(b) excludes the application of the Agreement from government services.[23] As previously mentioned the classification of services is generally defined by the CPC and W/120. The nature of the classification process and its application to water services will be discussed in depth in the following section.

The GATS document is divided into two key sections – the Framework Agreement containing the general rules and the accompanying schedules listing national commitments on specific domestic access for foreign suppliers.[24] GATS, like the GATT, contains a number of key provisions designed to promote equality between Member States, market access and non-differential treatment of like products.[25] These are:

---

[21] Ibid.
[22] GATS, art. I, Marrakesh Agreement, annex 1B.
[23] GATS, art I:3(c) defines a service supplied in the exercise of government authority to be 'any service which is supplied neither on a commercial basis, nor in competition with one or more service suppliers'.
[24] World Trade Organization, 'GATS: objectives, coverage and disciplines', n. 17 above.
[25] See Julien Chaisse, 'Assessing the Relevance of Multilateral Trade Law to Sovereign Investments: Sovereign Wealth Funds as "Investors" under the General Agreement on Trade in Services' (2015) 3(2) *International Review of Law* 30–54. See also Julien Chaisse and Mitsuo Matsushita, 'Maintaining the WTO's Supremacy in the International Trade Order – A Proposal to Refine and Revise the Role of the Trade Policy Review Mechanism' 18 *Journal of International Economic Law* 9 (2013).

The trade in water services 41

- Article II:1: Most Favoured Nation (MFN)
  - With respect to any measure covered in this Agreement, each Member shall accord immediately and unconditionally to service and service suppliers of any other Member treatment no less favourable than it accords to like services and service suppliers of any other country.
- Article III: Transparency
  - Each Member shall publish promptly and, except in emergency situations, at the latest by the time of their entry into force, all relevant measures of general application which pertain to or affect the operation of this Agreement. International agreements pertaining to or affecting trade in services to which a Member is a signatory shall also be published.
- Article XVI – Market Access
  - With respect to market access through the modes of supply identified in Article I, each Member shall accord services, and service suppliers of any other Member, treatment no less favourable than that provided for under the terms, limitations and conditions agreed and specified in its Schedule.
- Article XVII – National Treatment Obligation
  - In the sectors inscribed in its Schedule, and subject to any conditions and qualifications set out therein, each Member shall accord to services and service suppliers of any other Member, in respect of all measures affecting the supply of services, treatment no less favourable than that it accords to its own like services and service suppliers.

Despite the existence of similar principles in the two Agreements,[26] the GATT and GATS differ as a result of the mixed approach adopted by the GATS. This approach allows for the core provisions of Article XVII (National Treatment) and Article XVI (Market Access) only to apply to individual service sub-sectors nominated by the Member State for liberalisation whereas Articles II (Most Favoured Nation) and III (Transparency) apply 'horizontally' across all sectors in a similar manner to the GATT.[27] Consequently, Articles XVI and XVII will only apply in circumstances where a Member State has specifically nominated it for inclusion thus making GATS an 'opt in' Agreement. Therefore, Member States are

---

[26] Leroux, n 2 above.
[27] Hunter and others, n 17 above; WTO, n 17 above.

required to nominate their service sectors for liberalisation before Articles XVII and XVI have national application.[28]

Specifically, the MFN principle requires Member States to 'automatically and unconditionally' provide other Member States with treatment no less favourable than they would afford any other country. The concept of like services has not yet been fully explored by the WTO adjudication bodies. However, it did find in *Canada – Autos*[29] that 'manufacture beneficiaries' and 'non-manufacture beneficiaries' were like service suppliers 'regardless of whether they have production facilities in Canada'.[30] Members are required to afford this access without delay and to all WTO Members.[31] The GATS, however, allows a Member to 'maintain a measure inconsistent with [the MFN principle] provided that such a measure is listed in, and meets the conditions of, the Annex on Article II Exemptions', [32] thus enabling Members to exclude themselves from the operation of the provision for both legal and political reasons.[33] Similarly, Article II, the Market Access provision, requires Members wishing to liberalise a service sector to specifically nominate the sector for liberalisation and then enter into commitments under Articles XVI, XVII and XVIII.[34] Once nominated, the provision operates to restrict a Member from limiting the number of suppliers in the country, the value of services imported, the quantity of service output, the number of service operations, the number of persons employed, participation on foreign capital and certain forms of legal entities.[35] However, Article XVI.2 creates an exception to the rule allowing Members to meet its requirements 'according to services and service suppliers of any other Member, either formally identical treatment or formally different treatment to that it accords to its own like services and service suppliers'.[36] Mitsuo Matsushita, Thomas Schoenbaum and Petros Mavroidis argue that this allows Members to make exceptions through a number of means, including the use of population density tests

---

[28] WTO, ibid.
[29] *Canada — Certain Measures Affecting the Automotive Industry*, WTO Doc WT/DS139/R, WT/DS142/R (2000) (Report of the Panel); WTO Doc WT/DS139/AB/R, WT/DS142/AB/R, AB-2000-2 (2000) (Report of the Appellate Body) §§10.247–8.
[30] Matsushita and others, *The World Trade Organization: Law, Practice and Policy* (2nd ed., OUP 2006) 619.
[31] Ibid., 620.
[32] GATS, art II.2.
[33] Matsushita and others, n 30 above, 623.
[34] Ibid., 648.
[35] Ibid.
[36] GATS, art. XVI.2.

to determine the number of service suppliers permitted to operate or limiting the operation of foreign subsidiaries to a percentage of total domestic assets in an industry sector.[37] Consequently this provision, like the MFN principle does not apply to all Members in all circumstances. Finally, Article XVII, the National Treatment provision requires that members treat the 'like services' of Members in a manner no less favourable than their domestically produced 'like services'. This provision has a potentially large scope of operation, having the capacity to cover all GATS measures. However, in reality, its operation is limited to the areas affecting the trade in services excluding those already covered by Articles XVI and VI.[38] In *EC-Bananas III*,[39] the dispute resolution panel developed a four-pronged test to determine the inconsistency of a measure with the GATT National Treatment provision.[40] First, the test requires that the complainant establish that the Member had taken a 'specific commitment in the relevant sector and mode of supply'. Second, the Member must have adopted a measure that 'affected the supply of services in the sector and the mode of supply concerned'. Third, the disputed measure must have been 'applied to foreign and domestic like services and/or services suppliers' and finally, the measure must have accorded the foreign suppliers 'treatment less favourable than that accorded their domestic counterparts'.[41] However, it remains to be seen whether this approach will by applied by a dispute resolution panel with respect to the GATS National Treatment provision.

The GATS, however creates a number of general exceptions under Article XIV which provide for circumstances in which Members are allowed to take certain otherwise prohibited actions on a number of limited grounds in the same manner as Article XX of the GATT. These actions must not be applied in a discriminatory manner or act as a distinguished restriction on the trade in services.[42] Specifically of interest

---

[37] Matsushita and others, n 30 above, 468. See also the case of India, Debashis Chakraborty, Julien Chaisse, and Jaydeep Mukherjee, 'Deconstructing Services and Investment Negotiations – A Case Study of India at WTO GATS and Investment Fora' (2013) 14(1) *Journal of World Investment and Trade* 44–78.

[38] Matsushita and others, ibid., 659; GATS, art VI: (the Domestic Regulation provision) provides that in circumstances where a Member has made a GATS commitment, the Member must apply regulations that may affect the trade in services 'in a reasonable, objective and impartial manner'.

[39] *EC — Regime for the Importation, Sale and Distribution of Bananas*, WTO Doc WT/DS27/R/USA (1997) (Report of the Panel); WTO DocWT/DS27/AB/R (1997) (Report of the Appellate Body) (hereinafter '*EC — Bananas III*').

[40] Matsushita and others, n. 30 above, 662.

[41] Ibid.; *EC – Bananas III* n. 39 above.

[42] GATS, art. XIV.

with respect to the water services sector, the Article XIV(b) provides that: 'nothing in this Agreement shall be construed to prevent the adoption or enforcement by any Member of measures . . . necessary to protect human, animal or plant life or health'.[43]

Similarly, Article XIV also provides an exception for measures designed to protect 'public morals and public order'[44] however not as in the case of GATT Article XX(g) 'exhaustible' natural resources. Generally the Article XX/Article XIV case law has demonstrated a willingness of the WTO Panel and Appellate Body to accept the merits of trade restrictive measures in genuine circumstances, however, a general failing of the Member State is to construct the measures in a non-discriminatory manner.[45] An exception to this trend can be found in the recent Appellate Body decision, *EC – Seals Products*,[46] where measures adopted by the European Union (EU) to prohibit the importation and marketing of seal products were the subject of a complaint by Canada and Norway. In this dispute the EU justified the application of its measures under GATT XX(a), and the 'protection of public morals and public', on the basis that animal welfare concerns were of high importance to public morals in Europe.[47] These arguments were upheld by both the Panel and Appellate Body despite the measure being discriminatory under Articles I(i) and III(iv).[48] The finding in *EC – Seal Products* has the potential to inform the interpretation of Article XVI(a) as the 'protection of public morals' has been a key issue in the very limited case law to date.[49]

---

[43] GATS, art. XIV(b).
[44] GATS, art. XIV(a).
[45] See e.g. *US – Gambling*, n. 5 above; *United States – Import Prohibition of Certain Shrimp and Shrimp Products*, 12 October 1998, 38 ILM 118 (1999); Appellate Body Report *Reformulated Gasoline and Brazil: Measures Affecting Imports of Retread Tyres*, WT/DS332/AB/R, 3 December 2007.
[46] Appellate Body Report, *EC – Measures Prohibiting the Importation and Marketing of Seal Products*, WT/DS400/AB/R and WT/DS401/AB/R (May 22, 2014) ('*EC – Seal Products*').
[47] WTO, '*EC – Measures Prohibiting the Importation and Marketing of Seal Products*' < https://www.wto.org/english/tratop_e/dispu_e/cases_e/ds401_e.htm> accessed 4/8/15.
[48] *EC – Seal Products*, n. 46 above; Rob Howse and others, 'Sealing the Deal: The WTO's Appellate Body Report in EC – Seal Products' 18 *Insights* (2014) <www.asil.org/insights/volume/18/issue/12/sealing-deal-wto's-appellate-body-report-ec-–-seal-products> accessed 4/8/15.
[49] See, e.g. *US – Gambling* n. 5 above; Appellate Body Report, *China – Measures Affecting Trading Rights and Distribution Services for Certain Publications and Audiovisual Entertainment Products* - AB-2009-3 (2009).

GATS AND SERVICE CLASSIFICATION

The classification of water services is an area of relative uncertainty under the GATS. As previously mentioned, service classification within the Agreement is governed by the W/120 and CPC documents which categorise and define service areas and subcategories. The CPC agreement was created by the United Nations Statistical Office in 1991 with the goal of classifying goods and services in a 'comprehensive and mutually exclusive manner'.[50] Mireille Cossy notes that originally the document was created for statistical purposes but was later adopted by Member States as the guiding classification document following the Uruguay Round. Since 1991, the document has been revised twice but now, however, shares the responsibility for service classification with the Services Sectorial Classification List (W/120).[51] The W/120 was drafted by the GATT Secretariat in 1991 and creates 12 broad service sectors which are divided in 160 subsectors and as Cossy notes is generally viewed as a 'simplification' of the CPC. Member States are free to use either classification system or to adopt another of their choosing.[52] This 'freedom' has been a significant cause of the general uncertainty surrounding service classification as there are no definitive boundaries or groupings.[53] This, as will be discussed subsequently in relation to *US-Gambling*, presents challenges in terms of defining both the nature and boundaries of GATS commitments. Specifically, with respect to water services the issue of classification is particularly fraught. Neither the CPC or the W/120 contain a specific reference to water services, however the related areas of sanitation and sewage services are included within the environmental services category.[54]

The area of environmental services has seen a growth in the number of commitments made by Member States over recent years. The W/120 creates four subcategories within this sector namely 'sewage services', 'refuse disposal services', 'sanitation and similar' and 'other', which may include cleaning, noise abatement and landscaping.[55] At present there are over 60 commitments in the area, 54 of which are with respect to

---

[50] Mireille Cossy, 'Water Services at the WTO' in Edith Brown Weiss and others (eds) *Fresh Water and International Law* (OUP 2005) 117.
[51] Ibid., 122.
[52] Ibid.
[53] Ibid., 121.
[54] Ibid.
[55] W/120, n. 3 above.

sanitation.⁵⁶ This number is however minimal compared to the number of commitments made in other areas such as tourism and financial services.⁵⁷ The low level of commitments in this area can be partly explained by the operation of the public services exception in Article 1(3)(b) as many environmental services are state operated, and particularly in the case of sanitation, have monopolistic tendencies. There are however generally higher levels of community concern regarding the liberalisation of essential services, such as water and sanitation, which has made liberalisation in these areas more politically sensitive than other areas such as financial services.⁵⁸

The European Community submitted a proposal in 2000 to the WTO for greater clarity regarding the classification of environmental services. The proposal argued for the creation of seven new categories of 'purely environmental services' which it asserted would support the enhanced take up of commitments by Member States.⁵⁹ Importantly, it allocated a specific category for water services, 'water for human use and water management'.⁶⁰ This proposal while gathering a great degree of interest was not formally adopted by the WTO.⁶¹ As a result, water services are not specifically mentioned within the classification system and to date there has been no Member State commitment in this area.⁶² However, the increasing rate of commitments in the environmental services sector and in particular with respect to sewage and sanitation, raises the question of how long water services may remain outside the Agreement. Sewage and sanitation services both rely heavily upon water for their processes and clearly their water needs feed into water use and resource allocation. Water services therefore in their broadest meaning may be subject to GATS commitments while the specific area of drinking water may be unaffected. This fragmentation of water supply in terms of the Agreement may raise domestic challenges for water managers given the tendency of

---

⁵⁶ WTO/World Bank, Services Database < https://i-tip.wto.org/services/SearchResultGats.aspx > GATS Environmental Services Search accessed 4/8/15.

⁵⁷ WTO, n. 17 above.

⁵⁸ See for example World Trade Organization (WTO), *GATS – Fact and Fiction* at <www.wto.org/english/tratop_e/serv_e/gatsfacts1004_e.pdf> accessed 4/8/15; See also Shiva, n. 14 above, 97.

⁵⁹ WTO, 'Communication from the European Communities and their Member States' GATS 2000: Environmental Services S/CSS/W/38 (22 December 2000), 8, 6A.

⁶⁰ Ibid.

⁶¹ Cossy, n. 50 above, 123.

⁶² World Bank and WTO, I-TIP Services < http://i-tip.wto.org/services/(S(r2sl0omoomrwnqs4bwb02o4i))/default.aspx> accessed 4/8/15.

the sector towards a natural monopoly and require co-existence of public and private actors.[63] The complexity and uncertainty surrounding the classification of water services is a significant challenge for the WTO and the GATS Agreement. While the sector is currently outside the Agreement, the increasing activity in environmental services means that liberalisation may occur within some aspect of service. The likelihood of this occurring is enhanced by the interpretation of the Agreement in particular through the leading decision of *US – Gambling*.

US – GAMBLING AND UNINTENDED SERVICE LIBERALISATION?

There has been relatively little case law regarding the GATS within the Panel or Appellate Body level of the Dispute Settlement Unit (DSU). The GATS has only been considered by the Dispute Settlement Body in a handful of cases and in only two at the Appellate level.[64] The first of these decisions, *US – Gambling*[65] is a GATS specific dispute and one whose details will be considered subsequently. The second, *China – Publications and Audiovisual Products* (2009)[66] was a dispute between US and China over a number of Chinese measures which the US argued restricted the distribution of audiovisual and home entertainment products in China. This dispute considered both GATT and GATS provisions. The Chinese measures in this dispute were found to be inconsistent with Article XVII (national treatment) of the GATS. There are currently two disputes under consultation before the DSU,[67] including *EU and its Member States – Certain Measures Relating to the Energy Sector – Dispute between EU and Russian Federation* in relation to the EU's Third Energy Package.[68] If this dispute progresses it will be the first GATS decision in relation to natural resources and may provide some important insights in the area.

---

[63] See Budds and McGranahan, n. 11 above, 93; Peter Gleick and others, 'The New Economy of Water: The Risks and Benefits of Globalization and the Privatization of Fresh Water' (Pacific Institute for Studies in Development, Environment and Security 2002) 5.
[64] Leroux, n. 2 above, 750.
[65] *US – Gambling*, n. 5 above.
[66] *China – Measures Affecting Trading Rights* n. 49 above.
[67] WTO, Disputes by Agreement (GATS) <www.wto.org/english/tratop_e/dispu_e/dispu_agreements_index_e.htm?id=A8> accessed 4/8/15.
[68] WTO, *EU and its Member States – Certain Measures Relating to the Energy Sector* <www.wto.org/english/tratop_e/dispu_e/cases_e/ds476_e.htm> accessed 4/8/15.

This discussion will focus on *US – Gambling* as it contains significant implications for service classification and raises the prospect of what can be called 'unintended liberalisation'. In *US – Gambling,* Antigua and Barbuda (Antigua) claimed that the US had violated paragraphs 1 and 3 of Article VI, through a number of federal[69] and state measures[70] legislated in the US relating to the remote supply of gambling services.[71] Given the number of provisions, Antigua alleged the 'collective effect' of the state and federal measures amounted to a total prohibition on the cross-border supply of gaming services.[72] The GATS however only allows a Member to challenge the *effect* of *a measure* as opposed to the collective effect of a group of measures.[73] As a result, both the Panel and Appellate Body rejected Antigua's claim, focusing the failure of Antigua to structure its complaint in an appropriate form.[74]

Despite this technical outcome the Panel and Appellate took the opportunity to consider the nature of the US's GATS commitment to 'other recreational services (except sporting)' and whether the commitment included 'gambling and betting services' within its scope.[75] The Panel and Appellate Body found that the US had made a specific commitment with respect to gambling and betting services by applying the W/120[76] and 1993 Scheduling Guidelines[77] as a 'supplementary means of interpretation'

---

[69] United States Code (the 'Wire Act'), s. 1084 of Title 18; United States Code (the 'Travel Act'), s. 1952 Title 18; United States Code (the 'Illegal Gambling Business Act', or 'IGBA'), s. 1955 of Title 18.

[70] Colorado Revised Statutes ss 18-10-103; Louisiana Revised Statutes (Annotated) s. 14:90.3; Annotated Laws of Massachusetts s. 17A ch. 271; Minnesota Statutes (Annotated) s. 609.755(1) and subdiv. 2–3 of s. 609.75; New Jersey Constitution para. 2 of s. VII of art. 4, New Jersey Code s. 2A: 40–1; New York Constitution, art I, s. 9; New York General Obligations Law s. 5-401; South Dakota Codified Laws ss 22-25A-1–22-25A-15; Utah Code (Annotated) s. 76-10-1102: source US-Gambling, WTO/World Bank, Services Database http://i-tip.wto.org/services/(S(5s22pd0bjgendtvmrsulcamk))/SearchResultGats.aspx para. 4.

[71] Leroux, n. 2 above, 756; Panagiotis Delimatsis, 'Due Process and "Good" Regulation Embedded in the GATS – Disciplining Regulatory Behaviour in Services through Article VI of the GATS' (2006) 10 *Journal of International Economic Law* 13.

[72] Leroux, ibid.

[73] *US – Gambling*, n. 5 above, para 124–6.

[74] *US – Gambling*, ibid., 115 in Leroux, n. 2 above.

[75] Leroux, ibid., 762; *US – Gambling*, ibid.

[76] Uruguay Round, Group of Negotiations on Services, Service Sectors Classification List, MTN.GNS/W/120, 10 July 1991.

[77] Scheduling of Initial Commitments in Trade in Services: Explanatory Notes, MTN.GNS/W/164, 3 September 1993.

under Article 32 of the Vienna Convention.[78] The W/120 had been relied upon by the DSU[79] as a means of defining individual service sectors, while the Scheduling Guidelines was endorsed as a means of assisting Members achieve the 'greatest possible degree of clarity' when scheduling a specific commitment.[80] Consequently, both documents were deemed important by the bodies as a means of assisting them determine the scope and nature of the US's commitment.[81] The Appellate Body found that, even though the US commitment schedule did not specifically refer to the CPC[82] (and followed the W/120), both documents could be used as 'context' for the interpretation of specific Member commitments within the meaning of Article 32 of the Vienna Convention.[83] Consequently, the Appellate Body determined that the US GATS commitment to 'other recreational services (except sporting)' must be interpreted as including 'gambling and betting services' within its scope.[84] The Panel and Appellate Body also considered whether the US had acted inconsistently with paragraphs 1 and 2 of Article XVI. The Appellate Body upheld the decision of the Panel finding that the US had violated Article XVI on the basis that the disputed federal acts prohibited the cross-border supply of gambling services in circumstances where the US had made a specific GATS commitment in the area.[85] The Appellate body found that the federal acts in effect created a 'zero quota' which are prohibited under Article XVI:2(a) and (c) and were therefore invalid.[86] However, the Appellate Body reversed the Panel's decision with respect to the state laws as it found that Antigua had failed to establish a *prima facie* case.[87] Also with respect to Article XVI, the Appellate Body upheld the Panel's findings that the US laws had been designed 'to protect public morals or to maintain public order' within the meaning of Article XIV(a) and reversed that Panel's finding that the laws had been unnecessary.[88] However, the Appellate Body modified the Panel's decision with respect to the Article XIV determining that the US measures had not

---

[78] Vienna Convention on the Law of Treaties, 23 May 1969, 1155 UNTS 331; 8 ILM 679.
[79] Marrakesh Agreement, art. 23(3)(f)(ii) annex 2 (*Understanding on Rules and Procedures Governing the Settlement of Disputes*) ('DSU').
[80] Leroux, n. 2 above, 759.
[81] Ibid., 749.
[82] CPC, n. 4 above; Leroux, n. 2 above, 760.
[83] Leroux, ibid., 762; *US – Gambling*, n. 5 above.
[84] Leroux, ibid., 762, 761; *US – Gambling*, ibid.
[85] Matsushita and others, n. 30 above, 652; *US – Gambling*, ibid.
[86] *US – Gambling*, ibid.
[87] Ibid.
[88] Ibid.

satisfied its requirements as the prohibition on the remote supply of gambling had not been applied equally to domestic and foreign suppliers.[89]

The finding of the Appellate Body in *US – Gambling* raises a number of points of interest regarding the application of GATS to the liberalisation of services. The Panel and Appellate Body's readiness to accept the exception claimed by the US under Article XIV(a) illustrates a willingness on the part of the WTO to recognise claims made by countries under this provision. Thus, if a Member State legislates for a legitimate purpose within the scope of the Article XIV, there is a substantial likelihood that the measure will be held to be valid. This is particularly significant in light of the recent *EC – Seal Products* decision. With respect to any future cases involving water services, it would be hoped that Article XIV(b)[90] may be employed in a similar manner to protect non-discriminatory legislation aimed at protecting and promoting basic water access, quality and affordability as a means of promoting and protecting human health.

The Appellate Body's inclusion of gambling and betting services within the US's 'Other Recreational Services (except sporting)' commitment however also demonstrates the potential uncertainty with respect to GATS commitments. The decision demonstrates that the meaning and scope of a Member's commitment will ultimately be determined by the DSU in circumstances where a dispute arises.[91] Leroux argues that the *US – Gambling* decision illustrates a need for 'greater clarity, consistency, and precision in the scheduling of commitments under the GATS' and that this outcome should be pursued through negotiation between Members rather than dispute resolution outcomes.[92] However, for the present, it appears that the clarification of commitments will continue through dispute resolution channels as many Members fear that a clarification process may lead to a reduction in commitments.[93] With respect to Article VI, it is noteworthy, that despite the case's focus on domestic regulation, neither the Panel nor the Appellate Body considered the domestic regulation provision found in Article VI. Delimatsis argues that the Appellate Body highlighted the

---

[89] Ibid., the US had claimed an exception under GATS, art. XIV(a) which provides exemption for measures 'necessary to protect public morals or to maintain public order' provided they applied in a manner consistent with the chapeau of art. XIV. See also Leroux, n. 2 above, 787.

[90] GATS, art. XIV(b) provides an exception for measures 'necessary to protect human, animal or plant life or health'; on this exception, see Chaisse, n. 25 above.

[91] Leroux, n. 2 above, 766.

[92] Ibid.

[93] Ibid.

irrelevance of the provision when it asserted that '[i]t is neither necessary nor appropriate for us to draw, in the abstract, the line between quantitative and qualitative measures'.[94] Consequently, *US – Gambling* does not provide any insights into how Article VI will apply to domestic regulatory measures. This is unfortunate as Article VI has the potential to be a central GATS provision and therefore it is important to understand how the obligation to 'ensure that all measures of general application affecting trade in services are administered in a reasonable, objective and impartial manner' will be applied.[95]

Clearly, these aspects of the *Gambling* decision risk creating ambiguity for Member States regarding the scope of their commitments and a potential chilling effect as Members may be less willing in future to nominate a service sector for liberalisation. Moreover, given the uncertainty surrounding sector classification there exists a substantial risk that a commitment may be interpreted differently by different Member States and most importantly by the DSU. The issue of interpretative differences raises the prospect of a commitment being found to be wider than originally intended for liberalisation for the Member State. If a commitment includes an additional aspect or aspects of a service not envisioned to be included in the original classification, this may be said to be 'unintended'. This is not to say that entire service sectors will suddenly become the subject of an unintended GATS commitment, rather that related aspects of an existing service commitment may be interpreted to include related services not originally intended by the Member State for liberalisation. As a result of the *US – Gambling* decision it is clear that a Member's liberalisation commitment will only be fully defined after it has been considered by the DSU in the context of a dispute. This issue is now particularly important with respect to water services given the growth of commitments in the related areas of sanitation and sewage. As previously mentioned, water, sanitation and sewage services are interlinked and ultimately depend upon connected supply and infrastructure channels. The lack of a specific reference to water services under the classification documents and the likely expansive interpretation of any commitment by the DSU continues to raise practical questions as to how existing sanitation and sewage commitments may be interpreted and how a future water services commitment may function. As Cossy argues, the GATS can play an important role in supporting decisions regarding privatisation and private sector involvement in any of its service sectors however this is best achieved by providing a 'predictable

---

[94] *US – Gambling*, n. 5 above, para. 250; Delimatsis, n. 71 above, 14.
[95] Delimatsis, ibid.

legal framework' which will send a positive signal to investors and foster foreign direct investment.[96] Clearly, in this area greater certainty could still be achieved.

HOW CAN GREATER CERTAINTY BE ACHIEVED?

The application of the GATS to water services has always been a controversial issue in light of the associated 'threats' of enhanced privatisation and foreign control over water services. Many commentators, such as Shiva have raised concerns regarding the scope of the Agreement and its effect once in force.[97] In particular, Shiva has argued that once commercial activity or competition was introduced to a service area, there was a risk that this service area 'may be dragged into a free trade ambit' despite the lack of a specific commitment by a Member State. Moreover, she has also asserted that the ambiguity surrounding the meaning of 'commercial basis' in Article I:3(c) created uncertainty regarding the status of public services and that the inclusion of a service area under the GATS would allow companies to sue countries in circumstances where government restrictions prevent free market access.[98] Such concerns and a number of privatisation failures, unrelated to the Agreement,[99] resulted in significant public hostility to the role of the GATS and the wider liberalisation of water services. These concerns resulted in the publication of the 2001 document 'GATS: Fact or Fiction' by the WTO.[100] GATS: Fact or Fiction outlines the structure of the GATS Agreement and the benefits of service liberalisation. It clearly reflects a concerted attempt by the WTO to overcome the negative perceptions that were associated with the Agreement at the time. In particular one section of the document was devoted to the

---

[96] Cossy, n. 30 above, 141.
[97] Morrison and Gleick, n. 14 above, 5; Shiva, n. 14 above, 97; See also Maude Barlow and Barry Clarke, *Blue Gold: The Battle Against the Corporate Theft of the World's Water* (New Press 2002).
[98] Morrison and Gleick, ibid.; Shiva, ibid., 94.
[99] See for example Cochabamba and the resulting International Centre for Investment Disputes Case: *Aguas del Tunari v Bolivia* (Netherlands–Bolivia BIT) <https://icsid.worldbank.org/ICSID/FrontServlet?requestType=CasesRH&actionVal=showDoc&docId=DC629_En&caseId=C210> accessed 4/8/15; See also Eric Woodhouse, 'The "Guerra del Agua" and the Cochabamba Concession: Social Risk and Foreign Direct Investment in Public Infrastructure' (2003) 39 *Stanford Journal of International Law* 295.
[100] WTO Secretariat *GATS Fact or Fiction* <www.wto.org/english/tratop_e/serv_e/gatsfacts1004_e.pdf> accessed 4/8/15.

issue of water services entitled, 'The WTO is not after your water' which outlined the freedom of Member States to maintain a public or private owned monopoly service.[101] It is perhaps in this climate of distrust that the reforms to classification or additional commitments with respect to water services have remained off the agenda. To date the EC has been the only Member to have requested specific commitments with regards to water distribution. This proposal was, as previously mentioned, not adopted by the WTO.[102] More recent rounds of negotiations have also failed to touch upon the issue.[103] Therefore the central question remains, is the absence of a specific reference to water services from the classification schedules beneficial as it removes the pressure from governments to nominate their water sectors and separates the Agreement from this controversial area, or is the absence of a specific classification creating further uncertainty?

The nature of water supply presents significant difficulties with service liberalisation as it remains one of the only true natural monopolies. The private sector, as previously mentioned, now plays a significant role in the supply of water, however the creation of true competition remains a challenge. Water resources and networks are interconnected meaning that it can be difficult to fully separate water supplied for the purposes of household and commercial consumption, sanitation and sewage. In light of the existing commitments with respect to sanitation and sewage and the likely expansive interpretation of commitments by the DSU in the case of a dispute, it is clear that a specific services classification for 'water services including drinking' would benefit the overall operation of the agreement. This is not to say that Member States would therefore be required to nominate for liberalisation in this area, rather that the creation of the category would more clearly define the boundaries with respect to service.

CONCLUSION

The application of GATS to water services has been one of the more controversial topics within globalisation discourses since the adoption of the Agreement in 1995. The liberalisation of water services under the GATS is inherently linked to the processes of globalisation and privatisation, areas which have both been topics of significant public debate. These concerns have stemmed in part from the sector's traditional mode of public sector

---

[101] Ibid., 11.
[102] Cossy, n. 30 above, 140.
[103] Morrison and Gleick, n. 14; WTO, n. 17 above.

supply and also water's fundamental role in human health and survival. However, another contribution to these sentiments has been the challenge of reconciling the economisation of what has traditionally been viewed as a public good. Service liberalisation is not however a new process having been widely adopted within more traditionally commercial spheres such as banking and telecommunications. It has however struggled to make similar inroads within the environmental services sector and with respect to water services themselves.

Environmental services are a relatively new area of liberalisation activity under the GATS. The recent increase in commitments by Member States under the Agreement indicates a likely expansion of this area in coming years. However, the challenges regarding service classification present a number of difficulties in this area with respect to water services. The absence of a specific reference to the service area within the W/120 or the CPC means that a Member State is not able to specifically nominate their water services for liberalisation or in the alternate, not able to specifically exclude their water services from a liberalisation commitment. The interconnected nature of water supply and the growing number of commitments in the areas of sanitation and sewage raises the risk that part of Member State's water services may be included within a commitment. This uncertainty has been supported by the lack of GATS specific case law and the prospect of 'unintended liberalisation' raised by the Appellate Body decision, *US – Gambling*. The decision of *US – Gambling* demonstrates that the nature and content of a services commitment will ultimately be decided by the DSU in the context of a dispute. This was the case with respect to the US's commitment to 'recreational services (other than sporting)' which was found to include the remote supply of gaming services. Consequently, as a result of this decision it is clear that the exact boundaries of a commitment may be uncertain until adjudicated by the DSU. This raises particular challenges with respect to water services due to the interconnected nature supply and the lack of clarity regarding their classification.

The inclusion of a new subcategory specifically related to water service may support the overall operation of the Agreement in this area. The creation of such a category would allow the area of water services to be specifically included or excluded from a commitment and may also avoid 'commitment creep' in the case of existing sewage and sanitation commitments. Such an approach could facilitate greater certainty and enable a Member State to make a water services commitment if that was its intention. This is one circumstance where the voluntary nature of the GATS Agreement may prove to be highly beneficial to both Member States and to the overall functionality of the Agreement.

4. Virtual water: a global economic solution to a local environmental and political problem?

Paolo Turrini*

1. INTRODUCTION

Studies on water as a resource shared between two or more countries have changed a lot in the last decades. In the eighties and nineties of the last century the debate was focused on what also became known to a larger and lay audience as 'water wars': at that time, many scholars firmly believed that the water scarcity affecting some world regions would lead, someday, to outbursts of real violence between states.[1] The subsequent years swept this conviction away – at least in part: despite the previous forecasts, no war has apparently been waged in the last decades for causes primarily related to the control of water endowments. Thus, the scientific literature on the subject took different directions: some studies tried to explain why no regional armed conflict resulted from water scarcity;[2] others capitalized on the lessons learned and, building on them, provided suggestions on how to best avoid international tensions due to the use and possession of water resources; still others shifted the focus away from pathological episodes and began to pay more attention to the everyday management of water security.

---

* Post-doc Research Scholar, School of International Studies, University of Trento. This chapter is to be read as complementary to the article 'Virtual Water and International Law' that I wrote together with Prof. Marco Pertile and will appear in the 'Brill Research Perspective – International Water Law' series.

[1] See, e.g., the bibliography quoted in Philipp Stucki, *Water Wars or Water Peace? Rethinking the Nexus between Water Scarcity and Armed Conflict*, Programme for Strategic and International Security Studies occasional paper no. 3 (2005) 14 ff.

[2] An excellent – although now rather outdated – review of the literature on conflicts and the environment in general is Nils Petter Gleditsch, 'Armed Conflict and The Environment: A Critique of the Literature' (1998) 35 *Journal of Peace Research* 381.

As a consequence, a multi-headed hydra – *rectius*: a multi-headed hydro – rose from freshwater. A number of concepts were injected into the discourse on water security. The notion of hydro-politics emerged as an umbrella-term covering both conflicts and their antidote, i.e., co-operation, and more generally the issue of states 'taking place in shared international river basins'.[3] As in mythological tradition, cut one head off, two take its place. Hydro-politics can be split in hydro-governmentality on the one hand (that is, the way states regulate water-related matters, especially at the domestic level, as an expression of ordinary governmental powers)[4] and, on the other hand, hydro-diplomacy (being the way international tensions are prevented by diplomatic means,[5] which can sometimes lead to the conclusion of treaties[6]). Germane to the latter is the head named hydro-solidarity: under this label, the management and negotiations around transboundary watercourses are looked at with an emphasis on dialogue and ethics, which are realized through the creation of collaborative institutions, the promotion of stakeholder participation and the integration of water within its wider environmental, economic and social context.[7] Finally, hydro-hegemony reveals why political and legal arrangements among countries sharing a river basin, despite their being precious instruments to avoid open conflicts, sometimes merely crystallize the power relations existing in the group of riparians, and accord better treatment to the most influential of them.[8]

The above is a non-exhaustive list of the paths that have been trodden

---

[3] Anthony Turton, 'Hydropolitics: The Concept and its Limitations' in Anthony Turton and Roland Henwood (eds), *Hydropolitics in the Developing World: A Southern African Perspective* (African Water Issues Research Unit 2002); see also Arun P. Elhance, 'Hydropolitics: Grounds for Despair, Reasons for Hope' (2000) 5 *International Negotiation* 202.

[4] Chad Staddon and Nick James, 'Water Security: A Genealogy of Emerging Discourses' in Graciela Schneier-Madanes (ed.), *Globalized Water: A Question of Governance* (Springer 2014).

[5] Adelphi, 'The Rise of Hydro-diplomacy: Strengthening Foreign Policy for Transboundary Waters', Climate Diplomacy report (2014).

[6] See, using this term, Surya P. Subedi, 'Hydro-Diplomacy in South Asia: The Conclusion of the Mahakali and Ganges River Treaties' (1999) 93 *American Journal of International Law* 953, and, using the cognate 'hydro-politics', Salman M.A. Salman and Kishor Uprety, 'Hydro-Politics in South Asia: A Comparative Analysis of the Mahakali and the Ganges Treaties' (1999) 39 *Natural Resources Journal* 295.

[7] Andrea K. Gerlak, Robert G. Varady and Arin C. Haverland, 'Hydrosolidarity and International Water Governance' (2009) 14 *International Negotiation* 311.

[8] A very recent bibliography on this subject has been compiled by the Geneva Water Hub and can be found at <https://www.genevawaterhub.org/

by scholars dealing with water security, that is, with that particular concern of states that regards both the protection from water-related natural disasters (like droughts and floods) and the procurement of a sufficient amount of water, capable of satisfying the primary needs of any country (such as the provision of drinking water and sanitation services to the population as well as the production of food and energy).[9] The examples have been chosen less for their representativeness of the existing literature than for their names that betray a desire to coagulate the specialized studies around a number of approaches or ideas. Here, however, we will stare into the eyes of just one of the heads of the hydro, the one called hydro-centricity (that is, the failure to appreciate that there are options to provide food and water security that do not require an increased water abstraction from rivers and lakes),[10] and will do this by referring to the concept on which the fight against the hydro-centric perspective is hinged: the concept of virtual water.

This chapter is composed of three parts. In Section 2, the notion of virtual water is introduced and briefly discussed with a view to analyzing its possible applications. Section 3 aims at providing some hints on the role that law – and especially, international law – plays in actualizing policies inspired by the idea of virtual water. Subsequently, in Section 4 we wonder whether law serves water-security purposes more effectively outwith the realm of virtual water than within it. Finally, some conclusive remarks are offered.

2. VIRTUAL WATER: WHAT IT IS AND WHERE WE CAN CHANNEL IT

That water is needed for producing food is a truism. Crops can be impressively water-demanding, and animal husbandry is even more so.

---

sites/default/files/atoms/files/2015.03.14_biblio_hydrohegemony_en_0.pdf>, last accessed 15 February 2016.

[9] See Dan Tarlock, 'Water Security, Fear Mitigation and International Water Law' (2008) 31 *Hamline Law Review* 703, 715–16, and also John Briscoe, 'Water Security in a Changing World' (2015) 144 *Dædalus – Journal of the American Academy of Arts & Sciences* 27.

[10] Stephen Brichieri-Colombi, 'Hydrocentricity: A Limited Approach to Achieving Food and Water Security' (2004) 29 *Water International* 318; John Anthony Allan, 'Beyond the Watershed: Avoiding the Dangers of Hydro-Centricity and Informing Water Policy' in Hillel Shuval and Hassan Dweik (eds), *Water Resources in the Middle East: Israel-Palestinian Water Issues – From Conflict to Cooperation*, vol. 2 (Springer 2007).

Everyone knows that. Nonetheless, some of us – perhaps many of us – would be somewhat surprised in learning that, in the world, 'water scarcity tends to manifest itself as food shortages [...] rather than individual dehydration'.[11] The reason is that, on the basis of our (limited) personal experience, we are prone to erecting a wall between food security and water security: as food shortages cause death by starvation, thus – we think – water scarcity causes death by thirst. But we should not forget that the amount of water required for agricultural uses is 70 times more than the quantity required to meet basic household needs.[12] The same figure ('70') is of importance also because it represents the percentage of the total water withdrawals, at global level, that are due to agriculture.[13] The joint consideration of these data suffices, per se, to understand why water is to be regarded as a fundamental factor in food production, and water scarcity as a major cause of food insecurity.

If water is so strictly connected to food, then, water-scarce countries should be careful when they resort to their poor water endowments in order to produce food for their populations. On this very observation lies the intuition behind the idea of virtual water: all agricultural products need a certain amount of water to be produced (known as their virtual water content[14]), and water-scarce states would be better relying on food imports rather than depleting their own water resources. The oldest case study on this topic concerns Israel, in the mid-eighties it was noted by some economists for being an exporter of water-intensive agricultural goods (that is, goods whose production requires a lot of water) notwithstanding its chronic lack of water.[15] The British geographer Anthony Allan, in turn, noted these studies and built on them his now famous 'theory' of virtual water, predicating that global food security – and, with it, water security – can be better attained through the international trade in food.[16]

---

[11] Andrew Biro, 'Water Wars by Other Means: Virtual Water and Global Economic Restructuring' (2012) 12(4) *Global Environmental Politics* 86, 88.

[12] Ibid.

[13] It must be noted, however, that at regional level the ratio ranges from 20 per cent (Europe) to 80 per cent (Africa), due to climatic and economic reasons. These data are available at <http://www.fao.org/nr/water/aquastat/water_use/index.stm>, last accessed 15 February 2016.

[14] As opposed to the 'real' water content, which is the water a good is materially composed of.

[15] See, for instance, G. Fishelson, 'The Allocation and Marginal Value Product of Water in Israeli Agriculture' (1994) 58 *Studies in Environmental Science* 427.

[16] Allan's academic production is quite vast, thus I will cite here only what is perhaps the first of his contributions on this issue: John Anthony Allan, 'Fortunately there are substitutes for water otherwise our hydro-political futures

The word 'theory' is perhaps improper. After all, Allan's idea, in its first formulation, was not much more than a statement of good sense, suggesting that water-scarce states should stop growing water-intensive food (especially if they intend to export it) and buy it instead. It took about a decade for the idea of 'virtual water' to become popular in scientific circles, but, since the inception of the new millennium, it has turned into a real magnet for hydrologists and economists: while the former have focused on the calculation of the so-called water footprint of any kind of products and services (that is, their virtual water content),[17] the latter have devoted their efforts to investigating both the magnitude and the direction of international virtual water flows (i.e., how much water is needed to produce the goods that are exported, and which countries are net importers or exporters of virtual water).[18] Thus, in just a few years, a massive amount of numerical knowledge agglomerated around what was once the rough notion of virtual water, providing a detailed photograph of the quantity of water virtually embedded in agricultural goods as well as its movements across the globe.

This notwithstanding, the term 'theory' can still be used only as an approximation. The fact is that a theory, in order to be called such, has not only to be grounded on solid science, but also show a certain level of internal coherence, and be premised on an unequivocally identified

---

would be impossible' in ODA, *Priorities for Water Resources Allocation and Management* (1993). Other writings of his are, however, quoted in this chapter: see footnotes 10, 20, 24, 26, 28, 42, 46 and 51.

[17] Although it is agricultural products that mostly drew the attention of scientists dealing with such a topic (see, e.g., Mesfin M. Mekonnen and Arjen Y. Hoekstra, 'The Green, Blue and Grey Water Footprint of Crops and Derived Crop Products' (2011) 15 *Hydrology and Earth System Sciences* 157), it is possible to calculate the water footprint of virtually everything, from energy production (see, e.g., Mesfin M. Mekonnen, Winnie Gerbens-Leenes and Arjen Y. Hoekstra, 'The Consumptive Water Footprint of Electricity and Heat: A Global Assessment' (2015) 1 *Environmental Science: Water Research & Technology* 285) to building construction (see, e.g., Mengyao Han *et al.*, 'Virtual Water Accounting for a Building Construction Engineering Project with Nine Sub projects: A Case in E-town, Beijing' (2016) 112 *Journal of Cleaner Production* 4691).

[18] The literature on this subject is huge. Recent accounts of these flows and – more importantly – their drivers, are in Stefania Tamea *et al.*, 'Drivers of the Virtual Water Trade' (2014) 50 *Water Resources Research* 17, and Rosa Duarte, Vincente Pinilla and Ana Serrano, 'Understanding Agricultural Virtual Water Flows in the World from an Economic Perspective: A Long Term Study' (2016) 61 *Ecological Indicators* 980. See also the annexes to the volume by Arjen H. Hoekstra and Ashok K. Chapagain, *Globalization of Water: Sharing the Planet's Freshwater Resources* (Blackwell Publishing 2008).

problem – even though it may be silent as to its solution. However, though the academic discourse on virtual water is filled with references to a small number of suggested policy options (which we will see in Section 3), the question to which they try to find an answer is all but clear. And, actually, it is better to speak of 'questions', in the plural. Questions, and problems, that have ripened in the virtual water discourse – and the conceptual appendages that have grown out of it – in a way that we can easily define as disordered (even if some scholars have tried to sort them out rationally[19]). Nonetheless, understanding which problems the virtual water theory wants to tackle has, in principle, important legal consequences. The reasons are at least two: in the first place, because only by having a clear picture of the problems at hand, as well as their mutual relationship (e.g., trade-offs) and hierarchy, an effective legal strategy to solve them can be fine-tuned; and secondly, because some international legal rules (such as Art. XX of the General Agreement on Tariffs and Trades, allowing for some exceptions to the international trade legal regime, or para. 12 of Annex 2 to the Agreement on Agriculture, on lawful agricultural subsidies) explicitly make the legality of a measure contingent on the goal it aims at.

On account of this, I will now briefly review the purposes that a virtual water policy may serve. It is interesting to note that the trade in water-intensive goods, in Allan's opinion, is meant to have a double beneficial effect. On the one hand, it should alleviate a state's problems that are due to water scarcity and prevent irreversible deterioration of its water resources thanks to the outsourcing of food production, of which water-abundant states will take charge. On the other hand, it should stave off tensions among neighbouring countries in water-scarce regions by reducing the pressure on their common waters (indeed, trade in virtual water is seen by Allan as one of the reasons for the threat of water wars never

---

[19] See, e.g., Hong Yang and Alexander Zehnder, '"Virtual water": An unfolding concept in integrated water resources management' (2007) 43 *Water Resources Research* 1, 1–2 (according to whom the virtual water debate followed three paths: 'water resources availability and food trade relations', 'virtual water flows and water use efficiency' and 'the role of virtual water in conflict mitigation and national and regional scarcity managements'), and Arjen Y. Hoekstra, 'Virtual water: An Introduction' in Arjen Y. Hoekstra, *Virtual water trade: Proceedings of the International Expert Meeting on Virtual Water Trade*, Value of Water Research Report Series No. 12 (2003) 14 (who identifies a twofold practical value: 'Virtual water trade as an instrument to achieve water security and efficient water use' and 'Water footprints: making the link between consumption patterns and the impacts on water').

materializing).[20] Therefore, with the same stone a water-poor country kills two birds: it attains a *political* result by avoiding conflict with the states with which it shares its meager water resources, and ensures an *environmental* success by preserving the quantity and, especially, the quality of its waters.[21]

When hydrologists and economists embarked on the virtual water project, they introduced new concepts and ideas that mixed up the above scheme – and thus the potential objectives descending from the virtual water discourse increased. For example, it was highlighted that the water footprint of an agricultural good is not always the same, irrespective of the place or method of production. On the contrary, the water needed to grow food depends on, e.g., technological and climatic factors, meaning that countries with advanced technology and, even more significantly, favourable climatic conditions will require a lesser amount of water to produce the same good.[22] Moreover, the notions of green and blue water were also brought into the debate, the former referring to water 'embedded' in soil thanks to the rain, and the latter indicating water withdrawn from rivers and lakes. These two kinds of water may be virtually present in different proportions in the same good, once again for reasons linked to the place of production: as a consequence, a ton of food grown in a country with given natural and climatic conditions will make greater use of green water than blue water, while the same amount of food, if produced elsewhere, will show the opposite ratio.[23]

This additional information may help us figure out a couple of additional

---

[20] John Anthony Allan, 'Virtual Water – the Water, Food, and Trade Nexus. Useful Concept or Misleading Metaphor?' (2003) 28 *Water International* 107.

[21] It has been pointed out (see, e.g., Stephen Merrett, 'Virtual Water and Occam's Razor' (2003) 28 *Water International* 103, 104) that to speak of water consumption for agricultural purposes is wrong, since the water cycle is close, that is, the water used to grow crops is not lost but returns to the aquifer or comes back as rain after the plants have released it as vapour. For this reason, one could say that the environmental advantages arising from the abandonment of agriculture are all due to its polluting nature, not by the (non-)sparing of water. Nonetheless, the over-exploitation of water could also affect its quantity, if it is abstracted at an unsustainable rate (this is particularly so in the case of underground water basins, which replenish at a very slow pace and on which many farmers rely).

[22] See, e.g., the case of cotton with different water footprints because grown in different places: Ashok K. Chapagain, Arjen Y. Hoekstra, Hubert H.G. Savenjie and Rajani Gautam, 'The Water Footprint of Cotton Consumption: An Assessment of the Impact of Worldwide Consumption of Cotton Products on the Water Resources in the Cotton Producing Countries' (2005) 60 *Ecological Economics* 186, 190.

[23] Ibid.

purposes to be attached to the virtual water discourse – purposes that, in fact, have been considered with some interest by the 'father' of virtual water himself.[24] Rather than moving the production of water-intensive goods (a) from water-scarce states to water-rich ones, as originally envisaged, we could transfer it either (b) from a place where these goods need a lot of water to another where a smaller amount suffices,[25] or, alternatively, (c) from a place where they consume blue water in a greater proportion to another where green water is predominantly used (indeed, since blue water – differently from the green one – can be had recourse to for purposes that are not agricultural in nature, it has a greater opportunity cost).

I would like to comment on these different purposes in three respects. First, it is useful to note that they are not necessarily consistent with one another. For example, bringing the production of a certain good to a place where the proportion between green and blue water is more advantageous might correspond to bringing it to a place where more water is needed to grow that very good: in other words, blue water is spared at the expense of a greater amount of green water. Or, else, a good might be produced in a country in which it demands less water but that, at the same time, relies on smaller water endowments. Many such combinations are possible, and no optimal allocation can be established in principle due to the heterogeneous nature of the three purposes: something that can be explained through the second and third comments.

The former regards which types of goals are pursued in the three scenarios. Apparently, purposes (a) and (b) share an *environmental* objective, in that they both aim at consuming or polluting less water, whereas purpose (c) is meant to reach an *economic* target, as it privileges, *ceteris paribus*, the consumption of green water over blue water – that is, of the water with a lesser opportunity cost – irrespective of the environmental consequences.[26] Things, however, might not be that simple. One reason is that the desired

---

[24] Maite M. Aldaya, John Anthony Allan and Arjen Y. Hoekstra, 'Strategic Importance of Green Water in International Crop Trade' (2010) 69 *Ecological Economics* 887, 892.

[25] This is what actually happens at the global level, as international trade reduces water use in agriculture by 5 per cent with respect to the amount that would be needed if each country produced the food it consumes: see Arjen H. Hoekstra, 'The relation between international trade and water resources management' in Kevin P. Gallagher (ed.), *Handbook on Trade and the Environment* (Edward Elgar Publishing 2008), 122.

[26] Although Allan notes that relying on green water is more nature-friendly than (over-)using blue water (John Anthony Allan, 'Can improving returns to food–water in Africa meet African food needs and the needs of other consumers?' in Tony Allan *et al.*, *Handbook of Land and Water Grabs in Africa: Foreign Direct*

effect and the actual one do not necessarily overlap. If agricultural production is outsourced to a place where it is less water-demanding, but this place is also water-scarce, the final outcome might be, as a matter of fact, a zero-sum game or even environmentally detrimental. In addition, measures like (a) and (b) could be taken in conjunction with other, non-environment-friendly decisions – and, actually, it is likely to be so, given that a country renouncing the production of food will have to import it, and to this end it will probably use the spared resources to expand its industry and get the money it needs. In such a case, it is doubtful whether that state is acting in pursuit of an environmental purpose.[27]

My third comment concerns the scope of the three purposes suggested above. Purpose (a) is local in character, as it only purports to conserve a resource that is scarce at the *domestic* or, at most, the *regional* level: here, the virtual water policy addresses a problem that is of a single state, or of the group of states with which it shares its waters.[28] Both (b) and (c), however, look at the production of water-intensive goods from a *global* perspective, and their focus is not on the country that dismisses its agricultural production but, rather, on the one that takes charge of it – which could be anywhere in the world. In these cases, the objective is the most rational use of the planet's water resources and the most efficient allocation of food production at the universal level.[29]

Now that we have discovered the multi-faceted nature of virtual water, we can turn to the ways by which the law can tame it.

Investment and Food and Water Security (Routledge 2013) 5), this assessment should be made on a case-by-case basis.

[27] More precisely: in theory, a strategy like (a) can easily be described as environmental; however, it would not meet the requirements of rules that, like art. XX(g) of the General Agreement in Tariffs and Trade, consider the restriction to all domestic production based on an exhaustible natural resource as a necessary proof of the environmental character of a measure aimed at preserving that resource.

[28] As it is made clear by the title of John Anthony Allan, 'Virtual Water: A Strategic Resource – Global Solutions to Regional Deficits' (1998) 36 *Ground Water* 545.

[29] Limited exceptions exist as to the global reach of goals (b) and (c), since some vast countries encompassing different climatic and natural conditions (such as China and the US) could easily apply these strategies all within their borders.

3. BATHING THE LAW IN VIRTUAL WATERS

Some years ago, British water engineers made public their recommendations for attaining global water security. The first was about bringing the notions of virtual water and water footprint at international fora like those dealing with trade law and climate change law.[30] Curiously, this actually happened, as the only two references to virtual water I am aware of that have been made in official or informal documents of international organizations concern the relationship of water scarcity and international trade,[31] on the one hand, and climate change adaptation strategies, on the other.[32]

The bulk of academic literature that set forth policy suggestions, however, focused on the former topic only. This is hardly surprising. Given that the notion of virtual water revolves around the idea of buying and selling food on the international market, the relevance of the legal regime governing international trade relations cannot be overstated. Unfortunately, these pages do not provide enough room to conduct an in-depth analysis of how international trade law interacts with virtual water.[33] Thus, I will sum up here some of the legal solutions that have been proposed, discussing them in relation to two issues: the way agricultural production can be forced out of a country, and the way it can be transferred to another and fitter one.

With respect to the former aspect, in the writings on virtual water the belief that water is under-priced is almost a trope. Accordingly, an increase in the price of water is often suggested. Faced with a higher cost of their primary factor of production – it is said – many farmers will turn to other activities (hopefully less water-intensive ones). However, this cannot be taken for granted. Indeed, it is possible that only a fraction of the agricultural sector will be dismissed, while some farmers will convert to high-value crops: in the case these were more water-intensive, there could

---

[30] The Royal Academy of Engineering, 'Global Water Security – An Engineering Perspective' (2010) 8, 35.
[31] Arjen Y. Hoekstra, 'The relation between international trade and freshwater scarcity', World Trade Organization Staff Working Paper ERSD-2010-05 (2010).
[32] United Nations Economic Commission for Europe, *Guidance on Water and Adaptation to Climate Change* (2009) 92.
[33] Thorough accounts are in Edith Brown Weiss and Lydia Slobodian, 'Virtual Water, Water Scarcity, and International Trade Law' (2014) 17 *Journal of International Economic Law* 717, and Marco Pertile and Paolo Turrini, 'Virtual Water and International Law' (forthcoming in Brill Research Perspectives – International Water Law), Ch. 4.

be a net increase in water consumption rather than a drop[34] (and should they be export-oriented, the overall loss could be even larger[35]). Moreover, pricing schemes are numerous and not all equally effective. For instance, per-unit area fees can be good – and perhaps the only viable solution – in the case of green water (that is, rain), but do not lead to a decrease in the demand for water as far as blue water is concerned.[36] Such a scheme, however, is easier to administer, and the associated monitoring and enforcement costs are very low.[37] This is just an example, but an important and general lesson can be drawn out of it. States are complex human societies that do not always respond straightly and smoothly to changes in legal policies, and rarely do so without exacting a trade-off of some sort. Any ambitious social experiment like the one envisioned by the proponents of virtual water must take this into account. The concept has been summarized admirably by two scholars who are quite sceptical about the whole virtual water argument:

> The assumption is that populations can be unproblematically engineered into new agricultural practices (or be moved out of agriculture), relations to the market, and property rights regimes to land and water resources. However, in practice, much depends on the complex and often plural character of social and legal institutions and frameworks (law, property rights) that co-determine the (legitimate) access to natural resources in interaction with human behaviour.[38]

In practice, this means that the 'fetish' of water price could, in some or even many cases, be less effective than expected, and conducive to disappointing results. It might be necessary to adjust the legal strategy by resorting to a richer toolbox (e.g., water quotas, crop type limitations and quotas, modifications in the property rights regime, etc.), and be ready to change

---

[34] 'When high-value crops are also more water-intensive, higher prices may cause an increase in total demand for water': François Molle and Jeremy Berkoff, 'Water Pricing in Irrigation: Mapping the Debate in the Light of Experience' in François Molle and Jeremy Berkoff (eds), *Irrigation Water Pricing: The Gap Between Theory and Practice* (CABI 2007) 53.

[35] Please note that export quotas are generally prohibited in the World Trade Organization: see art. XI of the General Agreement on Tariffs and Trade.

[36] Yacov Tsur and Ariel Dinar, *Efficiency and Equity Considerations in Pricing and Allocating Irrigation Water*, World Bank Policy Research Working Paper No. 1460 (1995) 33–34.

[37] Ibid.

[38] See Dik Roth and Jeroen Warner, 'Food Security as Water Security: The Multilevel Governance of Virtual Water' in Otto Hospes and Irene Hadiprayitno (eds.), *Governing Food Security. Law, Politics and the Right to Food* (Wageningen Academic Publishers 2010) 313 (quotation) and, more generally, 309–14.

the course in the case the resistance of all those that have vested rights and interests in the *status quo* risked nullifying the success of the plan.[39]

When we turn to the possibility for a state to attract the food production that was outsourced by another one, the price of water is still there. Whichever the objective pursued, attributing a price to water can, at least in theory, help reach it. Indeed, if all countries raise this price so that it mirrors the scarcity of water at the domestic level, then the resource will be priced differently depending on how much water a country possess. As a consequence, water-intensive activities such as agriculture will tend to move towards water-rich countries, where the resource will – presumably – cost less (purpose (a)). At the same time, these activities will settle where water has a greater productivity, that is, where the same good can be produced with a lesser amount of water (purpose (b)). A third trend exists, that sees water-intensive activities going to states where agriculture is practised mostly by means of green water, which is seldom priced (purpose (c)). The price of water acts as a driver towards all these goals, and it should allocate food production in those places which are situated at the crossroads of the three trajectories, that is, the countries that simultaneously fit all conditions best (again, in an over-simplified theory). The only requisite is that the due economic value of water be recognized everywhere in the world: to this end, it has also been proposed that an international treaty on water pricing be stipulated.[40]

However, one might wonder whether it is possible to direct food production according to just one of our three purposes, or even other criteria.[41] If price is not a proper device, what can be done, for example, to relocate the production in water-rich countries only? Given that the idea of all water-scarce states dismantling their own agricultural sector is, for the reasons sketched above, highly implausible, encouraging food production in specific states by means of subsidies appears, *prima facie*, the most effective solution. Indeed, this has actually occurred, although

---

[39] For instance, it has been suggested that the rise in the price of water be substituted by a system facilitating the selling of water by farms to cities that are willing to pay it a substantial price: Dennis Wichelns, 'The Virtual Water Metaphor Enhances Policy Discussions Regarding Scarce Resources' (2005) 30 *Water International* 428, 435.

[40] Hoekstra *supra* footnote 31, 17–18.

[41] In fact, it cannot be taken for granted that the optimal allocation in an economic perspective is also the optimal one from the point of view of environmental sustainability. Moreover, country-specific legal and political situations (e.g., a conflict with neighbouring states due to contested rights to shared waters) may advise against the allocation of water-intensive activities on the basis of general principles or supposedly neutral means.

in an unplanned way, when water-poor countries like Egypt became net importers of virtual water thanks to the subsidies of their food providers, like the EU.[42] However, in international trade law, the regime governing agricultural subsidies is intricate to say the least: suffice to say that they are regulated by two long and very composite agreements establishing partly-conflicting legal frameworks whose relationship is still unclear.[43] In general, export subsidies – that is, subsidies that are contingent on export performances – should be reduced,[44] and the only ones that, under certain circumstances, are deemed as acceptable are those that do not distort trade flows:[45] a scenario that is not propitious to virtual water policies, which aim at warping current trade patterns by promoting food exports.[46]

Interestingly, these legal constraints perhaps make subsidies more apt to support the downsizing of a country's agricultural sector,[47] i.e., the reduction or abandonment of water-intensive activities where they are unsustainable, rather than their concentration where they are environmentally sound. Or, again, subsidies might be used to trigger changes in a country's domestic agricultural sector that do not distort international trade and are designed to convert the production to less water-intensive goods. This is one further purpose of virtual water: drawing public attention to the different water footprints of different agricultural products, and pushing for opting for those of them with a lesser virtual water content. What is likely to be the best instrument to attain this goal is food labelling, by means of which consumers may send a strong signal to producers by showing a preference for less water-intensive products.[48]

---

[42] Tony Allan, *Virtual Water: Tackling the Threat to Our Planet's Most Precious Resource* (I.B. Tauris 2011) 50.

[43] See Lorand Bartels, 'The Relationship between the WTO Agreement on Agriculture and the SCM Agreement: An Analysis of Hierarchy Rules in the WTO Legal System' (2016) 50 *Journal of World Trade* 7.

[44] Arts 8–10 of the Agreement on Agriculture.

[45] Ibid., Annex 2.

[46] Perhaps this is the reason why, still in 2010, Allan was convinced that the role of 'international public agents such as the WTO [was] not yet significant': John Anthony Allan, 'The role of those who produce food and trade it in using and "trading" embedded water: What are the impacts and who benefits?' in Arjen Y. Hoekstra, Maite M. Aldaya and Bernard Avril (eds), *Proceedings of the ESF Strategic Workshop on Accounting for water scarcity and pollution in the rules of international trade*, Value of Water Research Report Series No. 54 (2011) 43.

[47] See, e.g., Agreement on Agriculture, Annex 2, paras 9–11. These subsidies, too, should not distort trade, but since foreign production would be favoured, hardly any state would complain!

[48] On this, see Piotr Szwedo, 'Water Footprint and the Law of WTO' (2013) 47 *Journal of World Trade* 1259; Laura Manson and Tracey Epps, 'Water

The allocation of food production on the basis of water security reasons only is, evidently, both simplistic and unrealistic. A plethora of considerations should be made to render facts and prospects better on our canvas. For example, it is quite obvious that states can only afford to dismiss (part of) their agricultural sector if they are able to buy food abroad – something that the loss of income from agricultural activities makes even more difficult. Therefore, a swift restructuring of the economy towards new industrial activities would be necessary. We may wonder whether financial institutions like the World Bank – which are strangely neglected by the virtual water literature – could provide assistance in such a transformation of a country's economy. So far, their role in water security has been ambiguous: the funding of engineering projects such as dams must be read in conjunction with the imposition of a market economy that has not always gone in the direction of environmental sustainability. However, the 'necessarily planned virtual water markets'[49] cannot be implemented by the night-watchman state that financial institutions endorse.

But the most visible trade-off of a virtual water policy consists in getting water security at the expense of food security. The latter means not only being able to purchase enough food, but also to purchase it from many sources, since – as in investment – diversification is key to reducing risks (in particular in a world where 'nine countries account for 90 % of the world's wheat exports'[50] and a few companies dominate the international food market[51]). Such a strategy is likely to be disrupted by an allocation of food production primarily based on environmental concerns. This is even more so if we transfer production – as suggested – to countries who rely heavily on green water, in light of its lower productivity that entails higher prices for the agricultural output.[52] Although provisions addressing the issue of food security are disseminated across both the General

---

Footprint Labelling and WTO Rules' (2014) 23 *Review of European Community and International Environmental Law* 329.

[49] Eva Youkhana and Wolfram Laube, *Cultural, Socio-Economic and Political Constraints for Virtual Water Trade: Perspectives from the Volta Basin, West Africa*, Zentrum für Entwicklungsforschung Working Paper No. 13 (2006) 11.

[50] Gulbenkian Think Thank on Water and the Future of Humanity, *Water and the Future of Humanity: Revisiting Water Security* (Calouste Gulbenkian Foundation/Springer 2014) 136.

[51] Allan, *supra* footnote 46 and Suvi Sojamo, Martin Keulertz, Jeroen Warner and John Anthony Allan, 'Virtual Water Hegemony: The Role of Agribusiness in Global Water Governance' (2012) 37 *Water International* 169.

[52] Food and Agricultural Organization, *New Dimensions in Water Security: Water, Society and Ecosystem Services in the 21st Century*, Doc. No. AGL/MISC/25/2000 (2000) 7.

Agreement on Tariffs and Trade and the Agreement on Agriculture, it is a widespread conviction that the World Trade Organization does not offer a strong enough guarantee against famines due to a raise in prices. For this reason, the literature on water security has gone as far as to say that 'a country must also have guaranteed access to exportable quantities of food',[53] and to propose the establishment of an OFEC (Organization of Food Exporting Countries) so as to ensure that the global distribution of food will not be used as a political weapon.[54]

As we can see, law (and especially international law and its 'daughters', international organizations) can play a fundamental role in supporting policies following from the basic idea of virtual water, in many direct and indirect ways. What one should ask, however, is whether the virtual water 'theory' is the best solution to the problems it tries to solve.

4. THE VIEW FROM THE OTHER BANK OF THE RIVER

Originally born as a global answer to a local problem, the virtual water idea has been grafted with new concepts that have widened its scope of application. The quest for a better use of water at the global level has thus become one of the possible objectives of a policy inspired by the discourse on virtual water. The means by which this (mainly environmental) goal is to be attained is an economic one, as the allocation of food productions is organized through a regulated market: prices, customs duties, subsidies, property rights and the likes. But a question that should be of interest to both lawyers and policy-makers: are there any other promising legal strategies that may be useful in tackling the problem of global water security?

Generally speaking, international law has had not an astonishing record in addressing environmental affairs at world level. Only recently has the idea of commonality been affirmed in scholarship,[55] although its contours are still object of speculation, and just few years ago the planet's water crisis has been conceptualized by doctrine as a common

---

[53] Ibid. 11.
[54] See Herman Bower, 'Integrated Water Management: Emerging Issues and Challenges' (2000) 45 *Agricultural Water Management* 217, 226.
[55] For an overview, see Jutta Brunnée, 'Common Areas, Common Heritage and Common Concern' in Daniel Bodanski, Jutta Brunnée and Hellen Hey (eds), *Oxford Handbook of International Environmental Law* (Oxford University Press 2007).

concern of humankind.⁵⁶ States' duties in environmental as well as water security matters have been pushed by academics beyond states' borders by concocting bold legal strategies, such as the attribution of an extra-territorial reach to the obligations connected with the right to water,⁵⁷ or the linking of the notion of common concern to the idea of responsibility to protect.⁵⁸ This manifestation of legal inventiveness is commendable, but it is unlikely to engender concrete results in the near future. In this respect, therefore, we are tempted to say that the intuition behind virtual water proved to be insightful, as water security has been put in the tow of market forces. However, we have seen that the need for strong regulation at both the domestic level (in order to convince the society to move away from agriculture) and the international one (so as to guide food production towards the most suitable place) was underestimated. Reaching a consensus in the community of states on the necessity to price water higher, or amending the international trade regime so that agricultural subsidies are treated differently on the basis of environmental concerns, is the sole way to unleash the potential of virtual water – and at the same time this plan has no more chances of succeeding than the one that labels water as a common concern. After all, when some virtual water proponents say that 'a more water friendly international trade regime with equal access to global markets' must take into account 'both water productivity and blue/green water ratio in products',⁵⁹ they are ultimately treating water as a universal concern, something whose defence should inform the egotistical trade system irrespective of particularistic interests.

But states are selfish: they are more worried by what is 'their own' than what is 'common'. They tend to consider water issues in the light of the notion of sovereignty:

> Concerns about fresh water sovereignty typically arise in four contexts: the purchase of rural land with significant water resources attached, the purchase of water detached from land in areas [(such as Australia)] in which this is

---

⁵⁶ Edith Brown Weiss, 'The Coming Water Crisis: A Common Concern of Humankind' (2012) 1 *Transnational Environmental Law* 153.

⁵⁷ Takele Soboka Bulto, *The Extraterritorial Application of the Human Right to Water in Africa* (Cambridge University Press 2013).

⁵⁸ Krista Nakavukaren Schefer and Thomas Cottier, 'Responsibility to Protect (R2P) and the Emerging Principle of Common Concern' in Peter Hilpold (ed.), *Die Schutzverantwortung (R2P): Ein Paradigmenwechsel in der Entwicklung des Internationalen Rechts?* (Martinus Nijhoff Publishers 2013).

⁵⁹ Aldaya, Allan and Hoekstra, *supra* footnote 24, 892.

possible, the actual purchase of water for bulk export and the 'virtual export' of water through trade in commodities when water is itself a scarce commodity.[60]

The commodification of water reveals the proprietary logic with which this resource is dealt: who lacks it, purchases it from who has it in abundance; however, it is up to the latter to take charge of the environmental consequences of the sale. In the eyes of Arjen Y Hoekstra – one of the greatest contributors to the advancement of the virtual water discourse – this is part of the reason why a global perspective is needed: the growing number of transboundary water transfer projects and the growing importance of multinationals in the water sector, together with the joint effects of climate change and the global economy on local water resources, are phenomena that cannot be properly addressed at a national or regional level.[61] Though he may be right in principle, this does not make his proposal more feasible in practice.

However, there are those who disagree. Susanne Neubert maintains that, since water scarcity manifests itself only at the basin level, it is there that the solution must be brought.[62] Even if, apparently, this reasoning fails to grab the preventive character of the virtual water strategy (which claims that the worsening of water scarcity can be forestalled), the advantages and viability of the basin approach should be compared with those of the virtual water solution. Since, of course, this analysis cannot be conducted here, my words must be taken as a humble suggestion to academics to seriously walk this path. In the last lines of this chapter I will limit myself to just a couple of considerations in this regard.

We have seen above that the virtual water argument has two main purposes: an environmental one (the saving of water by non-use at local level) and, through this, a political one (the thwarting of regional conflicts that are due to, or exacerbated by, over-consumption of water).[63] How do water agreements stipulated at basin level perform with respect to these two goals? As to the latter, an interesting thesis has

---

[60] Francine Rochford, 'Water Sovereignty and Food Security' in Quentin Farmar-Bowers, Vaughan Higgins and Joanne Millar (eds), *Food Security in Australia: Challenges and Prospects for the Future* (Springer 2013) 241.

[61] Arjen Y. Hoekstra, 'The Global Dimension of Water Governance: Why the River Basin Approach Is No Longer Sufficient and Why Cooperative Action at Global Level Is Needed' (2011) 3 *Water* 21.

[62] Susanne Neubert, 'Strategic Virtual Water Trade – A Critical Analysis of the Debate' in Waltina Scheumann, Susanne Neubert and Martin Kipping (eds.), *Water Politics and Development Cooperation: Local Power Plays and Global Governance* (Springer 2008) 134.

[63] See *supra* footnotes 20 and 21 and related text.

been put forth. While it is true that hydro-hegemonic tendencies from the part of powerful countries often end up forging unequal treaties among riparians,[64] according to Ahmed Abukhater 'the perception of a particular treaty as being equitable and fair is mainly shaped by the negotiation process used to reach certain outcomes, rather than being determined mechanistically by the quantitative allocation of water to each party'.[65] In this sense, procedural fairness might be more important than substantial equity in defusing regional tensions. This insight sheds a new, perhaps more benevolent light on international law, which, given the objective difficulty of providing rules on water allocation other than open-ended ones, has more than once focused on the obligation to solve environmental and water disputes by good-faith negotiations:[66] from *Lake Lanoux* (1957) to *Pulp Mills* (2010), without forgetting *Gabčíkovo-Nagymaros* (1997).[67]

The vague water-sharing criteria are, nonetheless, relevant as far as the environmental objective is taken into consideration, as they determine the quantity of water a country can rely on and, as a consequence, the activities that can be carried on with it. But the process is circular: in turn, the activities of a state contribute to determine the quantity of water to which it is entitled.[68] Looking at this fact with virtual water in mind, a non-negligible problem arises. The dismissal of agricultural production in compliance with the virtual water argument risks compromising the position of a state vis-à-vis its co-riparians, since 'the law provides very little protection of downstream states who want to preserve some portion of the natural flow for either environmental reasons or future development':[69] a country that compresses its own

---

[64] See *supra* footnote 8.

[65] Ahmed Abukhater, *Water as a Catalyst for Peace: Transboundary Water Management and Conflict Resolution* (Routledge 2013) xv.

[66] In fact, 'equitable utilisation essentially involves a process of balancing the interests of states, rather than defining an outcome in advance': Tim Stephens, 'International Courts and Sustainable Development: Using Old Tools to Shape a New Discourse' in Brad Jessup and Kim Rubenstein (eds), *Environmental Discourses in Public and International Law* (Cambridge University Press 2012) 204–05.

[67] See, in chronological order, *Lake Lanoux Arbitration (France v. Spain)* (1957), 12 R.I.A.A. 281, 24 I.L.R. 101, *passim* (especially para. 22); *Gabčíkovo-Nagymaros Project (Hungary v. Slovakia)*, Judgment, I.C.J. Reports 1997 7, paras 109 ff., 142 ff.; *Pulp Mills on the River Uruguay (Argentina v. Uruguay)*, Judgment, I.C.J. Reports 2010 14, para. 145.

[68] See art. 6(b) of the 1997 Convention on the Law of the Non-navigational Uses of International Watercourses.

[69] Tarlock *supra* foootnote 9, 722.

water needs and, thus, its water entitlements is a country that cannot be sure whether it will be able to reclaim them in the future or not. Water basin treaties must be able to provide an answer to such problems, if they want to represent that legal sanctum that goes under the high-sounding names of 'community of interests'[70] or 'regional common concern'.[71]

My second and conclusive comment regards the extent to which the merging of the virtual water and the water basin approaches is possible. For many years now, experts have been advocating the so-called integrated water resources management as the only method capable of supporting a holistic understanding of the problems affecting a water basin. This means, on the one hand, that the whole basin (together with its hydrological cycle), rather than the mere river, is taken into account in the management of the shared water source. On the other hand, hydrological data are considered not in isolation, but jointly with more far-reaching environmental and also economic information. This should help incorporate virtual water-related considerations into the management of transboundary basins. For instance, Patricia Wouters wrote that

> [t]he rule of equitable and reasonable utilization invites the convergence of interdisciplinary water expertise. Premised on a review of all relevant factors, this mechanism readily facilitates consideration of more than just the blue water found in rivers and aquifers and includes *not only green water, but also ecosystemic and even virtual water.*[72]

A concept that further expands the idea of Malin Falkenmark, who argues that what has to be shared between upstream and downstream states is not only the water of the river, but also the rainfall (that is, green water) over the river basin.[73] Moreover, the planning of water and land use at basin level could benefit from extending water accounting

---

[70] Owen McIntyre, *Environmental Protection of International Watercourses under International Law* (Ashgate 2007) 28 ff.

[71] Bjørn-Oliver Magsig, 'Overcoming State-Centrism in International Water Law: "Regional Common Concern" as the Normative Foundation of Water Security' (2011) 3 *Goettingen Journal of International Law* 317.

[72] Patricia Wouters, 'The Relevance and Role of Water Law in the Sustainable Development of Freshwater: From "Hydrosovereignty" to "Hydrosolidarity"' (2000) 25 *Water International* 202, 204 (emphasis added).

[73] Malin Falkenmark, 'The Greatest Water Problem: The Inability to Link Environmental Security, Water Security and Food Security' (2001) 17 *International Journal of Water Resources Development* 539, 546.

to the water footprint methodology or other similar water accounting methods,[74] also because the calculation of the virtual water content of goods is highly sensitive to local climatic and management conditions, and doing it at a smaller scale would provide more meaningful information to policy-makers.[75]

Surely there are other potentially meaningful ways to incorporate virtual water in existing concepts of (international) law and policy, such as, e.g., the ever-developing right to water.[76] It is up to scholars to devise them and enrich a debate that, despite the fame acquired by the idea of virtual water, so far has remained impervious to legal disciplines.

5. CONCLUSIONS

It is more than 20 years since virtual water was onstage. The idea has grown stronger and stronger and, apparently, its momentum has not yet ceased. However, it is still unclear whether it can be had recourse to as a practicable policy option or if, on the contrary, it is more of a descriptive character. Certainly, the massive numeric knowledge that coagulated around this notion is a welcome progress of our comprehension of water security issues, and it will be crucial to factor it into the theoretical framework. But the impression is that this framework is still under-developed if compared to the hydrological and economic aspects of the matter. In this respect, in my opinion there are two urgent questions to be addressed. First, we should wonder whether the attention to the potential benefits of virtual water policies have eclipsed other solutions, less ambitious on paper but more feasible in practice. Secondly, the water–food nexus that lies at the basis of virtual water should be expanded both in width,

---

[74] Elena López-Gunn *et al.*, 'Rethinking Integrated Water Resources Management: Towards Water and Food Security Through Adaptive Management' in Bárbara A. Willaarts, Alberto Garrido and M. Ramón Llamas (eds), *Water for Food Security and Well-Being in Latin America and the Caribbean: Social and Environmental Implications for a Globalized Economy* (Routledge 2014) 393.

[75] Emily Kate Schendel *et al.*, 'Virtual Water: A Framework for Comparative Regional Resource Assessment' (2007) 9 *Journal of Environmental Assessment Policy and Management* 341.

[76] This was proposed, even though in vague terms, by Dinara Ziganshina, 'Rethinking the Concept of the Human Right to Water' (2008) 6 *Santa Clara Journal of International Law* 113, 124–25.

by absorbing other elements (such as energy),[77] and in depth, by taking connections (e.g., constraints of means as well as trade-offs among results) more seriously.[78] After all, the virtual water 'theory' entails a *vaste programme*, one that cannot be pursued without being aware of its real premises and consequences.

---

[77] See, e.g., Bassel T. Daher and Rabi H. Mohtar, 'Water–energy–food (WEF) Nexus Tool 2.0: Guiding Integrative Resource Planning and Decision-making' (2015) 40 *Water International* 748.
[78] Radoslav S. Dimitrov, 'Water, Conflict, and Security: A Conceptual Minefield' (2002) 15 *Society & Natural Resources* 677.

5. Foreign investment in water: privatization, globalization and the law

Julien Chaisse

INTRODUCTION

Water is often perceived as an infinite natural resource. Unfortunately, the tragic reality is it maintains a fixed invariable volume with less than 3 percent consisting of fresh water.[1] Never before has the inherent tension of scarcity versus overconsumption in the relationship between man and water reached such a critical point.[2] Accordingly, Earth's water supply faces many newfound demands and challenges in the years ahead.[3]

Water, Earth's "blue gold," is its most precious and essential commodity. It is fundamental to all aspects of drinking, eating, maintaining hygiene, and promoting population health. Water is basic to the preservation of most ecosystems and crucial to a safe and long-lasting environment. Moreover, it is critical to several types of businesses and industries.[4] It not only maintains social stability and environmental sustainability, but also fosters economic development across civilizations.[5] Consequently, access to clean water has been recognized by the United

---

[1] Amy K Miller, 'Blue Rush: Is an International Privatization Agreement a Viable Solution for Developing Countries in the Face of an Impending World Water Crisis' (2005) 16 IND INTL' & COMP L REV 217, 223.
[2] Ibid.
[3] *See generally* Edith Brown Weiss and Lydia Slobodian, 'Virtual Water, Water Scarcity, and International Trade Law' (2014) 17 J INT'l ECON L 717 (discussing the complicated connection between fresh water crisis and international trade law; analyzing in particular the impacts of trade negotiations on the problems of water scarcity).
[4] See ibid.
[5] Ibid., 721.

Nations (UN) as a basic human right that every government is obligated to provide.[6]

The world of water services changed significantly in the late 1990s due to an extraordinary boom in global population growth.[7] The sustained population increase sparked a need for water services' expansion.[8] Opportunities for investment in water services and sanitation infrastructure attracted tremendous support from a myriad of international financial institutions.[9] These institutions unlocked a host of new business opportunities for the water services and sanitation industry to address traditional problems ranging from fresh water scarcity to inadequate investment in sanitation infrastructure to the inability of many public authorities to meet coverage needs.[10] Coverage and accessibility, even at the most elementary level, necessitate functional water services and sanitation facilities.[11]

The inability of public authorities to provide coverage to their citizens prompted a rise in water-services privatization contracts between foreign investors and states, such that 10 percent of global consumers now receive their water from private companies.[12] Today, a growing number of businesses are engaging with the water services industry.[13] It is estimated that by 2025, annual spending on water infrastructure in OECD countries

---

[6] Sharmila L Murthy, 'The Human Right(s) to Water and Sanitation: History, Meaning, and the Controversy over Privatization' (2013) 31 BERKELEY J INT'L L 89.

[7] *See* United Nations, Department Of Economic and Social Affairs, Population division, *World Population Prospect: The 2006 Revision, Highlights,* (Working Papers ESA//P/WP 2007) 202 (explaining how in just over the past 50 years, the population on the planet has grown from 2.5 billion to over 6.5 billion and that it is expected to reach 9.2 billion people by 2050); Miller, n. 1 above, 217. See Rodrigez and others, 'Investing in Water Infrastructure: capital, operation and maintenance' *Water Paper* (The World Bank 2012) Figure 1 <http://water.worldbank.org/sites/water.worldbank.org/files/publication/water-investing-water-infrastructure-capital-operations-maintenance.pdf> (depicting the increased investment in water projects in developing countries starting in 1990).

[8] Miller, n. 1 above, 217.

[9] E.g., <www.worldbank.org/en/topic/water/projects> (providing examples of water projects and programs from the World Bank); see <www.worldbank.org/en/topic/water/projects> Part IV (discussing the influence of International Agencies to privatize developing countries water services).

[10] E.g., <www.worldbank.org/en/topic/water/projects/all>

[11] Jose Esteban Castro and Leo Heller, WATER AND SANITATION SERVICES: PUBLIC POLICY AND MANAGEMENT (Routledge 2009) 1.

[12] Murthy, n. 6 above. 90, 125 of the world's 400 largest cities are served by public sector. *See also* Miller, n. 1 above.

[13] *See* Miller, n. 1 above.

will exceed $1 trillion.[14] New technologies and the need for additional infrastructure investment will certainly increase demand in the market, potentially spawning billion dollar valuations. Such economic promise and opportunity largely explains why water has earned the moniker of Earth's "blue gold."

No international regime or regulatory body responsible for sanitation and water services exists. The growing execution of international investment treaties in the water sector, however, has resulted in the slow emergence of an international economic form of governance for cross-border sanitation and water services. This chapter explores the increased role of investment treaties in the context of disputes and investment arbitration and provides the first exhaustive analysis of this burgeoning water regime.

In Section A, this chapter provides the current international legal framework for foreign direct investment and how its wide variety of national and international rules and principles have inadvertently formed an emerging system of regulatory governance over the international water services sector.[15] In Section B, the chapter provides a definition for the term "investment" as it pertains to international investment treaties. In Section C, the chapter discusses the ever-evolving character of international investment treaties and how the expansion of cross-border investment in the sanitation and water services sector has amounted to a proliferation of "blue gold" disputes. Section D, discusses the privatization and concession contract related arbitration. Section E, delves further into the clean water distribution system, while Section F further clarifies the wastewater and sewerage services regime.

A. A RECALIBRATION OF CUSTOMARY INTERNATIONAL LAW TO INTERNATIONAL INVESTMENT TREATIES

Targeted investment host countries develop a tier one level of international investment regulation by promulgating national rules to ensure the realization of benefits of Foreign Direct Investment (FDI) and avoid

---

[14] Rodrigez and others, n, 7 above, 7–8, For developing countries alone, $103 billion per year was estimated to finance water, sanitation, and wastewater treatment. Rodrigez and others, 8,

[15] This chapter partially draws from the long Article by Julien Chaisse and Marine Polo 'Globalization of Water Privatization – Ramifications of Investor-State Disputes in the "Blue Gold" Economy' (2015) 38 BOSTON COLLEGE INTERNATIONAL & COMPARATIVE LAW REVIEW 1.

costs.[16] Due to pressure from competing countries, host states must generally combine economic openness with incentives for investors to attract FDIs.[17] Such incentives are comparable to "free zones," in which duty-free imports and exports are permitted along with direct subsidies and other financial incentives, and foreign investment guarantees such as, promises to stabilize domestic laws, provide tax relief, allow currency conversion, and the repatriation of sale proceeds and profits.[18]

On a secondary yet complementary tier, international investment law provides rules to ensure access for foreign investment to host country markets and to protect investments against risks, particularly political risks.[19] It creates a specific set of investment protection obligations on host countries, including protection against expropriation without compensation.[20] International investment law also provides access to financial compensation through investor–state arbitration if the host country breaches a protection obligation.[21]

International investment agreements (IIAs) serve as a source of international investment law.[22] IIAs include preferential trade and investment agreements (PTIAs) and bilateral investment treaties (BITs), both of which provide more comprehensive rules on investment.[23] Additionally, IIAs provide for investor–state dispute settlement.[24] IIAs strive to ensure

---

[16] See Jeswald W Salacuse, 'The Treatification of International Investment Law' (2007) 13 LAW AND BUSINESS REVIEW OF THE AMERICAS 155.

[17] UNCTAD, *The Role of International Investment Agreements in Attracting Foreign Direct Investment to Developing Countries* (UNCTAD 2009) 6.

[18] See Dorsati Madani, 'A Review of the Role and Impact of Export Processing Zones' (World Bank 1999) 5, <http://econ.worldbank.org.easyaccess1.lib.cuhk.edu.hk/docs/965.pdf> last visited Feb 6, 2015. Although Madani's report uses the term "export processing zones," the same scheme is referred to by many other names, including export free zones. See, e.g., Bureau for Multinational Enterprise Activities, International Labor Organization (ILO), 'Export Processing Zones: Addressing the Social and Labour Issues' <www.transnationale.org/pays/epz.htm last visited Feb 6, 2015.

[19] See Salacuse, n. 16 above.

[20] *See* Santiago Monitt, STATE LIABILITY IN INVESTMENT TREATY ARBITRATION: GLOBAL CONSTITUTIONAL LAW IN THE BIT GENERATION (Hart Publishing 2009) 83.

[21] See ibid., 125.

[22] See Anne Van Aaken, 'International Investment Law Between Commitment and Flexibility: A Contract Theory Analysis' (2009) 12 JOURNAL OF INTERNATIONAL ECONOMIC LAW 507.

[23] See UNCTAD 'Intellectual Property Provisions in International Investment Arrangements' 2 <http://unctad.org/en/docs/webiteiia20071_en.pdf>; Van Aaken, ibid.

[24] See Van Aaken, ibid.

a stable and predictable environment for investment through investor protection and arbitration mechanisms in cases of breach.[25]

This "treatification" shows the significant recalibration of international investment law over the last few years.[26] In his course delivered at the Academy of International Law in 1993, Prof. Juillard has synthesized this trend, noting that the evolution of the sources of law governing foreign investments has created losers—domestic law and customary international law—and winners—BITs and the general principles of international law.[27]

B. THE MEANING OF INVESTMENT: WHEN ARE SANITATION AND WATER SERVICES SUBJECT TO INVESTMENT TREATIES?

The most fundamental question in determining whether investment treaties are applicable to water services is whether a given sanitation or water service constitutes a "foreign investment" under international law. Indeed, this definition constantly changes as entrepreneurs, financiers and multinational companies develop innovative investment tools. IIAs tend to adopt a broad definition of "investment" that refers to "every kind of asset" of a foreign investor in a host country, suggesting the agreement covers anything of economic value.[28] Recent interpretations by the ICSID have determined that "investment" has an intrinsic meaning of contribution.[29] If it creates or generates "fruits and value," it deserves protection as an "investment."[30] In many IIAs, the oft-used asset-based definition typically includes an illustrative list of assets covered.[31] The categories of investments covered by most BITs remain substantially identical, namely: (a) movable and immovable property and other property rights; (b) interests in the property

---

[25] *See* Monitt, n. 20, 1; Van Aaken, ibid.
[26] See Salacuse, n. 16 above.
[27] Patrick Juillard, 'L'évolution des sources du droit des investissements' (1994) 250 COLLECTED COURSES OF THE HAGUE ACADEMY OF INTERNATIONAL LAW 208.
[28] See, e.g., Argentina – United Kingdom Bilateral Investment Treaty (BIT) signed on the 11th of December 1990; United Kingdom – Tanzania Bilateral Investment Treaty (BIT) signed on the 7th of January 1994.
[29] Céline Lévesque, 'Abaclat and Others v Argentine Republic: The Definition of Investment' (2012) 27 ICSID REVIEW 247.
[30] Ibid.
[31] See, e.g., Argentina – United Kingdom BIT.

of companies; (c) claims to money and claims to a performance; (d) intellectual property rights and (e) concession rights conferred by law or contract.[32]

Alternatively, some IIAs focus on foreign investment as an "enterprise" rather than as a variety of assets.[33] Those following the enterprise-based definition pay particular attention to the investor's objectives for establishing a long-term relationship with the economy of the host country—for example, the acquisition of a lasting interest in the ownership or management of an enterprise.[34]

Water sanitation is the process of cleaning water to make it safe for drinking, bathing, cooking, and other uses. Common methods of treating water include flocculation, filtration, absorption, ion-exchange, and disinfection.[35] Any and all of these methods require an individual in the host state to own and operate facilities (physical assets) with adequate technical expertise and proper technology to sufficiently purify the water.[36] Therefore, "investment," for purposes of water sanitation and water services in international law, typically encompasses both the facilities

---

[32] See, ibid.

[33] See Julien Chaisse and Puneeth Nagaraj, 'Changing Lanes: Intellectual Property Rights, Trade and Investment' (2014) 37 HASTINGS INT'L & COMP L REV 251.

[34] See Omar E García-Bolívar, 'G3 Agreement: A Comparison of its Investment Chapter with the Emerging International Law of Foreign Investment' (2004) 10 L & BUS REV AM 779. For instance, the G3, the free trade agreement between Colombia, Mexico and Venezuela, takes an enterprise-based approach to define investment.

> The G3's use of an enterprise-based definition is a positive one. By defining investment in terms of enterprise, the G3 grants protection to non-incorporated forms of FI as well as incorporated forms. The term "enterprise" is more general than the term "corporation," but the former comprises the latter. The drafters of the G3 differentiated between constituting an enterprise and organizing an enterprise. The agreement states that an enterprise will be any entity constituted, organized or protected under domestic laws. Such a provision opens the door for protection of non-incorporated forms of business organizations.

García-Bolívar, ibid., 790.

[35] See EPA, 'Drinking Water Treatment guidance' (EPA 816-F04-034 2004) <http://water.epa.gov/lawsregs/guidance/sdwa/upload/2009_08_28_sdwa_fs_30ann_treatment_web.pdf>.

[36] See Ohio River Valley Water Sanitation Comm'n, 'Strategic Plan for the Ohio River Valley Water Sanitation Commission' (2008) 1, <www.orsanco.org/images/stories/files/StratPlan2008.pdf>.

invested in by foreigners (tangible assets) as well as the research and development used to create new technologies (intangible assets).[37]

C. THE PROLIFERATION OF "BLUE GOLD" DISPUTES

Investors have increasingly exercised their arbitration rights under the IIAs in recent years. To date, 21 claims dealing with foreign investments in the water industry have been filed internationally. The proliferation of sanitation and water-related disputes before international investment tribunals serves as an important phenomenon.

The key feature of investment protection under BITs is that it allows foreign investors to challenge host governments' actions before an international arbitral tribunal.[38] This is imperative because domestic judicial systems may be biased against foreign interests. Moreover, national courts may be more likely to fall under pressure from other branches of government. The ability of foreign investors to bring their disputes to independent arbitrators provides assurances to the investors that domestic authorities will live up to their international obligations, thus ensuring a favorable and stable investment climate in the host country.[39]

This proliferation of water-services disputes must be further analysed by considering the type of services currently provided by foreign investors and the details of the claims and breaches of treaties, which often result in heavy compensation for the winning party.

D. PRIVATIZATION AND CONCESSION CONTRACT RELATED ARBITRATION

Privatization refers to a variety of partnerships between host states and private, foreign or local, investors or companies with different degrees of

---

[37] See Daniel Benoliel and Bruno Salama, 'Towards an Intellectual Property Bargaining Theory: The Post-WTO Era' (2010) 32 U PA J INT'L L 265, 275–90 and 25. See also Paul Krugman, 'A Model of Technology Transfer, and the World Distribution of Income' (1979) 87 J POL ECON 253.
[38] See Jeswald W Salacuse and Nicholas P Sullivan, 'Do BITs Really Work?: An Evaluation of Bilateral Investment Treaties and Their Grand Bargain' (2005) 46 HARV INT'L LJ 67.
[39] ibid.

ownership in management and operating services.[40] These partnerships have given the entire water sector management a brand new framework of policies.

Privatization is attractive as a means of improving, developing, and expanding the water service infrastructure by increasing its efficiency, distribution and technical expertise.[41] From 1990 to 1997, more than 90 projects in 35 different countries concerning water services had been undertaken by the private sector.[42] It is estimated that foreign investors are controlling almost 40 percent of the developing world's water market.[43] As a result, from 2003 5 percent of global water consumers are receiving services from private companies.

Privatization of public services can be formalized in various agreement configurations. A common and recurring practice can be clearly identified upon examination of the 21 water services disputes discussed later in this chapter. Indeed, in nearly every case, the same form of privatization and foreign investment was disputed: the concession agreement.

The most important foreign investments of the past 20 years were formalized with this special kind of privatization contract called a concession.[44] In 97 contracts with foreign investors, 48 of them were concessions, which accounts for almost 80 percent of total private investment in developing countries.[45] Concession contracts are popular with governments for several reasons. The different types of concession contracts allow governments to allocate risks and responsibility according to their needs.[46] For instance, in a concession contract, all associated risks, financial and commercial, are transferred to the investor.[47] Traditionally, a full concession agreement transfers the full operational and management responsibility of the entire water supply and sewerage system to the private contractor.[48] In other cases, a partial concession agreement option can stand, which defines the conditions and obligations transferred to the private contractor.[49] In these kinds of concessions, usually the assets remain controlled by the

---

[40] *See* Miller, n. 1 above.
[41] See Miller, ibid. And See George Mergos, 'Private Participation in the Water Sector: Recent Trends and Issues' (2005) 9/10 EUROPEAN WATER 59, 63.
[42] *See* ibid.
[43] See Carmen Arevalo-Correra and others, *Water Services and The Private sector in Developing Countries, Comparative Perceptions and Discussion Dynamics* (2012) 17.
[44] See Mergos, n. 41 above, 67.
[45] See ibid.
[46] See ibid.
[47] See Miller, n. 1 above.
[48] See ibid.
[49] See ibid.

government through monitoring and regulations and the water services are still public property.[50] The private company, however, has the complete responsibility for operation systems, maintenance and new investments, without actually owning the assets.[51] At the end of the contract, the assets are returned in good condition to the public authorities who had always retained title to the infrastructure.[52] All of these subtleties are actively negotiated between the two parties and differ on a case-by-case basis.

In any event, under a concession contract, the private company is generally responsible for everything including: the setting of customer and tariff rates, labor, expanding infrastructure, and new technologies.[53] These contracts are commonly made with a long duration, often over 25 years, to allow for more continuity and sustainability.[54] The specific conditions of the contract depend on the long-term and short-term goals of each party, such as the capacity of the government, how much private investors want to invest, and how much infrastructure construction is required.[55]

Furthermore, the contract can be thoroughly negotiated and regulated legally, constitutionally, and politically even before its application. If negotiated effectively, such contracts can ensure that water prices do not reach outrageous levels. Authorities can also reduce public outcry by arguing that they are actually not selling public assets nor placing them in the hands of foreign investors. It also reduces the regulatory burden on the government by using the concession contract itself as the regulatory mechanism.

Although every concession contract is unique, there is a classic pattern in the formation of such contracts.[56] The process begins with the launch of a public bid tender in order to find the best company to manage the water system.[57] Bidders compete to provide the best services at the lowest price, including the largest discounts to the public tariff among other considerations.[58] Generally, bids are awarded to companies with

---

[50] See ibid.
[51] See ibid.
[52] See ibid.
[53] See ibid.
[54] Judith A Rees, 'Regulation and Private Participation in the Water and Sanitation Sector' (1998) 22 NATURAL RESOURCE FORUM 95.
[55] See Miller, n. 1 above; See Rees, ibid., 99.
[56] See (book) CONTRACTING OUT WATER AND SANITATION SERVICES 5 <http://wedc.lboro.ac.uk/resources/books/Contracting_Out_Water_and_Sanitation_Services_-_Vol_2_-_Complete.pdf>.
[57] See Claude Crampes and Antonio Estache, 'Regulating Water Concession: Lessons from the Buenos Aires Concession' (1996) PUBLIC POLICY FOR THE PRIVATE SECTOR, Note No. 91, THE WORLD BANK GROUP 2.
[58] See ibid.

strong financial and technological capabilities.[59] The winner of the bid tender, called the concessionaire, will then be responsible for operating and maintaining the fixed assets, expanding coverage, and guaranteeing the water quality and developing the sewage system. Afterwards, negotiations begin, where the government describes what it expects from the investment such as goals, expansion plans, new technologies, coverage, overall efficiency, price and duration. The concessionaire identifies its investment priorities and responds to the government's requests by detailing the investor's methodology, costs, and expected duration. The government is usually responsible for paying a compensable monthly fee, set by a mathematical formula, to the concessionaire. The government is also responsible for setting the procedure of the collection of the bills from the customers. In addition, the government must assume the role of the regulator in the concession agreement.[60] In order to ensure that the customer is protected during the privatization process, different regulatory tasks must be observed by the state.[61] The regulatory tasks include, but are not limited to, control over unfair trading practice, safety net regulations, promoting water use efficiency, ensuring responsiveness to final customer needs.[62] In return, the concessionaire is responsible for providing services and guarantees of performance while meeting the standards set out in the concession contract. Generally, water and tariff prices are the first and most important points fixed in the agreement as these costs often involve several million of dollars.

The parties agree not only to a fixed general price at the outset of the concession, but also its maximum percentage augmentation for subsequent years. This enables the parties to identify their interests early in the process. The investor knows that it will be possible to make some profits and returns on investment in the following years, and the government is certain that only a maximum augmentation will be possible, maintaining an efficient access to water to its citizens. In this way, the concession is the key to expanding access to water and services rehabilitation, while minimizing the impact on tariffs. As it is impossible to predict the economic, environmental, social or even technological changes for such a long period of time, the tariff formula must be reviewed and renegotiated from time to time.[63] Annual adjustments can be made, for example, to accommodate an increase or decrease of water consumption or wastewater

---

[59] See ibid.
[60] See ibid.
[61] See Rees, n. 54, 100.
[62] See ibid.
[63] See ibid., 95.

treated, but every change must maintain the profit percentage initially fixed in favor of the investor.[64] This delicate point is often a key factor in the water investment arbitration cases.[65]

E. CLEAN WATER DISTRIBUTION SYSTEMS

Pure water, directly from rivers or oceans is rarely clean enough for direct human consumption. Therefore, water purification systems must be used to remove all possible bacteria and contaminants before it reaches consumers. This process commonly involves physical and chemical methods such as clarification and disinfection.[66]

Water quality refers to the biological, radiological, chemical and physical characteristics of the water.[67] In order to reach satisfactory quality, every water supply system must follow and reach certain national standards depending on the intended use of the water whether it be for industrial, agricultural, cleaning, or drinking use.[68] Certain considerations such as turbidity, color, taste (from dissolved organics), hardness (no mineral deposits), odor (from dissolved metals), and corrosiveness are considered when analysing water quality in order to avoid serious human health concerns resulting from procedural oversight.[69]

Access to water requires both a readily available water supply and a well-constructed distribution system.[70] Water distribution systems consist of a series of interconnected storage tanks, valves, pumps, hydrants and pipes that transport and provide water to consumers, both individuals and industries.[71] In other words, a water distribution system is the mechanism that carries water from the source to the consumer and is ultimately responsible for providing an uninterrupted supply of well-pressurized and safe drinking water.[72]

High-quality distribution systems are critical components in delivering

---

[64] See Crampes and Estache, n. 57, 3.
[65] See ibid.
[66] See WHO, Guidelines for Drinking-Water Quality (4th ed. WHO 2011) 4 <http://whqlibdoc.who.int/publications/2011/9789241548151_eng.pdf>.
[67] See ibid.
[68] Ibid., 2.
[69] See ibid., 6.
[70] See FEDERAL EMERGENCY MANAGEMENT AGENCY, U.S. FIRE ADMIN., *WATER SUPPLY SYSTEMS & EVALUATION METHODS* (FEMA 2008) 2 <www.usfa.fema.gov/downloads/pdf/publications/Water_Supply_Systems_Volume_I.pdf>.
[71] See ibid., 1.
[72] See ibid., 5.

safe drinking water in developing countries where too many people still receive poor water distribution services.[73] Water supply quality has three chief dimensions. First, it is necessary to have a certain continuity of supply and afford a sufficient quantity of water.[74] Second, the water must be provided with adequate pressure, sufficient for operating plumbing fixtures and firefighting equipment but at the same time not so high as to cause leaks or pipeline breaks that could result in waste of water.[75] It also allows the customer to have a sufficient amount of drinkable water available on a normal flow.[76] Third, and most important, high-quality service must provide high-quality, safe drinking water where needed.[77]

F. WASTEWATER AND SEWERAGE SERVICES

Yet another category of water related cases concerns "wastewater and sewerage services." Sewage services are the second phase in managing water resources. Once the water is used, either for domestic or industrial purposes, it is necessary to collect the wastewater and remove its impurities before it reaches its final destination for settlement or reuse.[78] Sewerage services consist of a network of pipes and pumps for the collection of wastewater.[79] Sewage is a complex and technical procedure. An understanding of three fundamental characteristics of sewage services is essential to mastering wastewater service goals.

The first fundamental characteristic is the collection and disposal of wastewater. It consists of a system of pipes, which transports the wastewater for treatment and ultimately for disposal.[80] There are many types of wastewater collection systems. Maintenance of these systems is

---

[73] *See* World Health Organization, *The Right To Water* (2003) (giving an overview of the discussion surrounding the right to water such as legal definition, implications and responsibilities) at 12–13. Available at http://www.who.int/water_sanitation_health/rtwrev.pdf.

[74] See FEDERAL EMERGENCY MANAGEMENT AGENCY, U.S. FIRE ADMIN., n. 70 above, 1.

[75] See ibid.

[76] See ibid.

[77] See ibid.

[78] See Sewage Treatment, Foundation for Water Research <www.euwfd.com/html/sewage_treatment.html>.

[79] 'Wastewater Disposal and Transport Options – Sewerage' World Bank Group <http://water.worldbank.org/shw-resource-guide/infrastructure/menu-technical-options/sewerage>.

[80] See Sewage Treatment, Foundation for Water Research, n. 78 above.

an integral component of the proper management of sewerage services and is critical for preventing illegal wastewater releases.[81] Moreover, disposal of the treated wastewater is a delicate procedure and difficult to achieve. The final destination of treated sewage water is largely earth's soil, oceans, and rivers.[82] Thus, many environmental and technical precautions must be made when discharging the treated water out of sewage systems including anaerobic digestion (bacterial process), composting, or sometimes, incineration.[83]

The second characteristic is disease potential. The sewage process is critical because it stops water-borne diseases such as cholera and typhoid. However, during the collection method, small amounts of chemicals or organic compounds from cleaning and disinfection operations are often discharged into sewers and into the air.[84] Sewage may also contain assorted chemicals and specialized disposables such as medical waste, microbiological pathogens, nitrates and oils. Therefore, without proper treatment, there is a significant risk of water-borne diseases such as infections, intestinal and lung problems, and fever that pose a danger to the public.

The third, and vital, characteristic of wastewater services is sewage treatment. In order to protect public health, sewage water must be purified to remove organics, destroy bacteria, and neutralize toxic and chemicals waste.[85] There are varying methods and processes to treat the wastewater, depending on the degree of contamination, local conditions, governmental regulations, and general industry standards.[86]

Although the wastewater treatment process is very technical, it can be summarized as follows. The primary treatment consists of physical operations to remove floating solids and sedimentation by gravity.[87] This removes approximately 60 percent of impurities.[88] The wastewater is passed through a screen to trap solid objects and sedimentation.[89] The rest of the liquid is then subject to the secondary treatment, which consists

---

[81] Wastewater Collection System Toolbox, n. 79 above.
[82] See http://water.worldbank.org/shw-resource-guide/infrastructure/menu-technical-options/final-disposal.
[83] *See* Sewage Treatment, Foundation for Water Research, n. 78 above.
[84] <www.who.int/water_sanitation_health/medicalwaste/130to134.pdf>.
[85] See World Bank group, 'Introduction to Wastewater Treatment processes' </water.worldbank.org/shw-resource-guide/infrastructure/menu-technical-options/wastewater-treatment> last accessed Oct 31, 2014.
[86] <www.euwfd.com/html/sewage_treatment.html.
[87] See World Bank Group, 'Introduction to Wastewater Treatment Processes', n. 85 above.
[88] See ibid.
[89] See ibid.

of a biological process that removes organic matter and approximately 85 percent of impurities.[90] After the biological methods, tertiary treatments are conducted to eliminate every other constituent left after the first and second treatment stages, removing approximately 99 percent of impurities.[91] The aim of the tertiary treatment is to significantly improve the quality of wastewater before discharging it into the environment and is often the most expensive stage for water companies.[92] Finally, the wastewater undergoes disinfection before it is discharged.[93] Chlorine is typically used to remove and destroy any remaining pathogens.[94] At this point, wastewater is ready for public use.

CONCLUSION

The last decade has witnessed a dramatic surge of investment disputes between foreign investors and host country governments. Arbitral panels have been charged with the task of applying the rules of IIAs in specific cases, a task that is not often straightforward given the broad and ambiguous terms of these agreements. This new phenomenon of investment litigation has resulted in a number of decisions from arbitral tribunals in the water service sector. Such decisions have contributed to the formation of an embryonic water service jurisprudence and the elucidation of key provisions, concepts, and definitions embedded in BITs and water service-related concession contracts. All of this has led to the emergence of a nascent framework for global economic regulation of the sanitation and water services industry.

The definition of investment is currently absorbing sanitation and water services, a rather new form of investment in the transnational scenario. Furthermore, international investment law is growing flexible enough to attract these specific types of highly sensitive disputes. International investment treaties and the tribunal in charge of applying these rules have contributed significantly to shaping the contours and substance of an international water service jurisprudence and the emerging international economic water services regime. The investment world fills a gap that no other organization has been able to address. While an investment tribunal's main task is to apply treaties, which protect foreign investors, the same

---

[90] See ibid.
[91] See ibid.
[92] See ibid.
[93] See ibid.
[94] See ibid.

tribunals may not be well equipped to consider non-economic issues, such as those essential to the water regulation industry.[95] Although the investment jurisprudence may be seen as progress towards the regulation of an important service, it also emphasizes the lack of a more global holistic approach to regulating water services and access to water. Future research will have to find a means of reconciling the great advances made in the area of investment with the urgent need of ensuring that the more nascent human right to water receives equal consideration in the coming years.

On a more practical level, water services are no longer solely under the purview of domestic regulation. IIAs apply by default, particularly in the absence of WHO standards. The international investment regime further contributes to the internationalization of the water services regime. Conversely, governments must design water related policies that comply with fair and equitable treatment (FET), expropriation regulation, and full protection and security (FPS), since not doing so can be costly and deter foreign investors from providing high quality services. Or, if policy makers do not agree with such a reality, they must re-design and re-engineer the applicable international law.

As this chapter has explained, most of the time, the term investment is defined very broadly. In order to avoid significant exposure to investment claims one must wonder whether investment should be required to contribute to development? The issue is whether the contribution to development in IIAs should be made an eligibility criterion. Host states would like to be sure that investments would be protected only if they contributed to development. Investors would likely see such a requirement as creating substantial uncertainty as to whether an investment would qualify for protection in the absence of a standard definition of development. If such a requirement were included in a treaty, a host state could claim that the tribunal did not have jurisdiction on the basis that the investment did not contribute to development.

Both the theoretical and practical conclusions drawn by this chapter anticipate future developments and assist in exposing the horizon of forthcoming research and debate in the global governance of sanitation and water services. The increasing need for water due, inter alia, to global warming and climate change and new technologies such as shale gas, means that foreign investments in water will increase and a sound public consists in anticipating this increase by strictly taking a more proactive approach in thinking and designing the international principles that should regulate water.

---

[95] See Bradley J Condon, 'Treaty Structure and Public Interest Regulation in International Economic Law' (2014) 333 17 J. INT'L ECON. L. 351.

6. The right of the host state to regulate water services
Catharine Titi

For many decades, the promotion and protection of international investment has been international investment agreements' (IIAs) principal function and the reason for their existence. States have concluded these agreements to offer investors safeguards, such as national treatment, most-favoured-nation treatment, protection in case of expropriation, fair and equitable treatment, full protection and security and free capital transfers, and have backed these up through offering access to investor-state dispute settlement (ISDS). The protection of international investment has been such a predominant concern, that, concluding these treaties, states restricted their policy space and their ability to adopt measures for public welfare objectives. Against this background, a new preoccupation was born, that of reserving the host state's right to regulate and narrow the interpretive leeway of arbitral tribunals by addressing states' right to pursue specific public policy goals.[1] A new generation of investment treaties, essentially launched with the US and Canadian Model bilateral investment treaties (BITs) of 2004, have started to provide contracting parties with a modicum of flexibility and the question is asked for the first time of the pertinence of this right to regulate for state measures relating to water services.

State regulation concerning water may adversely affect an investor not only where investment is made in water utilities, such as supply of drinking water and sanitation services, but also where the investor is engaged in water-intensive activities, such as in the agricultural, industrial, energy, and the mining and oil sectors. Water regulation may also become relevant where the investment 'pollutes or deteriorates the environment associated with water'.[2] Environmental legislation, and state measures for the

---

[1] Catharine Titi, *The Right to Regulate in International Investment Law* (Nomos and Hart Publishing 2014).
[2] Miguel Solanes and Andrei Jouravkev, *Revisiting Privatization, Foreign Investment, International Arbitration, and Water* (Santiago: UN 2007) 17.

protection of human, animal or plant life or health are two fields of public policymaking that may have a direct impact on a foreign investment and may overlay with investment protections afforded by the regulating state in an investment agreement. Given the potential for overlap between water, investment and state regulation, it is not astonishing that several water-related claims have been initiated, against both developing and industrialised countries. Among the numerous examples, one may cite the famous *Aguas del Tunari, S.A.* dispute against Bolivia and related events that became known as Bolivia's Water War, as well as the first *Vattenfall* dispute against Germany, involving local authorities' measures relating to compliance with cleanness of river water targets of EU legislation, earning Germany its first ever known investment arbitration. In light of the very large scope of cases that would fall within the range of what can qualify as a 'water dispute', the analysis will limit itself to the right of the host state to regulate directly related to water services, namely with reference to either water utility construction projects[3] or clean water distribution and sewage services,[4] the latter category attracting the overwhelming majority of investment disputes.[5]

If the question is asked on the host state's right to regulate water, this reflects both the fact that water regulation concerns the public interest, and therefore the centrality of the provision of clean drinking water and sanitation services, and the fact that these services come under the protective umbrella of international investment agreements. A wave of relatively recent privatisations of public services, including water, has resulted in 10 per cent of global consumers now receiving water from private companies.[6] Inevitably, some of these services have come into the hands of foreign investors and this is where the intersection is found between international investment law and water. At the time of writing, at least 21 claims relating to foreign investments with reference to water services

---

[3] E.g., *ATA Construction, Indus. and Trading Co v Jordan* [2010] (ICSID); *Salini Costruttori v Jordan* [2006] (ICSID).

[4] E.g., *Azurix Corp v Argentina* [2006] (ICSID); *Compañiá de Aguas del Aconquija SA and Vivendi Universal SA v Argentina* [2007] (ICSID); *Suez Sociedad General de Aguas de Barcelona SA and Vivendi Universal SA v Argentina* [2010] (ICSID); *Impregilo SpA v Argentina* [2011] (ICSID); *SAUR International v Argentina* [2014] (ICSID); *Biwater Gauff (Tan.) Ltd v United Republic of Tanzania* [2008] (ICSID).

[5] See Julien Chaisse and Marine Polo, 'Globalization of Water Privatization: Ramifications of Investor-State Disputes in the "Blue Gold" Economy' (2015) 38 *Boston College International & Comparative Law Review* 1, 18.

[6] Ibid., 3.

have been filed.⁷ Where investment protections, investor-state dispute settlement and regulation in the public interest overlay, we are clearly in the domain of the right to regulate.

The chapter will query into the extent to which a host state is able to adopt measures relating to water services affecting, inter alia, the economic value of an investment without violating its conventional international investment obligations and, significantly, without incurring a duty to compensate the foreign investor. It will commence with an overview of the concept of the right to regulate in international investment law. It will then enquire into the presence of water as a regulatory interest in international investment agreements. The following section will turn to arbitral investment jurisprudence on water, to examine the extent to which arbitral tribunals have recognised that states have a right to regulate water. The final section will conclude.

THE RIGHT TO REGULATE[8]

Definition and General Observations

The limiting effect of IIAs on host states, especially as evidenced in investment claims brought in front of arbitral tribunals, has led contracting parties to a search for policy space and for a rebalancing of international investment obligations. The right to regulate emerged prominently in recent negotiations, and notably in the exercise by the European Union (EU) of its competence over the conclusion of provisions on foreign direct investment. In the recent 'Concept Paper: Investment in TTIP and beyond – the path for reform', the EU stressed that its investment policy reaffirms the right of the parties to regulate and to achieve legitimate public policy goals, such as the protection of public health, safety and the environment[9] and underlined the further need for improvement in this area.[10]

In the narrow context of international investment law, the right to regulate is the 'legal right exceptionally permitting the host state to regulate in derogation of international commitments it has undertaken by means of an investment agreement without incurring a duty to compensate'.[11] The

---

[7] Ibid., 7.
[8] This section is based on Titi, n. 1 above.
[9] EU, *Concept Paper: Investment in TTIP and Beyond – the Path Ahead* (Brussels 2015) 3.
[10] Ibid., 7.
[11] Titi, n. 1 above, 33.

legal basis of this right is generally found, as we will see, in conventional law. Incidentally, jurisprudential doctrines and tribunal deference can increase host state regulatory freedom *ex post*, by conceding the state's 'legitimate regulatory interests'.[12] Although resembling a veritable right to regulate, the latter case involves a *legitimate* rather than a *legal* right of the host state to act as it does.[13]

Where the right to regulate is recognised, the state is exempt from compensating the aggrieved investor for the measures it has taken. The absence of the need to compensate is an important aspect of the concept. The latter would be deprived of meaning, if a subsisting requirement to compensate was recognised,[14] given that a state retains in any case the (*lato sensu*) right to regulate, so long as it accepts the consequences, viz. the obligation to compensate potentially affected investors.[15] The significance of compensation in this context was also underlined by the *Feldman* Tribunal which acknowledged that reasonable government regulation in the public interest cannot be carried out if adversely-affected businesses 'may seek compensation'.[16]

The Right to Regulate in Conventional Law

In the main, the right to regulate has its legal basis in conventional law, namely in investment treaty exceptions. Exceptions sometimes take the form of standard-specific clauses, such as provisions stating that the most-favoured-nation treatment shall not apply to privileges which either contracting party accords to investors of third states by virtue of agreements for the avoidance of double taxation;[17] or, they take

---

[12] Ibid., 33, 40.

[13] Ibid., 33.

[14] William W. Burke-White and Andreas von Staden, 'Investment Protection in Extraordinary Times: The Interpretation and Application of Non-Precluded Measures Provisions in Bilateral Investment Treaties' (2008) 48 *Va J Int'l L* 308, 338.

[15] Titi, n. 1 above, 34; L Markert, 'The Crucial Question of Future Investment Treaties: Balancing Investors' Rights and Regulatory Interests of Host States', in M Bungenberg, J Griebel and S Hindelang (eds), *EYIEL 2011, Special Issue: International Investment Law and EU Law* (Heidelberg: Springer 2011) 146, 150, 165.

[16] *Marvin Feldman v Mexico* [2002] (ICSID) (hereinafter Feldman Award), para. 103. See further Ruggie, J, 'Business and Human Rights: Towards Operationalizing the "protect, respect and remedy" Framework' (Report of the UNSRSG on human rights and transnational corporations and other business enterprises, A/HRC/11/13 2009) para. 30.

[17] E.g., art. 3(4) of the German Model BIT of 2009.

the form of general exceptions that apply to the whole treaty, such as general exceptions modelled on Article XX of the GATT or exceptions for essential security interests.[18] General exceptions, also described as provisions 'of comprehensive scope',[19] are drafted so as to indicate that *'nothing in the Agreement'* shall be construed to prevent the parties from adopting the measures specified therein. Their scope is larger than that of standard-specific exceptions, in that they are not limited to one investment protection but are applicable to any protection offered by the treaty

Apart from investment treaty exceptions, conventional law may also contain other complementary provisions that contain positive language on regulatory interests but which do not accord an independent right to regulate. Such positive language may establish 'soft' obligations – also called 'best efforts' commitments[20] – offering states an interpretative presumption that their interests will be taken into account.[21] Notably within this category fall provisions on the non-lowering of standards stipulating, for instance, that the parties consider that:

> it is inappropriate to encourage investment by relaxing domestic health, safety or environmental measures. Accordingly, a Party should not waive or otherwise derogate from, or offer to waive or otherwise derogate from, those measures to encourage the establishment, acquisition, expansion or retention in its territory of an investment of an investor. If a Party considers that the other Party has offered such an encouragement, it may request consultations with the other Party and the two Parties shall consult with a view to avoiding the encouragement.[22]

Another manifestation of positive language consists in the insertion of regulatory interests in the preamble. The source of a treaty's object and purpose,[23] the preamble does not establish independent legal rights or obligations[24] and therefore regulatory interests in the preamble do not provide a genuine right to regulate. However, they are still important for host state policy space, since, in accordance with the Vienna

---

[18] E.g., art. 18 of the US Model BIT of 2012.
[19] Burke-White and von Staden, n. 14 above, 331.
[20] UNCTAD, *Bilateral Investment Treaties 1995–2006: Trends in Investment Rulemaking* (New York and Geneva: UN 2007) 92.
[21] Titi, n. 1 above, 104.
[22] Article 15 of the Canadian Model BIT (version of 2012).
[23] R. Dolzer and M. Stevens, *Bilateral Investment Treaties* (The Hague: Martinus Nijhoff 1995) 20.
[24] A. Newcombe and L. Paradell, *Law and Practice of Investment Treaties* (Alphen aan den Rijn: Kluwer Law International 2009) 124; *Total SA v Argentina* [2010] (ICSID) para 116.

Convention on the Law of Treaties,[25] the preamble constitutes a tool for treaty interpretation.[26] Regulatory interests in the preamble may cover a large range of public policy objectives, including the protection of the environment,[27] health and safety,[28] human rights,[29] labour standards,[30] sustainable development,[31] corporate social responsibility,[32] culture,[33] and anti-corruption issues.[34]

The Right to Regulate Beyond Conventional Law

Beyond conventional law, and to a certain extent, states may be capable of pursuing some regulatory interests under public international law despite a *stricto sensu* violation of their investment treaty obligations.[35] Especially relevant for investment law is the case where a state 'breaches' the substantive provisions of a treaty where a customary law 'circumstance precluding wrongfulness' may be applicable. Customary law circumstances precluding wrongfulness are found contestably 'codified' in the International Law Commission's Articles on the responsibility of states for internationally wrongful acts (hereinafter ILC Articles on State Responsibility).[36] The ILC Articles on State Responsibility recognise six circumstances precluding wrongfulness: consent,[37] self-defence,[38] countermeasures in respect of an internationally wrongful act,[39] *force majeure*,[40] distress,[41] and necessity.[42] Not all circumstances are equally susceptible to become invoked in investment disputes.[43] In dealing with exigent or extraordinary

---

[25] Article 31 Vienna Convention on the Law of Treaties of 1969.
[26] Titi, n. 1 above, 115.
[27] Preamble to Japan-Papua New Guinea BIT (2011).
[28] Preamble to US-Croatia BIT (1996).
[29] Preamble to Canada-Colombia FTA (2008).
[30] Preamble to US-Chile FTA (2003).
[31] NAFTA Preamble.
[32] Preamble to Canada-Peru FTA (2008).
[33] Preamble to Canada-Costa Rica FTA (2001).
[34] Preamble to US-Peru TPA (2006).
[35] Titi, n. 1 above, 236.
[36] ILC Articles, Chapter V (arts 20–27).
[37] Ibid., art. 20.
[38] Ibid., art. 21.
[39] Ibid., art. 22.
[40] Ibid., art. 23.
[41] Ibid., art. 24.
[42] Ibid., art. 25.
[43] Titi, n. 1 above, 237.

circumstances, these customary law 'defences' relate to situations that fall outside the regular policymaking activity of governments.[44]

Necessity is the most important customary international law defence for investment disputes, as evidenced by its invocation in a number of recent awards and by its apparent proximity with the essential security interests exception.[45] Nonetheless, successful invocation of the necessity defence is subject to particularly strict criteria[46] and its very interpretation in investment disputes generally leaves a lot to be wished for;[47] in any case it fails to justify the assumption that customary international law guarantees a state its regulatory flexibility. The necessity defence has proved to be practically unattainable in the overwhelming majority of cases.[48] Further reasons, that have been discussed elsewhere, indicate that customary international law does not appear to safeguard the state's policy space.[49]

The last potential 'source' of regulatory flexibility that will be discussed here is the occasional attitude of deferent tribunals that lend an ear to the host state's legitimate public policy objectives. However, arbitral jurisprudence has interpreted investment treaty provisions sometimes with deference to the host state's policy decisions and sometimes with deference to the investor's complaints, and therefore it provides no *legal right* that permits the state *ex ante* to know that it can regulate as it wishes without having to compensate investors for such regulation.

---

[44] Ibid.
[45] Ibid., 237–8.
[46] Article 25 of the ILC Articles on State Responsibility provides:

1. Necessity may not be invoked by a State as a ground for precluding the wrongfulness of an act not in conformity with an international obligation of that State unless the act:

   (a) is the only way for the State to safeguard an essential interest against a grave and imminent peril;
   and
   (b) does not seriously impair an essential interest of the State or States towards which the obligation exists, or of the international community as a whole.

2. In any case, necessity may not be invoked by a State as a ground for precluding wrongfulness if:

   (a) the international obligation in question excludes the possibility of invoking necessity; or
   (b) the State has contributed to the situation of necessity.

[47] See Titi, n. 1 above, 240.
[48] Ibid., 254.
[49] Ibid., 255.

In short, it is essentially conventional law that can guarantee host states their right to regulate. For this reason, it is now opportune to briefly consider whether investment treaties contain a right to regulate specific to water.

NO GENERAL RIGHT TO REGULATE SPECIFIC TO WATER BUT AN *INDIRECT* RIGHT THROUGH CONVENTIONAL LAW

Provisions typically found in investment treaties and offering states regulatory flexibility do not relate *directly* to water. In other words, water does not figure as a standalone regulatory interest that needs to be protected under the umbrella of the host state's right to regulate. However, the question may be asked whether conventional law recognises an *indirect* right of the host state to regulate water.

The question of whether water constitutes a human right, regularly asked in the last decade, is in this context not directly relevant. On 28 July 2010, the United Nations General Assembly recognised, through Resolution 64/292, the 'right to safe and clean drinking water and sanitation as a human right that is essential for the full enjoyment of life and all human rights'.[50] Host states have the obligation to protect human rights but they also have the obligation to observe their investment commitments,[51] and so water as a human right does not help us resolve the question of whether the state can regulate it without incurring investment claims for compensation. A more relevant question for investment law is whether investment treaties contain provisions that can cover water.

Access to water and water services relates closely to human life or health; cleanness of water itself relates to human, animal and plant life or health and to the protection of the environment. Some exceptions found in IIAs may indeed be applicable to water and water services, although the latter are not explicitly named. For instance, exceptions for the protection of human, animal and plant life or health may be presumed to cover access to water and sanitation services. Exceptions or explanatory notes to the expropriation standard, specifying that except in rare circumstances, non-discriminatory regulatory measures taken by a party in order to protect 'legitimate public welfare objectives, such as public health, safety, and the environment' do not constitute an indirect expropriation[52] may likewise

---

[50] UN General Assembly Resolution 64/292.
[51] See also *Anglian Water Group (AWG) v Argentina* [2010] (UNCITRAL) para 262, Decision on Liability.
[52] Annex B(4)(b) of the US Model BIT of 2012.

be indirectly presumed to cover water. Every mention of environmental or health measures in a treaty may also concern different aspects of access to water, water services or water quality.

Finally, a minority of treaties may in fact reference water. Article X.08 on Rights and Obligations Relating to Water of the EU-Canada Comprehensive Economic and Trade Agreement's (CETA, version of 26 September 2014) chapter on Initial Provisions and General Definitions provides:

> 1. The Parties recognize that *water in its natural state*, such as water in lakes, rivers, reservoirs, aquifers and water basins, is not a good or a product and therefore, except for Chapter XX – Trade and Environment and Chapter XX – Sustainable Development, is not subject to the terms of this Agreement.
> 2. Each Party has the right to protect and preserve its natural water resources and *nothing in this Agreement obliges a Party to permit the commercial use of water* for any purpose, including its withdrawal, extraction or diversion for export in bulk.
> 3. *Where a Party permits the commercial use of a specific water source, it shall do so in a manner consistent with the Agreement.*[53]

The limitations of this article as a clause offering contracting parties a right to regulate is evident. The carve-out of the first paragraph relates only to 'water in its natural state' and the exception of the second paragraph is limited to not oblige a party to permit the commercial use of water. The third paragraph confirms that where a party has permitted the commercial use of water, it does not have a (*stricto sensu*) right to regulate. If it regulates water in a manner inconsistent with the agreement, claims and an obligation to compensate will likely ensue.

CASE-STUDIES FROM RECENT ARBITRAL CASES – NO RIGHT TO REGULATE WATER ABSENT TREATY EXCEPTIONS

A consideration of recent arbitral decisions involving water services is pertinent in this respect. A first example comes from the *SAUR* case.[54] The claim was registered following an alleged failure of the local government of the Argentine province of Mendoza to apply service tariff increases according to an agreement that existed between the foreign investor and the federal government in the wake of Argentina's economic crises of

---

[53] Emphasis added.
[54] *SAUR International SA v Argentina* [2012] (ICSID).

2001. The tribunal that was called to adjudicate the claim acknowledged that human rights in general, and what it recognised as the right to water in particular, constitute sources of law applicable to the dispute at hand.[55] The reason for that was not only that the host state safeguarded these as constitutional rights in its internal legal system but also because they form part of the general principles of international law.[56] The tribunal remarked that, from the point of view of the host state, access to drinkable water constitutes a vital public service and, from the point of view of the individual, a fundamental right.[57] In this respect, the arbitrators cited the Report of the United Nations High Commissioner for Human Rights on the scope and content of the relevant human rights' obligations concerning equitable access to safe drinking water and sanitation under international human rights instruments.[58]

The Report of the United Nations High Commissioner explains that although 'human rights treaties do not recognise access to safe drinking water and sanitation as a human right per se, specific obligations in relation to access to safe drinking water and sanitation have been increasingly and explicitly recognised in core human rights treaties, mainly as part of the right to an adequate standard of living and the right to health'.[59] The Report cites, inter alia, a 'right to sanitation', which is the 'right of everyone to have access to adequate and safe sanitation that is conducive to the protection of public health and the environment';[60] and adds that sanitation services and facilities should be 'physically accessible' and 'affordable'.[61]

The *SAUR* tribunal considered that the public authorities have to retain the legitimate functions of planning, supervision, police, sanction, intervention and even revocation of the contract in order to protect the public interest.[62] However, the tribunal stressed that these 'prerogatives' are compatible with investment protection; public authority powers over water are not exercised in an absolute manner but on the contrary in accordance with the protections offered foreign investors on the basis

---

[55] *SAUR International SA v Argentina* [2012] (ICSID) para 330, Decision on Jurisdiction and Liability.
[56] Ibid.
[57] Ibid.
[58] Ibid.
[59] A/HRC/6/3, 16 August 2007, para. 6.
[60] Ibid., para. 18.
[61] Ibid.
[62] *SAUR International SA v Argentina* (n 54 above) para. 330, Decision on Jurisdiction and Liability.

of the investment treaty.[63] If by consequence the investor is treated in a non-equitable manner or if the investor does not receive, for example, full protection and security, it acquires a right to compensation.[64] As already noted, the right of the host state to regulate *excludes* the obligation to compensate the aggrieved investor. In a next step, the tribunal set itself the task of 'counterbalancing' these competing interests.[65] The tribunal ended by finding that the host state had violated protections under the BIT in question. It must be noted that the latter, the 1991 bilateral investment treaty between Argentina and France, did not contain any particular exceptions that could apply in the discussed context or guarantee the host state its right to regulate water. The tribunal's findings may have been different under a new generation treaty more cognisant of the state's regulatory interests. However, it is also true that the tribunal's finding of a violation of the fair and equitable treatment standard, an investment protection considered to contain a balancing test, may have been unaltered even in the presence of a general exception covering it in principle.

Another investment dispute that may have involved the host state's right to regulate water is the *Aguas del Tunari, S.A.* dispute,[66] which arose out of the privatisation of water utilities in Cochabamba, Bolivia.[67] The privatisation itself had been encouraged by international financial institutions,[68] but it resulted in raising the cost of water sources and water treatment systems, and so ultimately in higher water prices for consumers.[69] The following observation is revealing: while according to the World Health Organization (WHO) an individual should not be spending more than 3–5 per cent of his or her income on water for it to be affordable, in the context of the *Aguas del Tunari, S.A.* dispute here households were spending in excess of 20 per cent of their income on water.[70] Following the onset of the privatization project, protests erupted among the population, in

---

[63] Ibid., para. 331.
[64] Ibid.
[65] Ibid., para. 332.
[66] *Aguas del Tunari, SA v Bolivia* (ICSID). On this dispute, see also Catharine Titi, 'Investment Arbitration in Latin America: The Uncertain Veracity of Preconceived Ideas' (2014) 30 *Arbitration International* 381.
[67] The way to privatisation opened through Ley N° 1544 (Bolivia) – Ley Marco de Capitalización of 21 March 1994.
[68] Erik B Bluemel, 'The Implications of Formulating a Human Right to Water' (2004) 31 *Ecology Law Quarterly* 957, 965.
[69] Ibid., 966. See also Titi, n. 66 above, 381.
[70] Bluemel, ibid., 966.

which about 100 persons were injured and one person was killed.[71] These violent protests marked what was later called Bolivia's Water War.[72] In the aftermath of the events, Bolivia ended the privatisation concession.[73] The investor initiated the claim before ICSID, but later settled out of arbitration; increasing public criticism at the international level and 'the civil unrest and the state of emergency' persuaded the investor to withdraw its claim.[74]

A further dispute that revolved around water utilities was the *AWG* case.[75] In the absence of a treaty exception applicable in that case (that would have been an essential security interest exception in relation to Argentina's economic and financial crisis of 2001), the *AWG* tribunal acknowledged Argentina's 'reasonable right to regulate'[76] and estimated that when interpreting the meaning of fair and equitable treatment, 'the Tribunal must balance the *legitimate* and *reasonable expectations* of the Claimants with Argentina's *right to regulate* the provision of a vital public service'.[77] The tribunal further determined that 'the legitimate and reasonable expectations of the investors [. . .] must have included the expectation that the Argentine government would exercise its legitimate regulatory interests with respect to the [Concession at issue] throughout the period of thirty years and in response to unpredictable circumstances that might arise during that time'.[78]

The following statement of the *AWG* tribunal is of interest.

---

[71] Mary H. Mourra (2008), 'The Conflicts and Controversies in Latin American Treaty-Based Disputes', in Mary H. Mourra and Thomas E. Carbonneau (eds), *Latin American Investment Treaty Arbitration – The Controversies and Conflicts*, Kluwer Law International, 60–61; Sébastien Manciaux, 'La Bolivie se retire du CIRDI' (2007) 5 *Transnational Dispute Management*, 1.

[72] B. Fernando Aguirre (2012), 'Bolivia', in Jonathan C. Hamilton, Omar E. García-Bolívar and Hernando Otero (eds) *Latin American Investment Protections: Comparative Perspectives on Laws, Treaties, and Disputes for Investors, States, and Counsel*, Leiden and Boston: Martinus Nijhoff, 68; Damon Vis-Dunbar and Luke Eric Peterson, *Bolivian Water Dispute Settled, Bechtel forgoes compensation* (Investment Treaty News 2006); 'Water war in Bolivia' *The Economist* (2000) <www.economist.com/node/280871>.

[73] Bluemel, n. 68 above, 966.

[74] Vis-Dunbar and Peterson, n.72 above; Manciaux, n. 71 above.

[75] See *Suez, Sociedad General de Aguas de Barcelona SA, and Vivendi Universal SA v Argentina* (ICSID) and *Anglian Water Group (AWG) v Argentina* [2010] (UNCITRAL), Decision on Liability.

[76] AWG Decision on Liability, *Anglian Water Group (AWG) v Argentina* [2010] (UNCITRAL) para. 236.

[77] Ibid., emphasis added.

[78] Ibid.

Argentina and the *amicus curiae* submissions received by the Tribunal suggest that Argentina's human rights obligations to assure its population the right to water somehow trumps its obligations under the BITs and that the existence of the human right to water also implicitly gives Argentina the authority to take actions in disregard of its BIT obligations. The Tribunal does not find a basis for such a conclusion either in the BITs or international law. Argentina is subject to both international obligations, *i.e.* human rights *and* treaty obligation[s], and must respect both of them equally.[79]

And added:

Had the Contracting Parties, after carefully negotiating a complex set of legal obligations to protect and promote investments, intended that such obligations would not apply in times of war, civil disturbance, or national emergency, they certainly would have so stated specifically. Indeed, in many other BITs, contracting parties have included exception provisions to provide for limited exemptions from BIT obligations in particular situations. The Contracting Parties of the BITs in question in these cases could also have done so if they had wished, but they did not.[80]

In other words the tribunal, despite mentioning Argentina's reasonable right to regulate, in the absence of specific exceptions, did not eventually concede this right.[81] The tribunal found that Argentina had violated the bilateral investment treaty in question.

Pac Rim Cayman v El Salvador is a further currently pending water-related dispute.[82] The dispute arose out of the government's refusal to issue licences for a gold mining project in El Salvador due to alleged environmental issues, including the large amounts of water and cyanide required in the extraction process.

CONCLUSION

Evidence on the basis of the analysis of the right to regulate in the public interest in general and in relation to water more particularly indicates that the state does not have a *stricto sensu* right to regulate water unless concrete conventional exceptions say it does. In all other cases, the state that regulates water and by doing so harms foreign investors protected

---

[79] Ibid., para. 262, emphasis in original.
[80] Ibid., para. 270.
[81] Titi, n. 1 above, 290–91.
[82] *Pac Rim Cayman LLC v El Salvador* (ICSID).

under an investment treaty is likely to be found liable to pay compensation to the aggrieved investors.

But the question of the host state's right to regulate water is not dissimilar to – indeed it is part of – the wider debate on policy space for contracting parties in respect of measures for the protection of the public interest. In some cases, it may be desirable that host states retain the capacity to regulate aspects of water and water services without having to compensate investors. These aspects will generally fall under the scope of treaty exceptions relating to the protection of animal, human, and plant life or health, or the protection of the environment. It is much less certain that contracting parties should be allowed to digress from their investment commitments for reasons that do not involve the public interest. A state that prematurely terminates a water concession contract because it has simply decided to grant the concession to a domestic investor or to another foreign investor should still be liable to pay compensation. New generation investment treaties are generally cognisant of host state regulatory freedom, and they often allow contracting parties to regulate in the public interest. These treaties will generally then also accord states a right to regulate 'important' aspects of water.

7. Regulation and protection of water in international law: terrestrial and marine perspectives
Virginie J.M. Tassin

Water surrounds us and creates favourable conditions for the development of ecosystems that can sustain life. Water covers 71 per cent of our blue or 'water' planet, with 97.2 per cent of its water located in oceans and seas. Among those blue spaces, 2.5 per cent is freshwater while 70 per cent of frozen freshwater is located in Polar Regions and glaciers, and only around 1 per cent of the world's freshwater is accessible for human uses. Solid, liquid or gas, water can take various forms and can be found in many locations, from the surface of Earth to the form of ice, rivers, lake, seas and oceans, and even within its underbelly with the recent discovery of the existence of huge freshwater reserves beneath the ocean floor.[1]

All these different forms of water are nevertheless under common threats, first, due to the acceleration of human activities on land and at sea, and, second, to the related effects of climate change.

The world demand pushes our society to slowly turn to the sea rich in promises of future treasures that remain undiscovered. Accessing these new resources and forms of energy requires developing new human activities, which will give a central role to water issues. These activities such as deep-sea mining, navigation in polar areas, bio-prospecting and the development of marine renewable energies will cause new types of

---

[1] Post VEA et al, 'Offshore fresh groundwater reserves as a global phenomenon', (2013) 504 *Nature* 71. According to scientists, it is estimated that half-a-million cubic kilometers of freshwater is buried below the continental shelves off the coasts of Australia, China, North America and South Africa. The authors declare that 'the widespread confirmation of the scale of offshore fresh and brackish groundwater reserves therefore provides opportunities for the relief of water scarcity in densely populated coastal regions. Offshore groundwater abstraction can help mitigate the adverse effects of onshore pumping, such as land subsidence and seawater intrusion. This provides another important impetus to shift the boundaries of hydrogeology into the offshore domain'.

disturbances and pollution to the marine environment, which might affect water quality.

In addition, the land-based human activities are affecting water quality. Litter, telluric ocean pollution, or pollution caused by a sewage treatment plant, wastewater or by the exploitation of shale gas are examples, among others, illustrating the many threats faced by water.[2]

The icing on the cake, climate change, is also threatening water quality, essential to the good health of biological resources and to our survival. Ocean acidification, melting ice caps and sea level rise already disrupt ecosystems, food webs and humans who have recently started to relocate to other countries claiming the status of climate change refugees.[3] Freshwater and drinking water are thus logically facing many issues such as floods and droughts, putting the development of countries in danger as well as the health and life of their inhabitants.

As our appetite for resources, food and energy rapidly grows, the relationship of our civilization to water is changing. Freshwater used for agriculture[4] and our everyday use is becoming scarce and its access is problematic in certain regions and countries – specifically in developing and emerging economies. On its side, the quantity of saltwater is not an issue, but its quality is greatly decreasing, leading some of us to claim that our oceans are 'broken'.[5]

Establishing a list of the various threats is not the point of this chapter. This non-exhaustive list should nevertheless be used as a way to remind us of the interconnectedness of all threats to all forms of water, be they on land or at sea.[6] Indeed, not only are the five different ocean basins interconnected, leading oceanographers often talk about *one single ocean*, but also all water from mountain freshwater to rivers and to the various

---

[2] 'Gaz de schiste: risque de pollution de l'eau potable par le méthane' Science et Avenir (Online Edition 12 2011).

[3] See *Ioane Teitiota v The Chief Executive of Ministry of Business, Innovation and Employment* (CA 50/2014 [2014] NZCA, 8 May 2014) 173.

[4] Agriculture uses around 87 per cent of the total water used globally.

[5] See e.g., G Ray, 'The ocean is broken', Sydney Morning Herald (Online Edition 2013) <www.smh.com.au/environment/the-ocean-is-broken-20131018-2vs7v.html> accessed in February 2016.

[6] The UNESCO Report 2014 recognised the 'interconnectedness of the water, energy and related domains' to call for a coherent policy which will require a certain number of actions such as ensuring the availability of finance, allowing markets and business to develop, supporting innovation and research into technological development and creating legal and institutional frameworks to promote this coherence. The United Nations World Water Development Report, 'Water and Energy: Executive Summary' (2014) 12.

seas are interconnected as well.⁷ A pollution or disturbance in one part of the water cycle will be likely impact the rest, which will then have an effect upon the related ecosystems relying on the various forms of water. Therefore, regulating and protecting only one small part of the water cycle, such as freshwater for example, will not, by far, guarantee an efficient protection of water and its role on Earth.

Reflecting on water issues, and more particularly on water regulation, requires bearing in mind that this topic is closely linked to the important challenges of our modern society such as environmental security, energy security, food security and health security.⁸ Hence, this topic is a highly sensitive one. Indeed, as it was clearly declared in 2014 by the Water Report of the UNESCO, the World Water Assessment Program and the United Nations Water:

> Water and energy supply and provision are interdependent. Choices made in one domain impact the other, for better or for worse. Policy-makers, planners and practitioners can take steps to overcome the barriers that exist between their respective domains. Innovative and pragmatic national policies can lead to more efficient and cost-effective provision of water and energy services . . . Water and Energy are both at the heart of sustainable development and need to be recognized as such.⁹

By shining light on the differences of regulation and protection of water on land (1) and at sea (2), this chapter wishes to raise awareness about the importance of a global water vision and a global water governance.

1. THE TERRESTRIAL APPROACH TO WATER REGULATION

Water is essential to our survival on Earth. Being composed of almost 60 per cent of water, we need to access clean and drinkable water in order to ensure our optimum health. Water resources able to provide safe and drinking water are nevertheless scarce, freshwater representing less than 2.5 per cent of water on Earth.

Since the boom of industrialization following the Second World War, the use of water has been considerably diversified. Besides domestic and

---

⁷ Emphasis added.
⁸ These challenges will not be addressed directly in this chapter but are discussed in Chs 1, 2 and 7 in this book.
⁹ UN Water 2014, The United Nations World Water Development Report 2014, Water and Energy : Executive Summary, 2.

personal uses, men are using very significant quantities of terrestrial water supplies for intensive agriculture and the building of secured livelihoods, consuming at a high speed.

The scarcity of water, which could be defined as the lack of water quantity and quality for human consumption, will increase at the pace of the development of our society and the demographic explosion. Education, hunger, health and poverty issues will then grow in line with the water crisis. This crisis, described sometimes as 'water wars',[10] will undoubtedly impact internal waters as well as transboundary waters. Therefore, States will need to articulate their actions internally and externally. It is beyond doubt that it will require new systems of protection and management of water resources.

Despite these numerous and changing factors of water stress, States have paid very little attention to global water issues. The human right to water, enshrined into a 'terrestrial and human' reading consisting of guaranteeing water for domestic and personal uses, faces many difficulties regarding its recognition and implementation. The link between the concept of a healthy environment and the regulation of water on land is thus becoming essential to the development of international norms regulating water use.

1.1 The Human Right to Water: a Concept More Than a Right

Human need is at the heart of the human right to water. Driven by their needs to secure freshwater supplies as well as an access to clean and safe water, States have built international norms and practices replying to their most direct concerns: ensuring the survival of their populations and the potential of development of their territories.

The importance of access and quality of water was first mentioned

---

[10] 'Water wars: a new reality for business and governments', The Guardian (Online Edition, 2014) <www.theguardian.com/sustainable-business/2014/oct/06/water-wars-business-governments-scarcity-pollution-access> accessed in February 2016. As highlighted in this chapter, water conflicts greatly impact companies' profits in the form of reputational damage, legal sanctions and regulatory obstacles. The non-sustainable use of transboundary water resources also significantly contributes to the escalation of disputes between States. According to the Global Risks Report 2016 of the World Economic Forum, water crises pose the highest risk for the next ten years. World Economic Forum (2016), 'Figure 1.2: The Top Five Global Risks of Highest Concern for the Next 18 Months and 10 Years', Global Risks Report, 13 <http://reports.weforum.org/global-risks-2016/part-1-title-tba/> accessed in February 2016.

in 1977 at the United Nations Conference organised in Mar del Plata, Argentina.[11]

Although not explicitly mentioned in article 6, paragraph 1, of the 1966 International Covenant on Civil and Political Rights[12] and article 11, paragraph 1, and article 12, paragraph 2, of the 1966 International Covenant on Economic, Social and Cultural Rights,[13] the access to water has been slowly recognised throughout the years as related to the socio-economic rights of these Conventions. Indeed, how could the right to life, recognised as a supreme human right,[14] or the right to food and to health exist without any water?[15]

Other Conventions mention explicitly the importance of access to drinking water, such as the 1979 Convention on the Elimination of All Forms of Discrimination Against Women,[16] the 1992 Convention on the Rights of the Child, the 1999 Protocol on Water and Health and the 1992 Convention on the Protection of Use of Transboundary Watercourses and International Lakes.[17] These Conventions are nevertheless classic examples of interstate instruments without any direct implementation possible for

---

[11] United Nations 1977, 'Report of the United Nations Conference on Water, Mar del Plata' 70/29 E/Conf. 14.

[12] Hereinafter 'ICCPR'.

[13] Hereinafter 'ICESCR'.

[14] The right to life enunciated in article 6 of the Covenant has been dealt with in all State reports. It is a supreme right from which no derogation is permitted even in time of public emergency, which threatens the life of the nation (art. 4). However, the Committee has noted that quite often the information given concerning article 6 was limited to only one or other aspect of this right. It is a right which should not be interpreted narrowly.

Human Rights Committee 1982, General Comment No. 6, Article 6 (Sixteenth Session), Compilation of General Comments and General Recommendations Adopted by Human Rights Treaty Bodies, UN Doc HRI/GEN/1/Rev.1, paras 1, 6.

[15] Indeed, '(a)ccess to basic supplies of safe water is a fundamental precondition for human sustenance. In that sense, the human right to water constitutes a "survival right" comparable to, e.g., the right to food. Therefore, from a conceptual point of view, a substantive "human right to water" would have to include as an essential component a positive entitlement to access to basic water supplies'. T Kieffer and C Brölmann, 'Beyond state sovereignty: the human right to water' 2005 5 *Non-State Actors and International Law* 184. See also art. 10, para. 2 of the United Nations Watercourses Convention, which highlights the role of watercourses as 'human vital needs'.

[16] Art. 14, para. 2 of the Watercourses Convention.

[17] 'Take into account specific water-quality requirements (raw water for drinking-water purposes, irrigation, etc)'. Annexe III, c), Convention on the Protection of Use of Transboundary Watercourses and International Lakes.

all humans.[18] The first important turning point was marked in 1992 when the international community consecrated the importance of the protection and quality of supply of freshwater resources in an international soft law instrument, 'Agenda 21', adopted at the United Nations Conference on Environment and Development in Rio de Janeiro. According to article 18, paragraph 2, of this Agenda 21:

> Water is needed in all aspects of life. The general objective is to make certain that adequate supplies of water of good quality are maintained for the entire population of this planet, while preserving the hydrological, biological and chemical functions of ecosystems, adapting human activities with the capacity limits of nature and combating vectors of water-related diseases (. . .).

The 'standalone' human-right-need emerged meanwhile quite late, in 1997, when the United Nations Sub-Commission on the Protection and Promotion of Human Rights appointed Mr. Guissé to investigate a right of access to drinking water and sanitation services for everyone. Subsequent to this investigation, Mr. Guissé was appointed as special rapporteur in charge of reporting on the topic of drinking water access in order to assist the United Nations to define, long term, right to water in relation to other human rights.[19]

Guided by the object and purpose of the Charter of the United Nations praising the promotion of social progress, better standards of life and the promotion of economic and social development of all people, the General Assembly of the United Nations affirmed in 2000 the right to clean water, along with the right to food, as 'fundamental human rights' promotion of which 'constitutes a moral imperative both for national Governments and for the international community'. It also adopted in the same year the Millennium Declaration recognising the collective responsibility of the head of States and governments to uphold the principles of human dignity, equality and equity at the global level.[20] This declaration invited States to halve, by year 2015, 'the proportion of the people who are unable to reach or to afford safe drinking water' and '(t)o stop the unsustainable exploitation of water resources by developing water management strategies at the regional, national

---

[18] Emphasis added.
[19] See Sub-Commission on the Protection and Promotion of Human Rights Resolution No 2001/2 and Decision No 2002/105.
[20] General Assembly 2000, 'United Nations Millennium Declaration', Resolution, A/RES/55/2, para 2.

and local levels, which promote both equitable access and adequate supplies'.²¹

Another turning point was reached in 2002 with the issuance of 'General Comment 15 on the Human right to water'²² by the United Nations Committee on Economic, Social and Cultural Rights.²³ By explicitly recognising the existence of a human right to water despite 'the widespread denial of the right to water in developing as well as developed countries',²⁴ General Comment 15 pointed to the main issue surrounding the recognition of a *human right* to water: its normative value.²⁵ Before engaging more deeply into this issue, some comments should be made regarding the content of General Comment 15.

First and foremost, General Comment 15 recognises water as a 'limited natural *resource* and a public good fundamental for life and health'.²⁶ As a consequence, the definition of the human right to water has been shaped broadly. It entitles everyone to 'sufficient, safe, acceptable, physically accessible and affordable water for *personal and domestic uses*. An adequate amount of safe water is necessary to prevent death from dehydration, to reduce the risk of water-related disease and to provide for consumption, cooking, personal and domestic hygienic requirements'.²⁷ By taking into consideration the various uses of water, General Comment 15 furthermore establishes a priority in the allocation of water in favour first, of personal and domestic uses and second, in favour of 'water resources required to prevent starvation and disease, as well as water required to meet the core obligations of each of the Covenant rights'.²⁸ The interconnectedness of water with health, food and housing issues (i.e., the direct needs of humans) is thus at the heart of this human right to water.

The CESR goes even beyond this priority list by stating that the human right to water is a 'prerequisite for the realization of other human rights'.²⁹

---

[21] General Assembly 2000, 'United Nations Millennium Declaration', Resolution, A/RES/55/2, paras 19 and 23.
[22] Economic and Social Council 2003, 'General Comment No 15 (2002) The right to water (arts 11 and 12 of the ICESCR)', E/C.12/2002/11. Hereinafter 'General Comment 15'.
[23] Hereinafter 'CESR'.
[24] Para. 1, General Comment 15.
[25] Emphasis added.
[26] Para. 1, General Comment 15. Emphasis added.
[27] Para. 2, General Comment 15. Emphasis added.
[28] Economic and Social Council 2003, 'General Comment No 15 (2002) The right to water (arts 11 and 12 of the International Covenant on Economic, Social and Cultural Rights)', E/C.12/2002/11, para. 6, 3.
[29] Ibid., 1.

Such a reference is clearly establishing a connection between the ICCPR and the ICESCR. It also gives a new dimension to this right to water. The right to water seems to become a *human* right, a right able to be claimed by individuals, therefore a 'direct right'.[30]

The difficulty around this *new* human right to water lies nonetheless very much in its normative content.[31] By pointing out the 'widespread denial of States', General Comment 15 confirms that this human right neither exists thanks to the consent of States, nor does it indicate a pre-existing or emerging customary rule. The Comment has thus no compulsory nature and is not a source of international law.[32] As a ricochet, individuals cannot claim this human right, which is lacking any binding nature. Hence, the situation is extremely delicate. How could individuals enforce this (soft law) human right to water? Indeed could it be enforced at all?

Mr. Guissé, himself highlighted this issue in 2003 after the publication of General Comment 15 by stating that, 'it would be necessary to establish a legal framework for this basic right since it would be impossible for individuals to call for this right without a legal text to support them'.[33]

One delicate point is the interpretation that could be given to articles 11 and 12 of the ICESCR. Without any direct reference to the right to water, and to any human right to water, could the right to water, and any human right to water, be taken into account when implementing articles 11 and 12? As stated in article 31 of the Vienna Convention on the Law of Treaties, the interpretation of a treaty shall be done in good faith 'in accordance with the ordinary meaning to be given to the terms of the treaty in their context and in the light of its object and purpose'. In the absence of any direct reference to the human right to water in the text of the treaty/convention and in the absence of any subsequent agreement of the States Parties relating to the ICESCR, the context and the object of the treaty could be assessed through the preparatory work of this Convention.

However, this preparatory work is not of any great help as water issues have been simply, and not surprisingly, omitted from the negotiating table. As it was the case for the regulation of water in Law of the Sea, this omission is not voluntary but results from a lack of awareness of States of the importance and various dimensions of water issues.

When studying the ICESCR, T.S. Bulto suggested that General

---

[30] Emphasis added.
[31] Emphasis added.
[32] Art. 38, para. 1, Statute of the International Court of Justice.
[33] Sub-Commission on the Promotion and Protection of Human Rights 2003, 'Sub-Commission Begins Consideration of Economic, Social and Cultural Rights', Press Release.

Comment 15 did not invent this human right to water; it discovered it through the object and purpose of the ICESCR. The right to water could be then apprehended as an autonomous right, 'a member of the illustrative list of art 11 of the ICESCR. Put differently, the human right to water can be treated as an independent right, deserving of equal protection just like the other more explicit rights listed under art 11 of the ICESCR'.[34]

In 2010, the General Assembly of the United Nations added an important stone in the recognition of a human right to water. Entitled 'The human right to water and sanitation',[35] Resolution 64/292 recognised the right to safe and clean water and sanitation as a human right 'that is essential for the full enjoyment of life and all human rights'. Conscious of the great economic challenges that such access would require, the General Assembly encouraged States and international organisations to provide resources, capacity building and technology transfer to implement this human right. Adopted by 122 voices and 41 abstentions, some States have however deplored the lack of consensus surrounding this Resolution. The United States noted that: 'This draft resolution describes a right to water and sanitation in a way that is not reflective of existing international law, as there is no right to water and sanitation in an international legal sense as described by the draft resolution'.[36]

Shortly after this Resolution, the Human Right Council of the United Nations adopted, by consensus, Resolution 15/L.14 on the basis of the work of Mrs. Albuquerque, independent expert in charge of examining the obligations of States with regard to a human right to water. This Resolution affirms that water and sanitation are 'derived from the right to an adequate standard of living and inextricably related to the right to the highest attainable standard of physical and mental health, as well as the right to life and human dignity'.[37] More importantly, the work of Mrs. Albuquerque clarifies the content of this human right to water. According to her, States are under an obligation to create an environment suitable for the realisation of all human rights, including the one of water. This obligation entails a duty to provide affordable water, which requires the establishment of affordable water services, so that *access* to safe water

---

[34] TS Bulto, 'The emergence of a human right to water: invention or discovery?' (2011) 12 *Melbourne Journal of International Law* 290, 303.
[35] General Assembly 2010, 'The Human Right to Water and Sanitation', A/RES/64/292, paras 1–2.
[36] General Assembly 2010, '108th Plenary Meeting' Official Records, A/64/PV108, 8.
[37] Human Rights Council 2010, 'Human rights and access to safe and drinking water and sanitation', A/HRC/15/L14.

and sanitation could be given to all humans. This human right to water is consequently, first, restricted to the personal and domestic water uses, and second, defined by its availability, quality, accessibility, affordability and acceptability.[38]

Obviously, this interpretation of the United Nations seeks to ensure and maximise the protection of the human right to water in international law. It nevertheless does not ease the great complexity surrounding its implementation.[39] Despite the growing recognition of this human right to water in a number of domestic case law and regional soft law instruments,[40] it is not a human right but a concept emerging through State practice.

Such State practice will be considerably reinforced through the years thanks to the adoption, in September 2015, of a new international agenda dedicated to sustainable development 'the Sustainable Development Goals',[41] which takes over the Millennium Development Goals[42] and complements it by adding new SDGs[43] and specific targets – 169 exactly – designed to guide their implementation. Among these goals, SDG 6 encourages States *and* stakeholders to ensure the availability and sustainable management of water and sanitation for all. The accuracy of SDG 6 gives a new momentum to these issues. Indeed, its associated targets set two deadlines: By 2020 States and stakeholders are expected to have protected and restored water-related ecosystems.[44] By 2030, States and stakeholders are expected to have achieved universal and equitable access to safe and affordable drinking water for all, achieved access to adequate and equitable sanitation and hygiene for all and end open defecation, improved water quality by reducing pollution, eliminated dumping and minimized release of hazardous chemicals and materials as well as halved the proportion of untreated wastewater and increased recycling and safe reuse globally. They are also expected to have implemented integrated water resources management at all levels and expanded international cooperation and capacity-building support to developing countries. SDG 6 is ambitious. By being much more precise than the one formulated in

---

[38] C de Albuquerque, 'Water and Sanitation as Human Rights' L'eau et son droit, Rapport public, Etudes et documents du Conseil d'Etat (2010) 483.
[39] L Caflisch, 'Le droit à l'eau - un droit de l'homme internationalement protégé?' L'eau en droit international, Colloque d'Orléans, Pedone, (Paris, 2011) 385.
[40] Bulto n. 34 above, 309–10, 312.
[41] General Assembly 2015, Transforming our world: the 2030 Agenda for Sustainable Development, Resolution A/RES/70/1. Hereinafter 'SDGs' or 'SDG'.
[42] Hereinafter 'MDGs'.
[43] The MDGs had eight goals against 17 for the SDGs.
[44] Target 6.6, General Assembly 2015, n 41 above, 18. It includes mountains, forests, wetlands, rivers, aquifers and lakes.

the MDGs, SDG 6 wishes to speed up a consistent States practice. It also subtly highlights the diversity of issues related to the implementation of a human right to water. [45]

Creating a human right to water as a treaty-based obligation will therefore not simply solve the issue of a human right to water. The main issue lies in the availability of water resources and the capacity of States to put in place water distribution networks and infrastructures replying to the needs of their populations and, more generally, the development of their territories. The *access* to clean and safe water is a priority, and a more realistic one than an ill-defined human right to water, which will require an extensive process before it can be precisely defined and implemented.

Guaranteeing the accessibility to water should thus be only one part of the answer. Indeed, the availability and accessibility of water for our direct use would be of no use if such water is not, as stated by General Comment No. 15, 'safe and acceptable'. In addition, the inextricable link between water access, water quality and a healthy environment should not be forgotten. As was stated in 2010 by the arbitral tribunal in *Suez, Sociedad General de Aguas de Barcelona, S.A. and Vivendi Universal S.A. v. Argentine Republic*: 'because of its importance to the life and health of the population . . . water cannot be treated as an ordinary commodity'.[46]

1.2 Water Regulation and Environmental Protection

By adapting and transforming their surroundings, humans have created amazing tools, structures and equipment in order to enhance the quality of their lives. The same actions have, however, often caused incalculable harm to the environment and to human beings.[47] Pesticides,[48] antibiotics, agricultural practices and increasing industrial activities are among the main causes of a deterioration in the availability and quality of water

---

[45] See also SDG 14 'Life Below Water: conserve and sustainably use the oceans, seas and marine resources for sustainable development' which is one of the new SDGs introduced into the sustainable agenda 2030. General Assembly 2015, n 41 above, 23–24.

[46] ICSID Case, *Suez, Sociedad General de Aguas de Barcelona, SA and Vivendi Universal SA v Argentine Republic,* Decision on Liability ARB/03/19, 30 July 2010, see paras 25 and 26.

[47] Declaration of the United Nations Conference on the Human Environment, Stockholm, 1972, para 3.

[48] See A Agrawal et al., 'Water pollution with special reference to pesticide contamination in India', (2010) 2 *Journal of Water Resource and Protection* 432. See also KH Bowmer, 'Ecosystem effects from nutrient and pesticide pollutants : catchment care as a solution', (2013) 2 *Resources* 439.

needed for humans' life and activities. As an example, in 2009, agriculture and the general industry were using respectively almost 70 per cent and 20 per cent, respectively, of the water used by humans.[49] This use of freshwater dramatically increases its pollution because much of this water is returned containing new nutrients and contaminants, threatening food security and more generally environmental management challenges. The expanding use of water (mainly freshwater) by the industry, be it as users or providers, encourages us to consider the issue of environmental and qualitative protection offered to this water by international law. How are we equipped to face what the United Nations Environment Programme describes as a 'global water quality crisis'?[50]

1.2.1 Identification of sources of international law applicable to environmental protection

Post World War II, the quality of water was one of the concerns of the international community because of barbarous acts committed in times of war. Inherited from various ancient war customs,[51] the First Protocol of the Geneva Conventions of 12 August 1948 codified the protection of water as a vital *resource* and an *environment*.[52] Article 54 provides that:

> It is prohibited to attack, destroy, remove or render useless objects indispensable to the survival of the civilian population, such as foodstuffs, agricultural areas for the production of foodstuffs, crops, livestock, drinking water installations and supplies and irrigation works, for the specific purpose of denying them for their sustenance value to the civilian population or to the adverse Party, whatever the motive, whether in order to starve out civilians, to cause them to move away, or for any other motive.[53]

---

[49] De Albuquerque, n. 38 above, 484.
[50] UNEP 2010, 'Sick water? The central role of wastewater management in sustainable development: A rapid response assessment' E Corcoran et al (eds), UNEP, UN-Habitat, GRID-Arendal 9.
[51] By reviewing different poisoning practices through history, P. Weckel describes how water could be used as a weapon in times of war in 'L'eau et le droit humanitaire' L'eau en droit international, Colloque d'Orléans, Pedone (Paris 2011) 371.
[52] Emphasis added.
[53] In 2005, the Eritrea-Ethiopia Claims Commission recognised that this provision reflects customary international humanitarian law.

> While the Protocol had not attained universal acceptance by the time these attacks occurred in 1999 and 2000, it had been very widely accepted. The Commission believes that, in those circumstances, a treaty provision of a compelling humanitarian nature that has not been questioned by any statements of reservation or interpretation and is not inconsistent with general State

As a complement, article 55 also protects water as a general component of the environment by prohibiting any widespread, long-term and severe damage intending or expecting to prejudice the health or survival of a population. The objective of this Protocol was therefore to avoid any poisoning of the water and any cuts of freshwater causing deaths by dehydration or starvation.

This 'no-harm' principle finds its roots in a broader custom, which guided the adoption of numerous norms and judicial decisions not relating only to water protection. Pursuant to this customary international law, States have a duty to prevent, control and reduce pollution arising from activities conducted within their territories and jurisdiction.

One of the key soft law instruments recognizing this custom is Principle 21 of the UN Stockholm Declaration of 1972:

> States have, in accordance with the Charter of the United Nations and the principles of international law, the sovereign right to exploit their own resources pursuant to their own environmental policies, and the responsibility to ensure that activities within their jurisdiction or control *do not cause damage* to the environment of other States or of areas beyond the limits of national jurisdiction.[54]

As the first and 'forward-looking'[55] soft law instrument dedicated to sustainable development and environmental law, the Stockholm Declaration stands out for its modern recommendations. By encouraging the preservation and enhancement of the 'human environment', this instrument calls for an *integrated approach of the environment* conciliating man's environment consisting of the 'natural and the man made'.[56] It invites States to cooperate and further develop 'the international law regarding liability and compensation for the victims pollution and other environmental damage caused by activities within the jurisdiction or control of such States to areas beyond their jurisdiction'.[57] In the context of this instrument, water is defined as part of the 'natural *resources* of the Earth',[58] which should

---

practice in the two decades since the conclusion of the treaty may reasonably be considered to have come to reflect customary international law.

Western Front, Aerial Bombardment and Related Claims Eritrea's Claims 1, 3, 5, 9–13, 14, 21, 25, 26, 19 September 2005, paras 105, 29.

[54] Principle 21, Stockholm Declaration. Emphasis added.
[55] See DA Wirth, 'The Rio Declaration on Environment and Development: two steps forward and one back, or vice versa' (1995) 29 *Georgia Law Review* 611.
[56] Para 1, Stockholm Declaration. Emphasis added.
[57] Principle 22, Stockholm Declaration.
[58] Emphasis added.

be 'safeguarded for the benefit of present and future generations through careful planning and management, as appropriate'. This duty of preservation is the backbone of a fundamental right of 'freedom, equality and adequate conditions of life, in a environment of a quality that permits a life of dignity and well-being (. . .)' that man *has*.[59]

The 1992 Rio Declaration continued to modernise sustainable development practices. It reaffirmed the no-harm principle in exactly the same way as the Stockholm Declaration but added a two-gear responsibility in case of 'damages to the environment' and in cases of 'significant adverse impact to the environment'. States are required to undertake, domestically, environmental impact assessment when it is likely that activities conducted will have a '*significant* adverse impact on the environment'.[60] In case of '*significant* adverse transboundary environmental effect', States shall also notify prior and in a timely manner the potentially affected State.[61] The Declaration further requires States to apply widely the precautionary approach defined as follows: 'When there are threats of serious or irreversible damage, lack of full scientific certainty shall not be used as a reason for postponing cost-effective measures to prevent environmental degradation.'[62]

On another level, The Rio Declaration deviates from the global environmental vision of the Stockholm Declaration by stating a right to development 'that must be fulfilled so as to equitably meet developmental and environmental needs of present and future generations'.[63] One explanation on the use of a 'right to development' rather than a 'right to environment' could be the practical implementation of such right. Praising a right to development seems easier to enforce than a right to environment. Indeed, declaring a right to environment would require defining its precise content (what is a healthy environment and what is an acceptable environment) and creating implementation mechanisms, which could raise innumerable difficulties considering the great diversities of ecosystems.

Logically, the no-harm principle has been integrated into a number of international conventions but with slight differences. The Preamble of the United Nations Framework Convention on Climate Change in 1992 makes direct references to the Stockholm Declaration of 1972 and recalls that States have 'the sovereign right to exploit their own resources

---

[59] Principle 1, Stockholm Declaration. Emphasis added to highlight the use of the present tense in Principle 1, which seems to indicate a pre-existing right.
[60] Emphasis added.
[61] Principle 19, Rio Declaration. Emphasis added.
[62] Principle 15, Rio Declaration.
[63] Principle 3, Rio Declaration.

pursuant to their own environmental and developmental policies, and the responsibility to ensure that activities within their jurisdiction or control *do not cause damage* to the environment of other States or of areas beyond the limits of national jurisdiction'.[64] This exact formula will be inserted in 1992 as the core principle of the Convention on Biological Diversity adopted at the United Nations Conference held in Rio de Janeiro.[65]

In 2001, the International Law Commission[66] decided to encourage the use of a criterion of harm, i.e., a risk of *'significant* transboundary harm'[67] as used by the Rio Declaration.[68] It adopted a very broad definition of harm by including the harm caused to persons, property or the environment[69] and highlighted the possible ambiguity of the term 'significant'. According to the ILC, this term should be understood as 'something detectable', which is not necessarily 'serious' or 'substantial'.[70] Hence, the risk[71] should not be assessed separately but at the same time as the evaluation of harm so as to combine the probability of the occurrence of an accident (the risk) and the magnitude of its impact (its significance). Article 1 of the ILC Draft Articles also interestingly included into this no-harm principle 'activities not prohibited by international law' in order to distinguish issues linked to the international liability of States from issues of States responsibility.

In 1996 and 2010, the International Court of Justice[72] added important stones to the crystallisation of these new principles of sustainable development and environmental law. The Advisory Opinion on the *Legality of the Threat or Use of Nuclear Weapons* in 1996 proved the existence of general obligations of States regarding the no-harm principle.[73] A few years

---

[64] Emphasis added.
[65] Article 3, Convention on Biological Diversity.
[66] Hereinafter 'ILC'.
[67] Emphasis added.
[68] 'The present articles apply to activities not prohibited by international law which involve a risk of causing significant transboundary harm through their physical consequences'. Article 1. ILC 2001, 'Draft Articles on Prevention of Transboundary Harm from Hazardous Activities', 149.
[69] The ILC highlights in this regard that 'the ecological unity of the planet does not correspond to political boundaries', ibid., para. 5, 152.
[70] Ibid., para. 4, 152.
[71] 'The notion of risk is thus to be taken objectively, as denoting an appreciation of possible harm resulting from an activity which a properly informed observer had or ought to have had', ibid., para. 14, 151.
[72] Hereinafter 'ICJ'.
[73] 'The existence of the general obligation of States to ensure that activities within their jurisdiction and control respect the environment of other States or of areas beyond national control is now part of the corpus of international law

later, the *Pulp Mills on the River Uruguay (Argentina v. Uruguay)* case went further *and* recognised environmental impact assessment as 'a requirement under general international law', required 'where there is *a risk* that the proposed industrial activity may have a *significant adverse impact* in a transboundary context, in particular, on a shared resource'.[74] The ICJ also described the planning and management duties of States. Composed of the duty of due diligence and the duty of vigilance and prevention, these planning and management duties 'would not be considered to have been exercised, if a party planning works liable to affect the regime of the river or the quality of its waters did not undertake an environmental impact assessment on the potential effects of such works'.[75]

Despite some differences among these various instruments, the no-harm principle has been inserted into a number of other conventions under a general duty which takes the form of a duty to 'prevent, reduce and control' pollution from its territory.[76] It should be noted that the measurement 'significant'[77] is not universally used, some treaties using the criteria of 'serious'[78] or 'substantial'[79] harm. That being said, the no-harm principle is nowadays a cornerstone of international environmental law and it has been declined through two key principles: the preventive and the precautionary principles.

However, uncertainties around the scope and content of the duty of prevention and precaution of States have raised many questions as to their legal status.[80] While their lack of fixed content can indeed cause issues when studying the practices of States, both of these principles are widely applied by States as 'a general practice accepted as law'.[81] In this way,

---

relating to the environment'. *Legality of the Threat or Use of Nuclear Weapons* (Advisory Opinion, ICJ Reports 1996), para. 29, 242.

[74] *Pulp Mills on the River Uruguay (Argentina v Uruguay)* (Judgment, ICJ Reports 2010) para. 204. Emphasis added.

[75] Ibid.

[76] See for example art. 194, para. 1, UNCLOS. See also section II, para. 1, Code of Conduct on Accidental Pollution of Transboundary Inland Waters 1990.

[77] See Article 4, para 2 of the Convention on the Regulation of Antarctic Mineral Resource Activities (not in force), which adopts a very broad definition of the no-harm principle.

[78] The Convention on the Law of the Non-navigational Uses of International Watercourses establishes two measurements: 'significant' harm and 'serious harm'. See arts 7 and 28.

[79] Art. 4, para. 2, Convention on the Regulation of Antarctic Mineral Resource Activities (not in force).

[80] P Sands and J Peel, *Principles of International Environmental Law* (3rd ed., Cambridge University Press 2012) 222.

[81] Art. 38(1)(b), ICJ Statute.

the diversity of means of enforcement by States does not preclude a wide application of these principles. Moreover, for the same reasons activities potentially causing damage to the environment have not been described in detail,[82] measures aiming at preventing and applying a duty of precaution cannot be listed as they adapt themselves at the pace of scientific discoveries and are in phase with the specificities of each single ecosystem.

In an Advisory Opinion dedicated to deep-sea mining, the Seabed Dispute Chamber of the International Tribunal for the Law of the Sea[83] addressed the issue of responsibility and liability of States sponsoring persons and entities with respect to activities conducted in the Area (the deep abyssal plains of oceans). This advisory opinion proclaimed the precautionary principle as part of customary international law[84] and reaffirmed the inextricable link between the precautionary principle and the duty of due diligence.[85] The Chamber also clarified the scope of the duty of prevention by declaring that a failure to take appropriate measures to prevent damage that may result from activities where there are plausible indications of potential risks 'would amount to a failure to comply with the precautionary approach'.[86]

Generally speaking this corpus of rules of international law is applicable to any type of environment and resources, including water, on

---

[82] 'It is, however, felt that specification of a list of activities in an annex to the articles is not without problems and functionally not essential. Any such list of activities is likely to be under inclusion and could become quickly dated from time to time in the light of fast evolving technology'. Art. 1(4), ILC 2001, n. 63 above, 149–50.

[83] Hereinafter the ' ITLOS'.

[84] *Responsibilities and Obligations of States with Respect to Activities in the Area, Advisory Opinion* (ITLOS Reports 2011 1 February 2011) para. 135.

[85] (I)t is appropriate to point out that the precautionary approach is also an integral part of the general obligation of due diligence of sponsoring States, which is applicable even outside the scope of the Regulations (of the International Seabed Authority).

The Chamber noted that ITLOS already established an implicit link between the two in the *Southern Bluefin Tuna Cases (New Zealand v Japan; Australia v Japan)*.

[86] *Responsibilities and Obligations of States*, n. 79 above, para. 131. On this topic see the special issue: INDEMER, 'Le régime juridique des grands fonds marins : Enjeux théoriques et pratiques à la lumière de l'avis consultatif du 1er février 2011' (2011) XVI Annuaire du droit de la mer, Pedone, Monaco 279. C Redgwell also describes this Advisory Opinion as 'what must currently rank as the high water mark of international judicial recognition of the precautionary approach'. Redgwell, *Transboundary Pollution: Principles, Policy and Practice'*, *Transboundary Pollution: Evolving Issues of International Law and Policy* (Edward Elgar Publishing 2015) 21.

shore and offshore. It sets out the relevant mechanisms to apply for the regulation and protection of water. These rules are however the minimum standards to apply and States are free to go further and create innovative instruments and mechanisms dedicated to the management and planning of water.

1.2.2 Water regulation in modern times: wastewater management challenges and access to safe water

Despite these well-recognised rules and principles, an immeasurable number of water challenges are posing issues today. Among them are wastewater management issues.

Wastewater management has become a pressing issue due to the increasing use of water by industries, agriculture and the growing demography worldwide, estimated to exceed 9 billion by 2050. Wastewater is in this context to be understood in a general way; it includes the collection, treatment, and re-use of water from domestic, agricultural, commercial and industrial activities but also water coming from storms and surface runoffs.

The challenge is considerable. A few numbers will set the picture quickly. According to the United Nations Environment Programme,[87] 80 per cent of wastewater is discharged untreated into water bodies with a number peaking at 90 per cent for developing countries where the discharge goes directly into lakes, rivers and/or oceans. Such human practice has therefore a considerable impact on ecosystem sustainability, the main effect being the phenomenon of eutrophication.[88] Long-term, eutrophic water will develop algal blooms. This high concentration of algae will gradually reduce the oxygen in the water and cause the death of all organisms, from the bottom to the surface of the water column. This is how 'dead zones' are created. This phenomenon is more common than one would think. In 2008, a study from Sweden's Göteborg University reported that the number of dead zones worldwide was around 405.[89] In 2012, The Global Environment Outlook Yearbook of the UNEP listed

---

[87] Hereinafter 'UNEP'.
[88] The phenomenon of eutrophication could be described as the excessive richness of nutrients in water, mainly due to run-off from the terrestrial land causing a dense growth of plant life and reducing therefore the depletion of dissolved oxygen.
[89] For more information, see D Perlman, 'Scientists alarmed by ocean deadzone growth', San Francisco Gate (Online Edition 2008). <www.sfgate.com/green/article/Scientists-alarmed-by-ocean-dead-zone-growth-3200041.php> accessed in February 2016.

415 eutrophic areas of which 169 have dead zones. The UNEP noted with concern the very little progress worldwide and the general trend of deterioration of freshwater and marine pollution control and reduction.[90]

The impact of wastewater on our health should also not be underestimated. In 2010, 1.8 million children under five years old died almost every 20 seconds from water-related diseases. To tackle this issue, 7, Target C, aimed to reduce by half the proportion of people without sustainable access to safe drinking water. In 2015, the international community has given greater importance to these water issues in granting them a specific goal – and not only a target – in the new 2030 sustainable agenda.[91] The scope of action of SDG 6, dedicated to improving the availability and sustainable management of water and sanitation for all, is therefore broader than the corresponding target of the Millennium Declaration of 2000.

Where are we now? In terms of drinking water access, the UNEP estimated in 2012 that 'the work will meet or even exceed the MDG drinking water target by 2015 if the current trend continues'. More precisely, around 90 per cent of the population in developing regions have supposedly gained an improved access to drinking safe water. A very recent research study published in February 2016 is nevertheless less optimistic. It reveals that previous assessments have been underestimating water scarcity by failing to capture the seasonal fluctuations in water consumption and availability. As a result, it concludes that 4 billion people (two-thirds of the world population) live under severe water scarcity at least one month per year.[92] These results are alarming. It proves that international law is not properly (or simply) implemented and could explain the ambitious targets of SDG 6.

Going back to wastewater management issues, it is important to keep in mind the snowball effect of water issues. Indeed, the more polluted the water, the greater the cost of treatment, and the more complicated it will be to reply to the growing 'safe and acceptable' water demand. In addition, health issues caused by unsafe water increase health care costs and contribute to the lost labour productivity. Therefore, water access and good wastewater management practices will improve human health and also food security and

---

[90] UNEP 2012, 'Global Environment Outlook – 5: Environment for the future we want', UNEP, Nairobi 111–12.
[91] See SDG 6. General Assembly 2015, n 41 above, 18–19.
[92] MM Mekonnen and A Hoekstra, 'Four billion people facing severe water scarcity' (2016:12) Science Advances (Online edition) < http://advances.sciencemag.org/content/2/2/e1500323> accessed in February 2016. The authors explain the difference of results of their research by the fact that previous researches have assessed water scarcity annually and not monthly.

economic development. The UNEP stated in this regard that: '(i)n terms of public spending on health issues, investing in improved wastewater management and the supply of safe water provides particularly high returns'.[93]

Improved wastewater management can further greatly improve food quality. Agricultural practices have long used wastewater as a resource. In 2012, around 20 million hectares of land were irrigated by wastewater worldwide.[94] This use of wastewater is nevertheless particularly risky considering its high proportion of contamination. For this reason, the World Health Organisation issued a guideline in 2008,[95] which is a code of good management practices to ensure good health practices in agriculture.

The ageing and under-dimensioned wastewater infrastructures worldwide is raising as well important challenges, requiring very important investments and funding mechanisms to be put in place. Research and Development and the private sectors are particularly dynamic, but a coordinated action and support of States' public administrations are missing.

To conclude, States have put in place well-recognised soft law and hard law instruments able to regulate and protect water, and protect humans from the various forms of pollution created by human activities. Although none of these rules are specifically dedicated to the regulation and protection of water, they set out (only) minimum standards and give States freedom to act as appropriate and in due time. They are however insufficiently implemented. As a result, water scarcity is spreading and water quality is degrading at high speed due to the acceleration and scaling-up of transformative human activities.[96]

We have the basic and minimum rules but implemented too timidly. Something is therefore missing. What?

One issue commonly identified is the lack of articulated economic and investment vision, short-term and long-term on water issues.[97] Facilitating private investments and clarifying the role of central, local authorities and

---

[93] UNEP 2010, 'Sick water?, n. 50 above, 11.

[94] Ibid., 32.

[95] For more information, see Mara D and Bos R, 'Risk Analysis and Epidemiology: the 2006 WHO Guidelines for the Safe Use of Wastewater in Agriculture', Wastewater Irrigation and Health: Assessing and Mitigating Risk in Law-Income Countries (International Water Management Institute, Earthscan, London 2010) 51.

[96] See generally UNEP 2011, 'Keeping Track of our changing environment: From Rio to Rio+20 (1992–2012)', UNEP, Nairobi 42.

[97] In 2001, the General Assembly identified, as a way to move forward in the implementation of the Millennium Development Goals, the promotion of increased investment in the water and sanitation sectors. General Assembly

the role of the private sector could be one solution to ensure the building of innovative and creative planning and management strategies around water issues.[98] A complementary and inescapable issue is the lack of any clear, coherent and forward-thinking water regulatory framework, able to conciliate human and economic issues as well the environment as a whole. Our following developments will therefore study the regulation of water by the Law of the Sea.

2. THE MARINE DIMENSION OF WATER REGULATION: LAW OF THE SEA & OTHER RELATED INTERNATIONAL INSTRUMENTS

Oceans mainly colour our blue planet. Their area is so important that they cover almost 80 per cent of the Southern hemisphere. It is therefore obvious to continue with the study of international law related to oceans and seas spaces in order to understand the regulation and protection of almost 97.5 per cent of the Earth's water.

2.1 Water as a Spatial Environment

The Law of the Sea is the one of oldest field of international law. It is the only field regulating, in one single Convention, the United Nations Convention on the Law of the Sea[99] of 1982, all oceans and seas spaces and their various uses. This field is nonetheless not well known and not mentioned often enough in legal studies dealing with the regulation and protection of water, most studies focusing mainly on human and land perspectives, and thus mainly on the issue of management and protection of freshwater. Although the direct use of freshwater by humans could obviously explain this focus, water issues are not limited to the management of this category of water. Before starting the study of the regulation and protection of water in Law of the Sea,

---

2001, Road map towards the implementation of the United Nations Millennium Declaration, 1/56/326, para. 92.

[98] As stated by the UN Water Report of 2014, 'The private sector can play a greater role in water and energy infrastructure investment, maintenance and operation. Private sector involvement and governmental support for research and development are crucial for developing alternative, renewable and less water intensive energy sources'. UN Water 2014, The United Nations World Water Development Report 2014, Water and Energy: Executive Summary, 2.

[99] Hereinafter the 'LOS Convention'.

some historical perspectives are necessary in order to understand how seawater, and therefore mainly saltwater, has been perceived by our civilization.

From the twelfth century, major European explorers like Marco Polo or Monte Carvino ventured to Asia in search of a spice trade route hitherto jealously guarded by the Arab world and Venice. Before water was perceived as falling off the Earth and being surrounded by monsters, the hostile environment of oceans and seas and became, during the Age of Discovery in the seventeenth century, spaces inviting discovery.

The changing perception of water pushed men to monitor and advocate the extension of their control offshore. The two main theories of Law of the Sea were developed in the seventeenth century. The first, formulated by Grotius, praised seas and oceans as free international areas, free for use by all Nations for seafaring trade. This Dutch theory '*De Jure Praedae*' known as '*Mare Liberum*' was countered a few years later by another theory – '*Mare Clausum, seu de Domino Maris Libri Duo*' – created by J Selden and supported by Portugal and Spain, praising a possible appropriation of sea space as a sovereign territory.

The momentum of the industrial revolutions, which gave men the means for their desires of discovery and conquest, and the end of the Second World War encouraged States to secure and protect more offshore spaces against any uncontrolled expansion. The first Conventions on the Law of the Sea, the Geneva Conventions of 1958,[100] were negotiated at this time, leading a few years later, in 1982, to the LOS Convention, extending greatly States' jurisdiction with the creation of the Exclusive Economic Zone covering up to 200 nautical miles[101] and the creation of the extended continental shelf which could extend up to 350 nautical miles on the seabed.[102]

States have thus gradually extended their control of the sea by slowly projecting their jurisdictions, gaining new spaces and, more importantly, access to the various resources that such space could offer. As it is the case for fisheries, mineral resources or even biological resources, States have been driven for centuries by a 'territorial obsession'[103] to find new resources to sustain their development and population.

---

[100] The Convention on the Territorial Sea and the Contiguous Zone; the Convention on the High Seas; the Convention on Fishing and Conservation of the Living Resources of the High Seas. Hereinafter 'Geneva Conventions'.
[101] Part V, LOS Convention.
[102] Art. 76, LOS Convention.
[103] G Scelle, 'Obsession du Territoire. Essai d'étude réaliste du droit international' (Symbolae, JHW Verzijl, Martinus Nijhoff, La Haye 1958) 347.

Water has thus always been considered as a space connecting territories, as doors of access to countless resources, and as a way to conquer new territories. It was an instrument for States' expansion.

The Conventions of 1958 and the LOS Convention give a specific significance to the term 'water'. Generally speaking, one could imagine that the term 'water' could be used to make a reference to its use, its quality or even to its various mechanisms of protection from the threats of human activities. Such regulation and protection could take the form, for example, of provisions highlighting the importance of water for a sustainable marine environment, the role of water in the marine environment and in transboundary ecosystem management, the various criteria of water quality, or requirements for water impact assessments. Nevertheless, the various Law of the Sea Conventions are using the term water in a completely different way.

The Convention on the Territorial Sea and the Contiguous Zone of 1958 uses the term 'water' as a way to refer to a space, the 'internal' waters,[104] or as a way to measure or delimit a space '(f)or the purpose of measurement, the area of an indentation is that lying between the low-water mark around the shore of the indentation and a line joining the low-water marks of its natural entrance points'.[105] The provision on the islands uses also the term 'water' as a way to describe a spatial environment – 'An island is a naturally formed area of land surrounded by water, which is above water at high-tide'.[106]

The Convention on the High Seas of 1958 does not mention 'water' when describing the High Seas and the Freedom of the High Seas. According to its article 2, the 'Freedom of the High Seas (...) comprises, inter alia, both for coastal and non-coastal States: (1) freedom of navigation; (2) freedom of fishing; (3) freedom to lay submarine cables and pipelines; (4) freedom to fly over the high seas'. This omission could be explained by the fact that the priority was not to describe the environment, its fragility and specificities, but more essentially to regulate human activities within a particular space.

Noting the fast development of modern techniques of exploitation, the Convention on Fishing and Conservation of the Living Resources of the High Seas of 1958 focused its provisions on the conservation of the living resources of the High Seas in order to prevent any over-exploitation. This instrument gathers 'an aggregate of the measures rendering possible the

---

[104] Art. 10, Convention on the Territorial Sea and the Contiguous Zone.
[105] Ibid., Art. 7, para. 3.
[106] Ibid., Art. 10.

optimum sustainable yield from those resources so as to secure a maximum supply to food and other marine products'.[107] The measures of conservation of living resources do not, however, take into consideration the marine environment as a whole, nor do they target seawater specifically. It therefore follows the focus of the Convention on the High Seas, which is the regulation of man's activities on the sea.

Water column,[108] archipelagic waters,[109] superjacent waters,[110] waters landmark,[111] depth of water,[112] the LOS Convention uses the term 'water' in the same way than the Geneva Conventions of 1958. It is used to measure, delimit and describe a spatial environment.[113]

When looking for ways of protecting and regulating water, we note, with interest, that the LOS Convention does not define the expression 'marine environment'. It nevertheless defines the term 'pollution of the marine environment' which consists of the direct or indirect introduction by man of substances or energy into the marine environment, including estuaries, and 'which results or is likely to result in such deleterious effects as harm to living resources and marine life, hazards to human health, hindrance to marine activities, including fishing and other legitimate uses of the sea, impairment of quality of use of sea water and reduction of amenities'.[114]

The absent definition of the 'marine environment' is surely strategic as it gives flexibility to the LOS Convention and enables it to go through the years and integrate gradually, in line with scientific discoveries, new resources and specificities of the marine environment. Furthermore, this lack of definition places heavier duties on Member States who are required, first, to generally protect and preserve the marine environment,[115] and second to take measures to prevent, reduce and control pollution of the marine environment.[116] As this obligation is focused on the impact of human activities, and not the activities, it imposes duties on Member States regardless of the activities or uses of seawater and the marine environment. Thus, when States are conducting activities that

---

[107] Art. 2, Convention on Fishing and Conservation of the Living Resources of the High Seas, 1958.
[108] Art. 257, LOS Convention.
[109] Ibid., art. 49.
[110] Ibid., arts 78 and 135.
[111] Ibid., art. 66.
[112] Ibid., art. 85.
[113] The LOS Convention also adds a new use of water: the production of energy from water currents. See ibid., art. 56.1(a).
[114] Ibid., art. 1(4).
[115] Ibid., art. 192.
[116] Ibid., art. 194.

are yet unregulated by international law, for example activities causing disturbances on the sea floor of the extended continental shelf that reduce the quality of seawater, States are still bound by their duty to protect and preserve the marine environment.

It is worth noting that none of these conventions define the term 'water' or 'seawater'. Nor do they describe what can be considered 'healthy seawater' nor a 'healthy marine environment'.[117] Seawater is therefore not considered as a resource that needs to be protected in the same way as any other non-living or living resources such as hydrothermal vents, minerals, biological or genetic resources. Seawater is perceived as the marine environment, present everywhere in various forms (liquid or solid) and with different specificities (warm, acidic . . .).

2.2 The Evolving Economic Dimension of the Marine Environment

The negotiations over the LOS Convention have been greatly marked by the areas of decolonisation. Emerging countries in Asia or Africa did not take part in the negotiations of the Geneva Conventions and thus did not recognize the values and rules established by the Western powers in these instruments. Consequently, they did not feature in the set of international laws that were implemented mainly by developed countries at that time.

Considering the Geneva Conventions as obstacles to their reconstruction and development, the majority of African Coastal States did not ratify them. This reflected a new political position aiming at prioritizing the protection of territories and national resources and was embraced by many emerging States, who declared their right of self-determination and urged a more equal international community.[118] This reformist attitude was echoed in the ideology of the 'New Economic Order', which consisted of promoting a more equitable economic cooperation among States, especially between new emerging States and developed States.[119]

---

[117] The only article of the LOS Convention mentioning the issue of water quality is art.1(4) dedicated to the definition of 'pollution of the marine environment'.

[118] The Declaration Granting Independence to Colonial Countries and Peoples and the Permanent Sovereignty Declaration on Natural Resources, 14 December 1960.

[119] The Establishment of a New International Economic Order (is) based on equity, sovereign equality, interdependence, common interest and cooperation among all States, irrespective of their economic and social systems which shall correct inequalities and redress existing injustices, making it possible to eliminate the widening gap between the developed and the developing countries and ensure steadily accelerating economic and social development and peace and justice for present and future generations.

As a result, the concept of marine environment inserted into the LOS Convention was shaped by an economic vision: oceans and seas were a space to claim in order to control and protect territories and resources.

The economic dimension of the marine environment is clearly illustrated by the original meaning of the concept of 'common heritage of mankind'.[120] This concept emerged as a way to soften and balance the great differences in technological, economic and national development among the negotiating States of the LOS Convention. Applied to a new ocean space, 'the Area'[121] situated beyond national jurisdiction, it was extremely innovative. Indeed, in order to protect and preserve the Area and its resources, the negotiating States created an Authority, the International Seabed Authority,[122] in charge of managing and supervising activities to benefit mankind.[123]

To reinforce this protection given by the common heritage of mankind, the LOS Convention prohibits any States from claiming or exercising sovereignty or sovereign rights over any part of the Area or its resources. Appropriation by any State as well as any natural or juridical person is also strictly prohibited.[124]

At the heart of the legal regime of the Area rests a key mechanism reflecting the true meaning of this concept. This mechanism is the one of 'profit sharing of production' aiming to restore the balance between countries with the financial and technological means to conduct activities in the Area, and

---

United Nations General Assembly 1974, Declaration for the Establishment of a New International Economic Order A/RES/S-6/3201

[120] Hereinafter 'UNCLOS III'.

[121] "Area" means the seabed and ocean floor and subsoil thereof, beyond the limits of national jurisdiction'. Art. 1.1(1), LOS Convention.

[122] Hereinafter 'ISA'.

[123] 1. Activities in the Area shall, as specifically provided for in this Part, be carried out for the benefit of mankind as a whole, irrespective of the geographical location of States, whether coastal or land-locked, and taking into particular consideration the interests and needs of developing States and of peoples who have not attained full independence or other self-governing status recognized by the United Nations in accordance with General Assembly resolution 1514 (XV) and other relevant General Assembly resolutions.

LOS Convention, art. 140, para. 1.

2. The Authority shall provide for the equitable sharing of financial and other economic benefits derived from activities in the Area through any appropriate mechanism, on a non-discriminatory basis, in accordance with article 160, paragraph 2(f)(i).

LOS Convention, art. 140, para. 2.

[124] Ibid., art. 137.

other less advanced countries. Cooperation among all States Parties is further strengthened through provisions requiring States to cooperate and promote technology transfer and scientific knowledge relating to activities in the Area. These cooperation and promotion measures aim first at facilitating the access to the relevant technology under fair and reasonable terms and conditions, and second to provide opportunities for training in marine science and technology, especially for developing States. As expressed by the representative of the UN Secretary General:

> It was a matter of understandable concern that the highly sophisticated technology necessary for exploiting the resources of the sea-bed was in the hands of a very limited group of countries and other even individual firms. The most effective way to ensure the transfer of technology to the developing and less advanced countries was through joint ventures organized either directly in association with the technologically advanced countries or by the International Authority.[125]

The marine environment of the Area is protected thanks to it belonging to the common heritage of mankind. The ISA plays a central role in this protection by adopting appropriate rules, regulations and procedure for the prevention, reduction and control of pollution and other hazards to the marine environment, and for the protection and conservation of natural resources of the Area.[126] Despite these mechanisms, the level of protection given by the common heritage of mankind remains unclear. This is mainly due to its scope. Indeed, even though it is solemnly affirmed in the LOS Convention, this concept is not defined in the legal regime of the Area, or in article 1 relating to the use of terms of the Convention. The scope of the 'common heritage of mankind' concept can nevertheless be understood by putting together a jigsaw puzzle.

The LOS Convention refers to the concept of a common heritage of mankind to promote a new equality of States 'irrespective of the geographical location of States, whether coastal or land-locked, and taking into consideration the interests and needs of developing States and of peoples who have not attained full independence'.[127] It guarantees 'the equitable sharing of financial and other economic benefits derived from activities in the Area, through any appropriate mechanism, on a non

---

[125] United Nations 1974, 'Third United Nations Conference on the Law of the Sea: Summary records of meetings of the First Committee' A/CONF62/C1/SR10, Vol II, 10th meeting, paras 26, 51.
[126] Art. 145, LOS Convention.
[127] Ibid., art. 144, para. 1.

discriminatory basis'.[128] Activities in the Area must also 'be carried out in such a manner as to foster healthy development of the world economy and balanced growth of international trade' and 'promote international cooperation for the over-all development from all countries, especially developing States'.[129]

In view of these provisions, the concept of common heritage of mankind is – fundamentally – an economic concept. Protection is understood in terms of regulation and control of activities in the Area and not in terms of ecological preservation. Resource protection and protection of the marine environment promote first and foremost the harmonious development of the world economy. The representative of Nepal declared in 1974:

> In as much as participation in the international area was open to all States, the extent of national jurisdiction should not jeopardize the *economic viability and potential of the area* covered by the international regime.[130]

2.3 Towards a More Ecological Apprehension of the Marine Environment

Since the ratification of the LOS Convention, the economic dimension of the common heritage of mankind has been softened by the adoption of the Agreement relating to the Implementation of Part XI of the United Nations Convention on the Law of the Sea of 10 December 1982.[131]

This 1994 Agreement, dedicated to the regulation of the Area, testifies to 'the importance of the LOS Convention for the protection and preservation of the marine environment and the growing concern for the global environment'.[132] It gives the ISA great responsibilities in the management and supervision of activities and gives it, more precisely, responsibilities to '(t)imely elaborate rules, regulations and procedures for exploitation, including those relating to the protection and preservation of the marine environment'.[133] Rules, regulations and procedures shall incorporate applicable standards for the protection and preservation of the marine environment.[134] Marine scientific research shall also be promoted and encouraged with a 'particular emphasis on research related to the

---

[128] Ibid., art. 144, para. 2.
[129] Ibid., art. 150.
[130] United Nations 1974, 'Third United Nations Conference on the Law of the Sea: Summary records of meetings of the First Committee', A/CONF62/C1/SR6, Vol II, 6th Meeting, paras 2, 24. Emphasis added.
[131] Hereinafter '1994 Agreement'.
[132] Ibid., Preamble.
[133] Ibid., s. 1, para. 5, k).
[134] Ibid., s. 1, para. 5, g).

environmental impact of activities'.[135] Scientific knowledge and monitoring of marine technology shall further be acquired 'in particular technology relating to the protection and preservation of the marine environment'.[136]

Embracing its mandate, the ISA has spent a significant amount of time during recent years negotiating, preparing and drafting rules, regulations and procedures, known under the name of the 'Mining Code', to regulate prospecting, exploration and exploitation of marine minerals in the Area. To date, the ISA has adopted different Regulations applicable to activities conducted on different types of resources (polymetallic nodules, polymetallic sulphides and ferromanganese crusts). Exploitation activities in the Area have not yet taken place; the ISA has thus mainly focused its attention on the rules, regulations and procedures dedicated to prospecting and exploration. It has nevertheless recently started to discuss the instruments dedicated to exploitation activities. One can therefore easily guess the impressive workload that this international body is facing.

Throughout the years, the ISA has been confronted with a growing number of environmental challenges due to the regulations of human activities on the seabed of the Area, all along the water column and at the surface. This codification work proves therefore to be more difficult than it might seem because of the lack of scientific knowledge surrounding deep-sea ecosystems and the impact of (new) activities in High Seas, especially at great depth.

The application of the precautionary principle, stated by Principle 15 of the Rio Declaration, as well as the ecosystem approach praised by the United Nations Convention on Biological Diversity of 1992, have greatly influenced the work methods of the ISA and the subsequent drafting of the Mining Code and of ISA Recommendations. To recall Principle 15:

> In order to protect the environment, the precautionary approach shall be widely applied by States according to their capabilities. Where there are threats of serious or irreversible damage, lack of full scientific certainty shall not be used as a reason for postponing cost-effective measures to prevent environmental degradation.

The first Advisory Opinion of the Seabed Dispute Chamber of the International Tribunal for the Law of the Sea in 2011[137] significantly

---

[135] Ibid., s. 1, para. 5, h).
[136] Ibid., s. 1, para. 5, i).
[137] *Responsibilities and Obligations of States*, n. 79 above, para. 135.

strengthened the implementation of the precautionary principle by recognising its incorporation into customary international law.

By looking closely at the various ISA instruments, their levels of details are striking. The precautionary principle, aiming at better protection and preservation of the marine environment, has been expertly distilled through the Mining Code. As a consequence, the environmental dimension of water has changed.

From a simple and implicit apprehension of water as a spatial environment, enabling human activities to furrow oceans and seas vertically and horizontally, water now becomes within the ISA instruments an explicit *component of the marine environment*.[138]

Its existence is testified first and foremost by the rather simple use of the term 'water' in many instruments of the ISA. This reference is an important step forward into a more ecological perception and regulation of the marine environment. Indeed, no mention whatsoever of water quality or of the importance of water was introduced in the United Nations Convention on Biological Diversity of 1992, nor in the Agenda 21, which nevertheless has a chapter on 'Oceans'.[139]

Facing an important legal gap, the ISA has therefore considerable responsibilities in the development of new regulations, which will foster an emerging State practice and could greatly influence the regulation of water in the years to come. The Regulations on Prospecting and Exploration for Cobalt-Rich Manganese Crusts and Polymetallic Sulphides[140] offer a good illustration of the new way of perceiving, but also monitoring and regulating water, in cases of activities conducted in the Area. According to these two regulations, water is now officially part of the definition of the marine environment: '"Marine environment" includes the physical, chemical, geological and biological components, conditions and factors which interact and determine the productivity, state, condition and quality of the marine ecosystem, *the waters of the seas and oceans* and the airspace above those waters, as well as the seabed and ocean floor and subsoil thereof'.[141]

Furthermore, water salinity is now recognised as an environmental

---

[138] Emphasis added.

[139] Agenda 21 is a non-binding instrument dedicated to sustainable development, which was adopted in 1992 at the United Nations Conference on Environment and Development held in Rio de Janeiro, Brazil. For more information on this instrument, see Part 1 of this chapter.

[140] ISBA/18/A/11 and ISBA/16/1/12/Rev1.

[141] Regulation 1, para. 3, d), ISBA/18/A/11, and reg. 1, para. 3, c), ISBA/16/1/12/Rev1. Emphasis added.

parameter of the marine environment.[142] In order to better guide contractors in the implementation of these Regulations, the ISA issues recommendations. The Ferromanganese Crusts Recommendations for the guidance of contractors for the assessment of the possible environmental impact arising from exploration for marine minerals in the Area[143] provides contractors with a high level of technical details on all methods and criteria used in the Ferromanganese Crusts Regulation. It recommends contractors to conduct a three-step approach consisting first of the conduct of environmental baseline studies, second, on monitoring to ensure that no serious harm is caused to the marine environment from activities, and third, on monitoring during and after testing of collecting systems and equipment.[144] These studies seek to prevent potential effects of mineral slurry at the sea surface, which will bring back to the surface a large quantity of cold water and nutrient-rich and particle-laden water that could significantly alter and damage the water quality and general marine environment.

When detailing how to conduct 'environmental impact assessment and best environmental practices', the Recommendation of 2013 invites contractors to collect information on the oceanographic conditions including the current, temperature and torpidity regimes along the entire water column[145] *before* conducting any activities.[146] Measurement programmes are also encouraged to be adapted to the regional hydrodynamic activities at the sea surface[147] and a study of the particle concentrations and composition to record distribution along the water column is recommended.[148] Contractors are also invited to conduct observations and measurements *after* the performance of activities of prospecting and exploration.[149] These observations and measurements concern, among others, the changes in the characteristics of the water at all the level of the discharge plume during the mining test, and changes in the behaviour of the fauna at and below

---

[142] Section II, para. 19, ii), Annex II, ISBA/18/A/11 and Section II, para. 19, ii), Annex 2, ISBA/16/A/12/Rev.1.
[143] International Seabed Authority 2013, 'Recommendations for the guidance of contractors for the assessment of the possible environmental impacts arising from exploration for marine minerals in the Area', ISBA/19/LTC/8. Hereinafter the 'Recommendation of 2013'.
[144] Ibid., para. 11.
[145] Ibid., para. 15, a), i).
[146] Emphasis added.
[147] Para. 15, a), iii), Recommendation of 2013.
[148] Ibid., para. 15, a), v).
[149] Emphasis added.

the discharge plume.¹⁵⁰ Besides this, they concern the changes in the water currents and the response of organisms to changes in water circulation.¹⁵¹

The technicality of these provisions could easily prevent anyone from understanding a simple but yet important evolution slowly taking place in State practice. The great environmental pressures caused by the explosion of activities at sea encourage States and international bodies such as the ISA to develop new monitoring and regulations recognising the interconnectedness of the marine environment, including the role of water quality. Although the mechanisms are still at an early stage, these new rules and regulations prove that the international community of States is slowly changing its appreciation and perception of seawater. Water is longer apprehended as an environment. It is not as yet a resource to protect as is the case of terrestrial land, but water is an indicator of the good health of the marine environment: it is now an explicit component of it.

2.4 Ocean Health and Climate Change

As the primary lungs of our planet, oceans and seas have an important role to play as climate regulators, food reservoirs and more generally as a life source. From the air that we breathe, to the fish that we eat, the medicine we take, the makeup we buy, and the recreational activities we have, oceans and seas have such a multi-faceted role in our everyday lives that we tend to forget how important this space is for our planet and for ourselves.

Recent disasters such as the explosion of the platform *Deepwater Horizon*¹⁵² or the continuous release into the Pacific Ocean of radiation-contaminated water off the nuclear power plants of *Fukushima*¹⁵³ are examples of the diverse threats facing this seawater. Those examples are nonetheless accidents and do not reflect the common, creeping, subtle and everyday effects of the climate change on our oceans and seas.

Since they are able to absorb one-third of excessive carbon dioxide

---

¹⁵⁰ Para. 30, e), Recommendation of 2013.
¹⁵¹ Ibid., para. 30, j).
¹⁵² G Wilson, 'Deepwater Horizon and the Law of the Sea: was the cure worse than the disease?' (2014) 41 *Boston College Environmental Affairs Law Review* 63.
¹⁵³ On this topic see G Darian, 'There's something in the water: the inadequacy of international anti-dumping laws as applied to the Fukushima Daiichi radioactive water discharge' (2012) 27 *American University International Law Review* 473. See also A Hutchinson, 'Is radioactive water worth worrying about?' *The New Yorker* (Online Edition 2015) <www.newyorker.com/tech/elements/is-radioactive-water-from-fukushima-worth-worrying-about> accessed in February 2016.

emissions from the atmosphere (22 million tons a day),[154] oceans and seas, and therefore seawater, happen to be a very precious resource and spatial environment for the fight against climate change. This capacity of absorption, often mentioned in favour of the development of marine protected areas,[155] however has a side effect. The more oceans and seas absorb carbon dioxide, the faster they warm up and the faster their acidity will increase.[156] Since the pre-industrial revolution, water acidity has increased of 30 per cent. As Dr. Richard Feely, senior scientist at the US National Oceanic and Atmospheric Administration, explains: 'if our blood – which is pH neutral – became .3 more acidic, our bodies would cease to function'.[157]

The impact of these important changes is not certain. In this regard, JP Barry et al have recently declared that 'it remains unknown whether ocean acidification will drive species to extinction but it is possible (. . .) (considering) the sensitivity and performance of marine organisms under future high-CO_2 conditions'.[158] Water acidification will nonetheless inevitably provoke changes of biodiversity and function of marine ecosystems.[159]

In 2010, the General Assembly of the United Nations expressed concerns 'over the current and projected adverse effects of climate change on food security and the sustainability of fisheries'. It also urged States to intensify efforts, directly or through sub-regional, regional or global

---

[154] National Geographic, 'Ocean Acidification' Pristine Seas (Online Edition) <http://ocean.nationalgeographic.com/ocean/explore/pristine-seas/critical-issues-ocean-acidification/> accessed in February 2016.

[155] For more information on the challenges of water quality in marine protected areas, see E McLeod et al., 'Designing marine protected area networks to address the impacts of climate change' (2009) 7 *Frontiers in Ecology and the Environment* 362 <www.reefresilience.org/pdf/McLeod_etal_2009.pdf> accessed in February 2016.

[156] B Walsh, 'Ocean acidification will make climate change worse' *The Time* (Online Edition 2013) <http://science.time.com/2013/08/26/ocean-acidification-will-make-climate-change-worse/> accessed in February 2016.

[157] B Waymouth, 'How to protect the ocean from us?' *The Huffington Post* (Online Edition, 2015) <www.huffingtonpost.com/belinda-waymouth/how-to-protect-the-ocean-_b_8067698.html> accessed in February 2016.

[158] JP Barry et al., 'Effects of ocean acidification on marine biodiversity and ecosystem function' in JP Gattuso and L Hansson (eds), *Ocean acidification* (Oxford University Press, 2012) 195–6.

[159] The Ad Hoc Open-ended Informal Working Group to the President of the General Assembly of the United Nations recognised in 2010 that 'human pressures on the marine environment were increasing and impacting the long-term health, resilience and productivity of marine ecosystems and marine biodiversity, including as a result of climate change'. General Assembly 2010, 'Letter dated 16 March from the Co-Chairpersons of the Ad Hoc Open-ended Informal Working Group to the President of the General Assembly' A/65/68, para 28, 5.

organisations or arrangements, to assess and address the impacts of global climate change on the sustainability of fish stocks and the habitats that support them.[160]

These modifications, more or less important depending on the various scientific scenarios, will inevitably impact not only resources but also the services we rely on. Such effects on industries have been highlighted by a report of the World Ocean Council in 2010:

> On a global scale, the impacts of climate change on oceans, such as the increased frequency and severity of storms and the rise in ocean acidification, should be of concern to all ocean industries. Increased instability and unpredictability in the marine environment increase the difficulty, risk and cost of doing business, and adds to the potential for environmental impacts from accidents.[161]

Ocean acidification and the impact of climate change on oceans and seas – and thus water quality – have not been yet regulated by international law. However, pursuant to the LOS Convention, States are under an obligation to protect and preserve the marine environment.[162] This core general and customary principle is listed in a number of other provisions of the LOS Convention dedicated to the prevention, reduction and control of pollution of the marine environment. Pollution from the atmosphere is also regulated. The LOS Convention establishes a difference between actions that States shall take in areas within national jurisdiction or on vessels and aircrafts under their control and the general actions that States can take in areas beyond national jurisdiction. Under the first category of actions, States shall adopt laws and regulations 'to prevent, reduce and control pollution of the marine environment from or through the atmosphere'. Under the second category of (broader) actions, States shall take 'other measures as may be necessary to prevent, reduce and control such pollution' and 'shall endeavour to establish global and regional rules, standards and recommended practices and procedures to prevent, reduce

---

[160] General Assembly 2010, 'Sustainable fisheries, including through the 1995 Agreement for the Implementation of the Provisions of the United Nations Convention on the Law of the Sea of 10 December 1982 relating to the Conservation and Management of Straddling Fish Stocks and Highly Migratory Fish Stocks, and related instruments', A/RES/64/72, para 3, 5.

[161] World Ocean Council, 'Conference Report' (Belfast, United Kingdom 2010) 5.

[162] Article 192, LOS Convention.

and control such pollution' especially through competent international organization or diplomatic conference.[163]

The 21st Session of the Conference of the Parties to the United Nations Framework Convention on Climate Change (COP21/CMP11) organised in Paris in November–December 2015 aimed at elaborating a new development model able to conciliate sustainable and economic development and at adopting a binding instrument fighting climate change. Emergency, Hope and Ambition, keywords expressed by the special envoy of Earth protection to the President of the French Republic, Nicolas Hulot, reflects very well the difficulties, challenges and expectations around this Conference. Indeed, one might think that this event was offering an opportunity to States to better implement their duties of protection and preservation of the marine environment by establishing global and regional rules preventing reducing and controlling atmospheric pollution, pursuant to article 212 of the LOS Convention. However, ocean issues, such as impact of atmospheric pollution on the marine environment, the acidification and the warming of water or the regulation of pollution from shipping activities, have been excluded from the preparatory work.[164]

In an attempt to remedy this situation, the United Nations Educational, Scientific and Cultural Organization (UNESCO) and the Intergovernmental Oceanographic Commission organised on Ocean Day, 8 June 2015, a conference on Ocean and Climate to raise awareness of the impact of humans actions on the ocean and to mobilise actors behind a project of sustainable global ocean management. This conference did not address in detail any specific political proposals but addressed in a subtle way the important gaps between, first, the mobilisation of civil society and of the private

---

[163] Ibid., art. 212. See also SDG 14 inviting States and stakeholders to 'Enhance the conservation and sustainable use of oceans and their resources by implementing international law as reflected in UNCLOS, which provides the legal framework for the conservation and sustainable use of oceans and their resources, as recalled in para 158 of The Future We Want'. General Assembly 2015, n 41 above, para. 14 c), 24. Such inclusion of ocean issues in the new sustainable development agenda is in line with the implementation road map of the Millennium Declaration which already stated in 2001 that '(a) focused agenda should foster discussion of findings in particular environmental sectors (forests, oceans, climate, energy, fresh water, etc.) as well as cross-sector areas, such as economic instruments, new technologies and globalization.' General Assembly 2001, Road map towards the implementation of the United Nations Millennium Declaration, 1/56/326, para. 170.

[164] See: 'COP21: on a oublié d'inviter l'océan' CNRS Le Journal and Libération (Online Edition, 2015). See also A Milman, 'Ocean warming and acidification needs more attention, argues US' *The Guardian* (Online Edition, 2015).

sector, and second, the political will of States omitting ocean issues from the negotiations and States' actions.

The resulting instrument, the Paris Agreement, unanimously adopted, excluded seawater issues from the final core text. However, it included a (too-simple) reference to oceans, and thus seawater issues, by declaring in its Preamble:

> *Noting* the importance of ensuring the integrity of all ecosystems, *including oceans*, and the protection of biodiversity, recognized by some cultures as Mother Earth, and noting the importance for some of the concept of 'climate justice', when taking action to address climate change.[165]

The inclusion of 'oceans' in the Preamble proves that States have heard the call of some dynamic States, experts, and organizations, pointing out the importance of seawater in climate change regulation. This simple mention is a good step, but a small one. It leaves the door open for future negotiations, the development of implementation measures and new legal instruments, which could be dedicated to climate change and ocean/seawater issues. Nevertheless, this rather simple mention of ocean issues reveals once again the current inability of States to *create new* regulatory mechanisms that can, concretely, conciliate the terrestrial and marine dimensions and respect the concept of 'integrity of all ecosystems' that has gradually withered away the last 40 years.[166]

The trend of the moment is to encourage the gathering of more scientific data to measure and understand the complex effects of climate change on oceans' health. While more data are needed, we should be careful with this scientific obsession. Indeed, more data will require more studies and more professionals as well as new models of analysis able to interpret and treat these data throughout the years.[167] Furthermore, this data obsession should be tempered by recalling the complementary role of science and the humanities that seems to have slowly faded away over the last years.[168]

---

[165] Paris Agreement, FCCC/CP/2015/L.9/Rev.1, 12 December 2015. Hereinafter the 'Paris Agreement'.

[166] Emphasis added. Reference is made to para. 3 of the Stockholm Declaration.

[167] EO Wilson: 'there is no guarantee that the exploration of Earth's biodiversity can be completed before the twenty-third century. The problem is a severe shortage of expert researchers. Technology without science is like an automobile without wheels and a road map'. Wilson, *Half Earth: Our Planet's Fight for Life* (Liveright 2016) 163.

[168] 'To understand the meaning of life, to know that we know and how and why we know, is the premier driving force of all science and the humanities'. Ibid., 50.

Indeed, the regulation and protection of water should not be mainly based on (available) scientific proof. With the current transformation of our international society and the growing will of individuals to influence/participate in meaningful international decision making, regulations have and will have, even more in the years to come, a profound humanistic and social resonance. The balance between science and the humanities should thus be seriously taken into consideration before and while creating any water regulatory framework.[169]

CONCLUSION

States have proved their capacity to develop innovative legal instruments applicable to water issues. However, protecting and regulating the various uses of water, as a resource and as an environment, does not only require having new legal instruments. It requires the implementation of the regulations already in place. In this regard, economic policy instruments could have an important role to play. As was encouraged by the UNEP, investments, planning and management measures 'must transcend the entire water supply and disposal chain involving ecosystem management (including coastal waters) agricultural efficiency and production and treatment of wastewater and a stronger focus on urban planning'.[170]

Indeed, whenever we talk about food, health, environment or urban development and even climate change, water is central and omnipresent.[171] For that reason, water regulation is heavily fragmented and sectorized. But water is a vital element for which we have no *global vision*.[172] Such lack of vision, inherited the twentieth century, is still shaping our present. Coastal waters are separate from their general ecosystems; we had and we

---

[169] Ibid.:

There is greatness in understanding the basic elements of human evolution and wisely taking upon the way they are linked. The form it is taking can be expressed succinctly as follows: the biosphere gave rise to the human rise, the evolved mind gave rise to culture, and culture will find the way to save the biosphere.

[170] UNEP 2010, 'Sick water?, n 45 above, 11.
[171] World Economic Forum (2016), 'Global Risk Interconnections Map', Global Risks Report. <http://reports.weforum.org/global-risks-2016/shareable-infographics/> accessed in February 2016.
[172] The term of vision is used wisely to highlight the need of developing planning and management tools against future long-term scenarios.

are still willing to separate the various uses and forms of water depending on their direct use and benefits for humans.[173]

This excessive tendency of fragmenting, cutting and pasting uses and spaces, be they terrestrial or marine, is greatly affecting the understanding and the regulation of water issues. Indeed, saltwater and freshwater are facing common issues such as acidification, eutrophication, and the discharge of wastewater putting in danger the ecosystem they support and their function and use on the planet. As a result, there are no 'waters' but a global water system combining human water security and natural system security.

This lack of understanding is reflected by the recent climate change negotiations, which have excluded ocean issues from the core provisions of the Paris Agreement. This is worrying because the stakes are high. How could we regulate the global water crisis, organise the 'global response to the threat of climate change'[174] and stimulate the global economy by omitting 71 per cent of the Earth's environment and one of its main components, water?

---

[173] See in this regard the UNEP inviting to recognise wastewater as a resource. Sick water: The central role of wastewater management in sustainable development: A rapid response assessment UNEP 2010, 'Sick water?, n. 45 above, 54–5.

[174] Art. 2, para. 1, Paris Agreement.

PART II

Ethical, legal and social issues

8. Is investment arbitration inimical to the human right to water? The re-examination of arbitral decisions on water services

Miharu Hirano and Shotaro Hamamoto

INTRODUCTION

In June 2015, a group of UN experts, including the Special Rapporteur on the human right to safe drinking water and sanitation, voiced concerns over the adverse impact of investment agreements on the human rights.[1] The issued statement draws attention to 'the potential detrimental impact these treaties and agreements may have on the enjoyment of human rights', including the human right to water. This statement goes one step further than the 2007 Report of the High Commissioner for Human Rights (UNHCHR) on equitable access to safe drinking water, which had simply noted that further analysis was needed in this field.[2] The 2015 statement was quoted by a parliamentary member in the Japanese Diet to criticise Japan's policy on investment treaties and the Minister for Foreign Affairs replied that Japan would design and negotiate investment treaties so that no detrimental impact would be exerted upon the enjoyment of human rights.[3]

In fact, urban water utility has now become rather a common sector

---

[1] 'UN experts voice concern over adverse impact of free trade and investment agreements on human rights' (2015) <www.ohchr.org/EN/NewsEvents/Pages/DisplayNews.aspx?NewsID=16031&LangID=E>.

[2] 'Report of the United Nations High Commissioner for Human Rights on the scope and content of the relevant human rights obligations related to equitable access to safe drinking water and sanitation under international human rights instruments', A/HRC/6/3 (2007), paras 63–64.

[3] Mr Keiji Kokuta (Japanese Communist Party) and Mr Fumio Kishida (Minister for Foreign Affairs, Liberal Democratic Party), House of Representatives, Committee on Foreign Affairs (28 August 2015) 21.

in the list of publicly available investment arbitration proceedings.[4] As of January 2016, 13 disputes have been brought to investment arbitration, all under the ICSID Convention;[5] so far seven of them have reached merits,[6] four have been discontinued[7] and two are still pending.[8] Among them, nine cases were brought against Argentina, but other countries like Algeria, Bolivia, Estonia and Tanzania also became respondents. A major privatisation project in Jakarta faces ongoing controversies,[9] and a last resort to seek arbitration is open for investors.[10]

---

[4] For instance, see the column of the subject of dispute by economic sector on the ICSID's website <https://icsid.worldbank.org/apps/ICSIDWEB/cases/Pages/AdvancedSearch.aspx> or the column in the UNCTAD's Investment Dispute Settlement Navigator <http://investmentpolicyhub.unctad.org/ISDS/FilterByEconomicSector>.

[5] 'Convention on the Settlement of Investment Disputes between States and Nationals of Other States' (1965) 575 United Nations Treaty Series 159. Note that *AWG Group Ltd v. Argentina* was brought under the UNCITRAL Arbitration Rules but was administered by the ICSID together with ICSID case of *Aguas Argentinas, SA, Suez, Sociedad General de Aguas de Barcelona SA, and Vivendi Universal SA v. Argentina*, on the basis of the parties' agreement.

[6] In chronological order, *Azurix Corp v. Argentina*, ICSID Case No. ARB/01/12, Award (2006); *Compañiá de Aguas del Aconquija SA and Vivendi Universal SA v. Argentina*, ICSID Case No. ARB/97/3, Award (2007); *Biwater Gauff (Tanzania) Ltd v. Tanzania*, ICSID Case No ARB/05/22, Award (2008); *Suez, Sociedad General de Aguas de Barcelona SA, and InterAguas Servicios Integrales del Agua SA v. Argentina*, ICSID Case No ARB/03/17, Decision on Liability (2010); *Suez, Sociedad General de Aguas de Barcelona, SA and Vivendi Universal, SA v. Argentina*, ICSID Case No. ARB/03/19, Decision on Liability (2010); *Impregilo SpA v. Argentina*, ICSID Case No. ARB/07/17, Award (2011); *SAUR International SA v. Argentina*, ICSID Case No ARB/04/4, Décision sur la competence et sur la responsabilité (2012).

[7] *Aguas Cordobesas SA, Suez, and Sociedad General de Aguas de Barcelona SA v. Argentina*, ICSID Case No ARB/03/18; *Aguas del Tunari SA v. Bolivia*, ICSID Case No ARB/02/3; *Azurix Corp v. Argentina*, ICSID Case No ARB/03/30; *Gelsenwasser AG v. Algeria*, ICSID Case No ARB/12/32.

[8] *Urbaser SA and Consorcio de Aguas Bilbao Bizkaia, Bilbao Biskaia Ur Partzuergoa v. Argentina*, ICSID Case No ARB/07/26; *United Utilities (Tallinn) BV and Aktsiaselts Tallinna Vesi v. Estonia*, ICSID Case No ARB/14/24.

[9] The Central Jakarta District Court annulled the concession contracts for water services on 24 March 2015. The order has not been executed as concessionaires appealed. *See* 'Suez will fight to keep its Jakarta water contract' Reuters. (2015) <www.reuters.com/article/2015/04/10/us-suez-jakarta-idUSKBN0N126W20150410>.

[10] One of the two concessions is owned by France-based Suez, whose investment may be protected under the France-Indonesia BIT. Despite the termination of the BIT in 28 April 2015, the protection continues to be effective for the investments covered by the treaty and admitted by the Contracting Party prior to the

Is investment arbitration inimical to the human right to water? 147

These arbitral proceedings aroused the interest of academics and practitioners as shown in the growing number of literature discussing the relationship between the human right to water and investment arbitration.[11] Their often critical examination of the arbitral decisions tend to conclude that tribunals failed to take into account the human right to water as a

---

notification of termination of the present Agreement. *See* art. 10, 'Agreement on the Encouragement and Protection of French Investments in Indonesia' (1973) 985 United Nations Treaty Series 258.

[11] E Cadeau and F Duhautoy, 'Le droit à l'eau, soluble dans le droit international de l'investissement?', (2013) 216 Droit de l'environnement 338; J Chaisse and M Polo, 'Globalization of water privatization: Ramifications of investor-state disputes in the "blue gold" economy', (2015) 38 *Boston College International and Comparative Law Review* 1; H Chen, 'The human right to water and foreign investment: Friends or foes?', (2015) 40 *Water International* 297; J Echaide, 'Sobre el derecho humano al agua y la fragmentación del derecho internacional: El régimen internacional de protección de inversiones vis-a-vis las obligaciones *erga omnes* en materia de derechos humanos', (2014) VIII *Revista Electrónica del Instituto de Investigaciones* 140; B Farrugia, 'The human right to water: defences to investment treaty violations', (2015) 31 *Arbitration International* 261; R Greco, 'The impact of the human right to water on investment disputes', (2015) 98 *Rivista di Diritto Internazionale* 444; F Marrella, 'On the changing structure of international investment law: The human right to water and ICSID arbitration', (2010) 12 *International Community Law Review* 335; P Mayer, 'Les arbitrages CIRDI en matière d'eau', in SFDI, *L'eau en droit international : Colloque d'Orléans* (Paris: Pedone 2011) 163; O McIntyre, 'Emergence of the human right to water in an era of globalization and its implications for international investment law', in JF Addicott, MJH Bhuiyan and TMR Chowdhury (eds), *Globalization, International Law, and Human Rights* (Oxford; Oxford University Press 2011) 175; T Meshel, 'Human rights in investor-state arbitration: The human right to water and beyond', (2015) 6 *Journal of International Dispute Settlement* 277; K Miles, 'Blue oil: Water resources, social justice and the international law on foreign investment', in S Alam, N Klein and J Overland (eds), *Globalisation and the quest for social and environmental justice* (London: Routledge 2012) 53; W Schreiber, 'Realizing the right to water in international investment law: An interdisciplinary approach to BIT obligations', (2008) 48 *Natural Resources Journal* 473; A Tanzi, 'Public interest concerns in international investment arbitration in the water services sector: Problems and prospects for an integrated approach', in T Treves, F Seatzu and S Trevisanut (eds), *Foreign Investment, International Law and Common Concerns*, (Abingdon: Routledge 2014) 333; P Thielbörger, 'The human right to water versus investor rights: Double-dilemma or pseudo-conflict?', in PM Dupuy, F Francioni and EU Petersmann (eds), *Human Rights in International Investment Law and Arbitration* (Oxford: Oxford University Press 2009) 487; E Truswell, 'Thirst for profit: Water privatisation, investment law and a human right to water', in C Brown and K Miles (eds), *Evolution in Investment Treaty Law and Arbitration*, (Cambridge; Cambridge University Press 2011) 570; JE Viñuales, *Foreign Investment and the Environment in International Law* (Cambridge: Cambridge University Press 2012) Ch 7.

relevant applicable international rule.[12] The human right to water is viewed as a potential legal tool for 'defence' to safeguard the regulatory policy of the host State to protect and promote universal access of safe drinking water against a foreign investor's claim that seek compensation for such measure.

Such line of argumentation is, however, counterintuitive for a lawyer since human rights law is structured, like investment law, to impose obligations on States, i.e. human rights obligations are designed to limit, rather than to broaden, the autonomy of sovereign States. Further, it seems unrealistic to suppose that 'persons of high moral character and recognised competence in the fields of law, commerce, industry or finance, who may be relied upon to exercise independent judgment'[13] consider it to be 'fair and equitable' to look only at the economic interests of investors while ignoring human rights. It is indeed often argued that international investment law and human rights law are not in conflict but converging.[14] Nevertheless, there certainly exists the perception that investment arbitrations are natural foes to the host State's regulatory powers, including the right/duty to regulate water services and to protect and promote the human right to water (1). In order to verify whether this perception reflects the reality, the chapter clarifies the source and scope of the human right to water (2) and conducts analytical research of arbitral decisions (3). The chapter concludes by expressing doubt as to whether investment arbitration exerts any negative impact upon the human right to water, at least at the current stage.[15]

---

[12] Art. 31(3)(c), 'Vienna Convention on the Law of Treaties' (1965) 1155 United Nations Treaty Series 331.

[13] Art. 13(1), ICSID Convention.

[14] E.g. A Pellet, 'Notes sur la "fragmentation" du droit international: Droit des investissements internationaux et droits de l'homme', *Unity and Diversity in International Law. Essays in honour of professor Pierre-Marie Dupuy* (Martinus Nijhoff Publishers 2014) 757. *See also* SL Karamanian, 'Human rights dimensions of investment law', in E De Wet and J Vidmar (eds), *Hierarchy in international law: The place of human rights*, (Oxford: Oxford University Press, 2012) 236.

[15] The social scientific research on a chilling effect demonstrates that such a concern is not necessarily shared among domestic public officials. *See*, C Côté, 'A chilling effect? The impact of international investment agreements on national regulatory autonomy in the areas of health, safety and the environment' (PhD Thesis, The London School of Economics and Political Science (LSE) 2014).

1. CONCERNS OVER THE 'POTENTIAL' IMPACT OF INVESTMENT ARBITRAL DECISIONS ON THE HOST STATE'S RIGHT/DUTY TO REGULATE WATER SERVICES

The structure of the treaty-based arbitration allows a foreign private entity to bring a case on the conformity of governmental measures with the applicable investment treaty, even when such measures concern public interests. It is now a widely shared view that treaty-based arbitration may have certain impacts on the domestic regulatory authority.[16] Such concerns arise from the application of substantive obligations contained in the applicable treaty to domestic measures of the host State, accompanied by the enforcement mechanism of the awarded compensation.

Domestic measures of the host State are subject to the substantive obligations provided under investment agreements, such as the expropriation clause or the fair and equitable treatment (FET) standard. Moreover, given the similarities of the languages used under investment agreements, tribunals tend to refer to, and quite often rely on, the reasoning given by other tribunals whether or not based on the same treaty. For this reason, the implications of an award may extend to the States that are not parties to the treaty on the basis of which the arbitral decision was made.[17] A State thus needs to be attentive not only to arbitral decisions rendered on the basis of the treaties to which it is a party but also to the general trend of the evolution of arbitral jurisprudence.[18]

Arbitral decisions rendered by arbitrators appointed by the disputing parties are enforceable, including in third States.[19] This mechanism gives

---

[16] See e.g. G Van Harten and M Loughlin, 'Investment treaty arbitration as a species of global administrative law', (2006) 17 *European Journal of International Law* 121.

[17] On the formation of 'case law', see C Titi, 'The Arbitrator as a lawmaker: Jurisgenerative processes in investment arbitration', (2013) 14 *Journal of World Investment and Trade* 829; M Audit 'La jurisprudence arbitrale comme source du droit international des investissements', in C Leben (ed.), *Droit international des investissements et de l'arbitrage transnational* (Paris: Pedone 2015) 119.

[18] B Kingsbury and S Schill, 'Investor-state arbitration as governance: Fair and equitable treatment, proportionality and the emerging global administrative law', (2009) 146 New York University Public Law and Legal Theory Working Papers 1, <http://lsr.nellco.org/nyu_plltwp/146>.

[19] E.g. Art. 54(1), ICSID Convention. See T Mizushima, 'The role of the State after an award is rendered in investor-State arbitration', in S Lalani and RP Lazo (eds) *The Role of the State in investor-State arbitration* (Leiden: Nijhoff 2015) 274; L Achtouk-Spivak and A Ben Mansour, 'Reconnaissance et execution des sentences arbitrales en matière d'investissement', in C Leben (ed.), n. 17 above, 999.

investment arbitration the distinguished characteristics of a coercive force.[20]

It is, however, one thing to argue that the structure of the investment arbitration may *potentially* influence domestic water policy, and another to find such an effect. The UN experts' statement cited in the introduction seem to suggest that investment arbitration has already influenced domestic authority. They stated that the bilateral and multilateral investment treaties might aggregate the problem of extreme poverty and affect the right of persons leaving in vulnerable situations due to 'the "chilling effect" that intrusive [arbitral] awards *have had*, when States *have been penalized* for adopting regulations.'[21] Unfortunately, the UN experts provide no concrete examples of such detrimental effects of investment arbitration. One can merely speculate that the statement, which refers to the reduction of smoking, had in mind the then ongoing *Philip Morris v. Australia*[22] on tobacco packaging regulation.

The former independent expert on the issue of human rights obligations related to access to safe drinking water and sanitation also indirectly raised the issue in her 2010 Report.[23] While noting that the investment law may potentially conflict with human rights law by limiting the regulatory space of the host State, the Report put this issue outside its scope. Nevertheless, it cited the Report of the Special Representative of the Secretary-General on the issue of human rights and transnational corporations and other business enterprises,[24] which stressed the constraints on the policy discretion of States to pursue legitimate public interest objectives,[25] and referred to then ongoing *Piero Foresti v. South Africa*,[26] in which investors raised the question of legality of measures taken under the Black Economic Empowerment Act.

The Report on human rights, trade and investment submitted by the

---

[20] Van Harten and Loughlin, n. 16 above, 135.

[21] 'UN experts voice concern', n. 1 above, [emphasis added].

[22] *Philip Morris Asia Ltd v Australia*, UNCITRAL (1976), PCA Case No. 2012–12. The tribunal denied its jurisdiction in its decision on 17 December 2015. The text of the decision was not publicly available at the time of writing this chapter.

[23] C de Albuquerque, 'Report of the independent expert on the issue of human rights obligations related to access to safe drinking water and sanitation', A/HRC/15/31 (2010), fn 2.

[24] Ibid., para. 35.

[25] J Ruggie, 'Report of the Special Representative of the Secretary-General on the issue of human rights and transnational corporations and other business enterprises', A/HRC/14/27 (2010), para. 22.

[26] *Piero Foresti, Laura de Carli & Others v South Africa*, ICSID Case No. ARB(AF)/07/01. The proceedings were later discontinued and the tribunal only rendered an award on costs. *See* Award, 4 August 2010.

UNHCHR to the General Assembly in 2003[27] studied two cases concerning regulation of the treatment of the materials potentially harmful to the environment, namely *Ethyl Corporation v. Canada*[28] and *Metalclad v. Mexico*,[29] and stated that '[t]o the extent that broad interpretations of expropriation provisions could affect States' willingness or capacity to introduce new measures to promote and protect human rights, then the use and interpretation of expropriation provisions is a cause of concern'.[30]

In comparison to the above-cited statements, the language used in the 2007 Report of the UNHCHR specifically dedicated to the issue of equitable access to safe drinking water was considerably nuanced. While noting that some submissions received during the consultation process raised concerns over the potential impact of investment arbitration on the duty of States to regulate companies in the context of private water provision, the Report did not take any position regarding this matter, and stated that '[i]t remains unclear whether and how the obligations of Governments under international human rights instruments will be taken into account in ICSID judgements' and concluded that 'this research would benefit from further analysis'.[31] It seems that this reserved position was prompted by the *Aguas Argentinas/Suez v. Argentina*[32] tribunal's decision to receive *amicus curiae* briefs from civil society organisations.[33]

None of the documents mentioned in this section provides any concrete

---

[27] High Commissioner for Human Rights, 'Human rights, trade and investment', E/CN.4/Sub.2/2003/9 (2003), para. 35.
[28] *Ethyl Corporation v Canada*, UNCITRAL (1976).
[29] *Metalclad Corporation v Mexico*, ICSID Case No. ARB(AF)/97/1.
[30] It added that 'it will be important to avoid a situation where the threat of litigation on the basis of broadly interpreted expropriation provisions has a "chilling effect" on government regulatory capacity, conditioning State action to promote human rights and a healthy environment by the commercial concerns of foreign investors.'
[31] Report of the United Nations High Commissioner for Human Rights, n. 2 above, paras 63–64.
[32] *Aguas Argentinas, S.A., Suez, Sociedad General de Aguas de Barcelona, S.A. and Vivendi Universal, S.A. v. Argentina*, ICSID Case No. ARB/03/19, Order in Response to a Petition for Transparency and Participation as *Amicus Curiae* (2005). As Aguas Argentinas withdrew as a claimant after the Order on *amicus* participation. For this reason, the decision on liability could be referred to as '*Suez*'. However, for the sake of consistency, the name '*Aguas Argentinas/Suez*' will be used. For the same reason, *Aguas Provinciales de Santa Fe S.A, Suez, Sociedad General de Aguas de Barcelona, S.A., and InterAguas Servicios Integrales del Agua, S.A. v. Argentina*, ICSID Case No. ARB/03/17 is referred to as '*Aguas Provinciales de Santa Fe/Suez*'.
[33] On *amicus curiae*, see footnotes 50–55 and the accompanying text of this chapter.

example of investment arbitration that had actual negative impact on domestic public policy. Most of the cases referred to were ongoing at that time (*Aguas Argentinas/Suez*, *Philip Morris*, *Piero Foresti*) or had been settled outside the proceeding (*Ethyl*). Whether *Metalclad* is appropriate to claim 'chilling effect' of investment arbitration is questionable, as this chapter examines later.

2. THE SCOPE OF THE HUMAN RIGHT TO WATER

In assessing the impact of investment arbitral decisions on the enjoyment of the human right to water, there is evidently a need to clarify its source and scope.

Despite the silence of the universal human rights treaties,[34] its status under such treaties has already been established. The Committee on Economic, Social and Cultural Rights published the well-known General Comment No. 15 on the right to water in 2002, which proclaimed that '[t]he right to water clearly falls within the category of guarantees essential for securing an adequate standard of living, particularly since it is one of the most fundamental conditions for survival'.[35] Accordingly, the most widely accepted legal basis of the human right to water is article 11 of the International Covenant on Economic, Social and Cultural Rights (ICESCR).[36] General Assembly Resolution 68/157 adopted in 2013 without a vote educes such an interpretation.[37]

---

[34] Riedel points out that water was not a major concern when the International Covenants were negotiated. EH Riedel, 'The human right to water and General Comment No. 15', in EH Riedel and P Rothen (eds), *The Human Right to Water*, (Berlin: BWV, Berliner Wissenschafts-Verlag, 2006), fn. 19.

[35] Committee on Economic, Social and Cultural Rights, 'General Comment No 15: The right to water', UN Doc E/C12/2002/11 (2002), para 3.

[36] International Covenant on Economic, Social and Cultural Rights, (1966) 993 United Nations Treaty Series 3. Article 11(1) provides that '[t]he States Parties to the present Covenant recognize the right of everyone to an adequate standard of living for himself and his family, including adequate food, clothing and housing, and to the continuous improvement of living conditions'.

[37] It was recalled in the preamble that 'the human right to water and sanitation is derived from the right to an adequate standard of living and is inextricably related to the right to the highest attainable standard of physical and mental health, as well as to the right to life and human dignity'. *See* 'The human right to safe drinking water and sanitation', A/RES/68/157 (2013). *See also* 'The human rights to safe drinking water and sanitation', A/RES/70/169 (2015). Note that in the latter resolution, the right to water and the right to sanitation are recognised as separate rights so that they are collectively referred to in the plural.

On the other hand, its customary status remains controversial. Unlike Resolution 68/157, Resolution 64/292 adopted in 2010 does not state that the right to water is derived from any other rights.[38] Hence, one may argue that the resolution thus recognised the right to water as an independent right under customary international law. However, it is to be noted that it was adopted with 41 States abstaining.[39]

However, when turning to the decisions of investment arbitrations, no arbitral tribunal has negated the human right to water. They are rather willing to recognise the human right to water under international law. Some of the tribunals dealing with water services explicitly referred to human rights considerations.[40] The most recent *SAUR* tribunal takes a more audacious position, stating that:

> [e]n réalité, les droits de l'homme en général, et le droit à l'eau en particulier, constituent des diverses sources que le Tribunal devra prendre en compte pour résoudre le différend car ces droits sont élevés au sein du système juridique argentin au rang de droits constitutionnels, et de plus, ils font partie des principes généraux du droit international.[41]

Hence in practice, it is not so relevant to determine the source of the human right to water, be it a right under treaties, customary international law, or general principles of (international) law.[42]

When turning to the scope of the human right to water, it is important to remind ourselves, in the first place, that States are under the obligation to achieve 'progressively the full realization' of the economic and

---

[38] UN General Assembly, 'The human right to water and sanitation', A/RES/64/292 (2010), para. 1.

[39] For a cautious approach, see T Winkler, *The Human Right to Water: Significance, Legal Status and Implications for Water Allocation* (Oxford: Hart Publishing 2012) 79. Some States in fact expressed reserved positions. *See* A/64/PV.108 (2010).

[40] *Azurix Corp v. Argentina*, ICSID Case No. ARB/01/12, Award (2006), para. 261; *Suez,* Case No ARB/03/17, n. 6 above, para. 240; *Suez,* Case No. ARB/03/19, n. 6 above, para. 262.

[41] *SAUR International*, n. 6 above, para. 330, citing the 2007 Report of the UNHCHR. The tribunal delivered this sweeping statement although it would have been sufficient for it to mention that the host State (Argentina) was a party to the ICESCR.

[42] One of the arbitrators of the SAUR tribunal stated in one of his publications: 'it may be argued that when explicitly referring to customary rules lawyers are in fact moving within the purview of general principles of international law, where the element of practice holds only a marginal position'. C Tomuschat, *Human Rights: Between Idealism and Realism* (3rd ed, New York: Oxford University Press 2014) 43.

social rights.[43] In addition, as far as the necessary steps are taken, the human right to water is neutral on the mode of water service provision.[44] Private sector participation in drinking water supply causes no problem so long as the State guarantees the fulfilment of the right to water.[45] Therefore, privatisation as such is in no way incompatible with the right to water.

Amongst the elements of the human right to water,[46] the most relevant and important for our analysis is the concept of affordability. The UNHCHR explains this principle in the following manner:

> Affordability requires that direct and indirect costs related to water and sanitation should not prevent a person from accessing safe drinking water and should not compromise his or her ability to enjoy other rights, such as the right to food, housing, health and education. [. . .] The human rights framework does not imply, therefore, a right to free water and sanitation but highlights the fact that nobody should be deprived of access because of an inability to pay. [. . .] Consequently, the affordability requirement is not incompatible with the principle of cost recovery for water and sanitation services, which is also recognized in several international declarations. However, it defines limits to cost recovery and highlights the fact that it should not become a barrier to access to safe drinking water and sanitation, notably by the poor.[47]

Affordability under the human right to water thus wields two imperatives; the cost recovery of water services and the universal access to safe drinking water for citizens. Nevertheless, an attempt to strike a right balance in financial viability of water service and affordable pricing for the poor

---

[43] Art. 2(1), ICESCR.
[44] Committee on Economic, Social and Cultural Rights, 'General Comment No. 3: The Nature of States Parties' Obligations', UN Doc E/1991/23, annex III (1991), para. 8.
[45] de Albuquerque, n. 23 above, para. 15.
[46] The 2013 Human Rights Council Resolution states that the human right to water is composed of the following elements: sufficiency, safety, acceptability, physical accessibility, affordability and non-discrimination. 'The human right to safe drinking water and sanitation', A/HRC/RES/24/18 (2013). *See also* Committee on Economic, Social and Cultural Rights, n. 35 above, para. 2.
[47] 'Report of the United Nations High Commissioner for Human Rights', n. 2 above, para. 28.
Broader and somewhat different definition is provided by the General Comment No. 15: 'Any payment for water services has to be based on the principle of equity, ensuring that these services, whether privately or publicly provided, are affordable for all, including socially disadvantaged groups. Equity demands that poorer households should not be disproportionately burdened with water expenses as compared to richer households.' (para 25).

frequently causes tensions in the real world.[48] As the Special Rapporteur on the human rights to safe drinking water and sanitation states, '[t]he affordability of water and sanitation services is highly contextual, and States should therefore determine affordability standards at the national and/or local level'.[49] Hence, the more a case becomes contested, the less the human right to water would give a concrete baseline. What role, then, does the right to water play in investment arbitration?

3. HUMAN RIGHTS CONSIDERATIONS IN THE ARBITRAL DECISIONS

Arbitral tribunals do refer to human rights considerations.[50] However, they take account of such considerations as elements of the public interests of the host State (3.1). As in cases where public interests of the host State are involved, tribunals recognise the host State's regulatory powers in cases where the application and implementation of the human right to water is put into question (3.2). When a tribunal finds the host State's abusive use of such powers, it also finds a violation of the applicable investment treaty (3.3).

3.1 Human Rights Considerations Subsumed in Those of Public Interests

The human right to water was referred to for the first time by an investment arbitration in an order in which it decided to receive *amicus curiae* submission.[51] When faced with requests from civil society

---

[48] UNDP, *Human Development Report 2006 - Beyond Scarcity: Power, Poverty and the Global Water Crisis* (New York, 2006) 97; SL Murthy, 'The human right(s) to water and sanitation: History, meaning, and the controversy over-privatization', (2013) 31 *Berkeley Journal of International Law* 89, 134.

[49] L Heller, 'Report of the Special Rapporteur on the human right to safe drinking water and sanitation', A/HRC/30/39 (2015), para. 28. For comparative analyses, *see* OECD, *Pricing Water Resources and Water and Sanitation Services* (Paris 2010) 26; H Smet, *De l'eau potable à un prix abordable* (Paris: Editions Johanet 2007).

[50] N Rubins, The notion of investment in international investment arbitration, *in* N Horn (ed.) *Arbitrating Foreign Investment Disputes* (Kluwer 2004), 283, 291–2. *See also* J Chaisse, 'Exploring the confines of international investment and domestic health protections—is a general exceptions clause a forced perspective?' (2013) 39 *American Journal of Law & Medicine* 332.

[51] *Amicus curiae*, while its official definition is absent under international law, is generally understood as an entity interested in a judicial proceeding, as a non-party to it, that submits an unsolicited written brief, or even makes an oral

organisations to submit *amicus* brief, the tribunal in *Aguas Argentinas/ Suez v. Argentina* based its decision to accept such a brief on public interests involved in water-related issues. Human rights considerations were considered to be part of such public interests. The tribunal stated that:

> The factor that gives this case particular *public interest* is that the investment dispute centers around the water distribution and sewage systems of a large metropolitan area, the city of Buenos Aires and surrounding municipalities. Those systems provide basic public services to millions of people and as a result may raise a variety of complex public and international law questions, *including human rights considerations*. Any decision rendered in this case, whether in favor of the Claimants or the Respondent, has the potential to affect the operation of those systems and thereby the public they serve.[52]

This passage was later cited in *Biwater v. Tanzania*,[53] which added that receiving *amicus* submission in the proceedings was 'an important element in the overall discharge of the Arbitral Tribunal's mandate'.[54]

---

statement in some cases, on law or fact before an international court or tribunal. For its definition, historical background and the way it functions at different international litigations, *see* L Crema, 'Tracking the origins and testing the fairness of the instruments of fairness: Amici curiae in international litigation', *Jean Monnet Working Paper* (2012), <http://jeanmonnetprogram.org/>; *see also* L Bartholomeusz, 'The *amicus curiae* before international courts and tribunals', (2005) 5 *Non-State Actors and International Law* 209.

Because of its common law origin, more neutral terminology is preferred in recent days, such as 'non-disputing party' (ICSID, 'Rules of Procedure for Arbitration Proceedings' (2006)) or 'third person(s)' ('UNCITRAL Rules on Transparency in Treaty-based Investor-State Arbitration' (2014)). However, this chapter uses the traditional term *amicus curiae*, which is used by the tribunals mentioned here, in order to avoid useless confusion.

[52] *Aguas Argentinas, SA, Suez, Sociedad General de Aguas de Barcelona, SA and Vivendi Universal, SA v. Argentina*, ICSID Case No ARB/03/19, Order in Response to a Petition for Transparency and Participation as *Amicus Curiae* (2005), para. 19 [emphasis added].

[53] *Biwater Gauff (Tanzania) Ltd v. Tanzania*, ICSID Case No ARB/05/22, Procedural Order No 5 (2007), para. 52.

One can question the hesitant rejection of receiving *amicus curiae* submission in *Aguas del Tunari v. Bolivia*. However, the tribunal refers to the respondent State's rejection as well as the unnecessity of such submission in jurisdictional phase. *See* Letter responding to the petition from the President of the Tribunal in the matter of *Aguas del Tunari SA v. Bolivia*, ICSID Case No ARB/02/3 (2003). For other possible explanations, consult B Stern, 'Civil society's voice in the settlement of international economic disputes', (2007) 22 *ICSID Review* 280.

[54] *Biwater Gauff*, n. 53 above, para. 50.

These tribunals' willingness to accept *amicus* participation by virtue of their discretionary power[55] seem to imply their broader concern over the legitimacy of international arbitration dealing with domestic matters.[56]

Human rights considerations are not only raised by civil society organisations but often by the host State. However, no tribunals have been influenced by such arguments in their interpretation of relevant provisions of applicable investment treaties.

In *Azurix v. Argentina*, the respondent raised the issue of the compatibility of the applicable investment treaty (Argentina-US BIT) with human rights treaties to which Argentina was a party. However, the tribunal summarily responded that this matter 'has not been fully argued and the Tribunal fails to understand the incompatibility *in the specifics of the instant case*'.[57] Similarly, the tribunal in *Aguas Argentinas/Suez v. Argentina* and *Aguas Provinciales de Santa Fe/Suez v. Argentina* stated that '[u]nder the circumstances of these cases, Argentina's human rights obligations and its investment treaty obligations are not inconsistent, contradictory, or mutually exclusive'.[58] In the tribunal's opinion, 'Argentina is subject to both international obligations, i.e. human rights *and* treaty obligation, and must respect both of them equally'.[59]

This compatibility question was further elaborated in *SAUR v. Argentina*. As mentioned above, the tribunal observed that these human rights constituted one of the various sources that it would have to take into account to settle the dispute.[60] It then transformed the human rights discourse into that of public interest. According to the tribunal:

---

[55] *Methanex Corporation v USA*, UNCITRAL (1976), Decision of the Tribunal on Petitions from Third Persons to Intervene as 'amici curiae' (2001), para. 47.

[56] J Harrison, 'Human rights arguments in *amicus curiae* submissions: Promoting social justice?' in PM Dupuy, F Francioni and EU Petersmann (eds), *Human Rights in International Investment Law and Arbitration* (Oxford: Oxford University Press 2009) 396.

[57] *Azurix Corp v. Argentina*, ICSID Case No ARB/01/12, Award (2006), para. 261 [emphasis added].

[58] *Suez*, Case No ARB/03/19, n. 6 above, para. 262; *see also Suez*, Case No ARB/03/17, n. 6 above, para. 240.

[59] *Suez*, Case No ARB/03/19, ibid.; *see also Suez*, ICSID Case No ARB/03/17, ibid. [emphasis in original].

[60] *SAUR International*, n. 6 above, para. 330.

> L'accès à l'eau potable constitue, du point de vue de l'État, un service public de première nécessité et, du point de vue du citoyen, un droit fondamental. Pour ce motif, en cette matière, l'ordre juridique peut et doit réserver à l'Autorité publique des fonctions légitimes de planification, de supervision, de police, de sanction, d'intervention et même de résiliation, afin de protéger l'intérêt général.[61]

The text of the tribunal's decision suggests that human rights are not taken into account as such but as elements of the public interests involved in the dispute. It follows that we have to broaden our perspective so as to consider how the tribunals take account of public interests in the host State when interpreting and applying relevant provisions of applicable investment treaties.

3.2 Public Interest Formula

Non-water cases

Recent arbitral decisions on environmental protection, archaeological site protection or banking regulation resort to the concept of public interest to recognise the host State's regulatory authority over these matters when applying various provisions of investment treaties, inter alia, those on expropriation and the FET standard.

To begin with the expropriation clause, it usually provides that investments shall not be expropriated or subject to a measure having equivalent effect to expropriation except where such measure is: (a) for the public purpose; (b) not discriminatory; (c) carried out under due process of law; and (d) accompanied by appropriate compensation. Under this formula, if it is literally interpreted, any regulatory measure which would have the effect equivalent to expropriation on an investor or investment would require compensation, even where the property is not seized and the legal title to the property is not affected (indirect expropriation[62]). However, this so-called 'sole effects doctrine' has never been applied in investment

---

[61] Ibid. Elsewhere in the decision, the tribunal similarly recognised that:

> les agissements irresponsables d'une entreprise prestataire de services de première nécessité peuvent dégénérer en risques graves pour la santé et le bien-être de la citoyenneté. Dans [ce] cas, il est justifié que l'État se réserve, pour la protection et au bénéfice de la société dans son ensemble, des pouvoirs spéciaux lui permettant d'interférer dans la gestion, voire même dans la propriété, de l'entreprise privée afin de redresser la situation. (para. 396)

[62] For the terminology, *see* OECD, '"Indirect expropriation" and the "right to regulate" in international investment law', *Working Papers on International Investment* (2004) 3.

arbitration, despite the lip service given by some tribunals to this doctrine, so that the provision on indirect expropriation has lost its *raison d'être* particularly when the applicable investment treaty contains the FET clause.[63] For example, *Metalclad v. Mexico* has frequently been criticized for considering environmental measures taken by the local authorities to constitute an indirect expropriation requiring compensation, thus undermining environmental considerations. However, a closer look at its reasoning reveals that the tribunal did not find that the measures allegedly taken to protect the environment constituted an expropriation simply because they deprived the investor of the use of its investment. It held, in fact, that the measures in question constituted an expropriation because they not only interfered with the use of the investment but also amounted to unfair and inequitable treatment.[64]

When applying the provision on indirect expropriation, tribunals always recognise the host State's regulatory space to protect and promote public interests.[65] The representative example is *Methanex v. United States*[66] and there are many such decisions.[67] As long as the host State takes *bona fide* and proportionate regulatory measures respecting the investor's legitimate expectations, the tribunal finds no indirect expropriation.[68]

---

[63] S Hamamoto, 'Requiem for indirect expropriation: On the theoretical and practical uselessness of a contested concept', *Private International Law as Global Governance (PILAGG) e-series*, IA-1, (2013), http://blogs.sciences-po.fr/pilagg/pilagg-e-series/ also at http://ssrn.com/abstract=2666836/.

[64] *Metalclad Corporation v. Mexico*, ICSID Case No. ARB(AF)/97/1, Award (2000), paras 103–104. As for the decision of the British Columbia Supreme Court partially setting aside the Metalclad award, see S. Hamamoto, 'Domestic review of treaty-based international investment awards: effects of the Metalclad judgment of the British Columbia Supreme Court', in M. Kanetake and A. Nollkaemper eds., *The Rule of Law at the National and International Levels* (Oxford: Hart Publishing 2016) 99–113.

[65] For the detailed study of this subject, *see e.g.* S Robert-Cuendet, *Droits de l'investisseur étranger et protection de l'environnement* (Leiden: Nijhoff 2010).

[66] *Methanex Corporation v. USA*, UNCITRAL (1976), Final Award of the Tribunal on Jurisdiction and Merits (2005), Part IV, Chapter D, para. 7.

[67] E.g. *Saluka Investments BV v. The Czech Republic*, UNCITRAL (1976), Partial Award (2006), para. 262; *Chemtura v. Canada*, UNCITRAL (1976), Award, 2 August 2010, para 266.

[68] T Waelde and A Kolo, 'Environmental regulation, investment protection and "regulatory taking" in international law', (2001) 50 *International and Comparative Law Quarterly* 846; R Moloo and J Jacinto, 'Environmental and health regulation: Assessing liability under investment treaties', (2011) 29 *Berkeley Journal of International Law* 33; LY Fortier and SL Drymer, 'Indirect expropriation in the law of international investment: I know it when I see it, or *caveat investor*', (2004) 19 *ICSID Review* 306.

The tribunals always recognise the host State's regulatory space to protect and promote its public interests also when applying the FET clause.[69] The FET obligation is generally considered to include the obligation of the host State to protect the investor's legitimate expectations.[70] It is not difficult to understand that the investor cannot legitimately expect that the host State will refrain from exercising its *bona fide* and proportionate regulatory measures to protect foreign investments and investors.[71]

Cases concerning water services

This general formula is also applied by the tribunals in cases on water and sanitation sector.

In the assessment of direct expropriation, the tribunal of *SAUR* accepted that '[l]e fait que l'exercice *légitime* de pouvoirs de police ne constitue pas un délit international est un principe général du droit international coutumier, communément accepté, et dans certains cas, il est même justifié que la personne lésée ne soit pas indemnisée'.[72] This statement reminds us of the finding by *Methanex* and *Saluka* tribunals.

Similarly, the tribunals dealing with water-related disputes follow the general trend when interpreting and applying the FET clause. In *Aguas Argentinas/Suez* and *Aguas Provinciales de Santa Fe/Suez v. Argentina*, the tribunal bore in mind that 'the Concession by its terms was subject to the regulatory authority of the Argentine State, which had a reasonable right to regulate'. Thus, under the FET standard, 'the Tribunal must balance the legitimate and reasonable expectations of the Claimants with Argentina's right to regulate the provision of a vital public service'.[73] Similarly, the tribunal of *Impregilo v. Argentina* also reserved the regulatory

---

[69] E.g. *Enron v. Argentina*, ICSID Case No ARB/01/3, Award, 22 May 2007, para. 261; *Ulysseas v. Ecuador*, UNCITRAL (1976), Final Award, 12 June 2012, para. 248.

[70] See S Hamamoto, 'Protection of the investor's legitimate expectations: Intersection of a treaty obligation and a general principle of law', in W Shan and J Su (eds), *China and International Investment Law* (Leiden: Brill/Nijhoff 2014) 141; M Potestà, 'Legitimate expectations in investment treaty law: Understanding the roots and the limits of a controversial concept', (2013) 28 *ICSID Review* 88.

[71] E.g. *Parkerings-Compagniet AS v. Lithuania*, ICSID Case No ARB/05/8, Award (2007), para. 332; *EDF v. Romania*, ICSID Case No ARB/05/13, Award (2009), para. 217; *El Paso v. Argentina*, ICSID Case No. ARB/03/15, Award (2011), paras 366–369.

[72] *SAUR International*, n. 6 above, para. 398 [emphasis in original].

[73] *Suez*, Case No ARB/03/17, n. 6 above, para. 216; *Suez*, Case No. ARB/03/19, n. 6 above, para. 236.

space of the host State, holding that '[t]he legitimate expectations of foreign investors cannot be that the State will never modify the legal framework, especially in times of crisis'.[74]

3.3 Measures Found to be in Violation by Tribunals

While the tribunals state that they recognise the host State's regulatory power, all arbitral decisions rendered thus far on urban water utilities have found that the host State had breached its treaty obligations.[75] Should we point out a contradiction between the tribunals' words and its actions? We do not think so, for the following reasons.

In some of the water-related cases, the host State intervened in the investor's activities in an unjustifiable manner, which cannot be considered a legitimate exercise of its regulatory powers. *Vivendi v. Argentina* arose out of a troubled privatisation of the water and sewage services in a province of the host State.[76] After the privatisation process had started, an anti-privatisation group won a local election and formed a new local government.[77] This new government was found to have been mounting, improperly and without justification, an illegitimate 'campaign' against the concession from the moment it took office, aimed either at reversing the privatisation or forcing the concessionaire to renegotiate lower tariffs.[78] The provincial independent water regulator became politicised and directed by the executive branch of the provincial government including the Governor and Ministers.[79] The tribunal stated that the claimants 'had every reason to expect that their privatisation partner, the Province, would not mount an illegitimate campaign to force them, on threat of rescission, to renegotiate a lower tariff',[80] and concluded that 'measured by a "do no harm" standard, Respondent directly undermined Claimants' legitimate expectations of their investment and breached its Treaty commitments'.[81]

---

[74] *Impregilo SpA v. Argentina*, ICSID Case No. ARB/07/17, Award (2011), para. 291.
[75] *Biwater* case is peculiar in that it did not award any pecuniary compensation to the claimant. This was because at the time the internationally wrongful act was taken by the host State, the fair market value of the local company was nil. *Biwater*, n. 6 above, para. 797.
[76] *Compañiá de Aguas*, n. 6 above, para. 4.2.1.
[77] Ibid., paras 4.8.1, 4.8.5.
[78] Ibid., para. 7.4.19.
[79] Ibid., paras. 4.6.10, 4.13.18, 7.4.22, 7.4.29, 7.4.36, 7.4.37, 7.4.44.
[80] Ibid., fn 355.
[81] Ibid., para 7.4.42 [footnote omitted].

The acts of the local government in question in *Azurix* were no better than in *Vivendi*. The tariff regime set in the course of general privatisation was politicised by the competing party. The repeated calls of the new provincial governor and other officials for non-payment of bills by customers were found to be on the verge of bad faith.[82] After the concessionaire was forced to leave without successfully renegotiating raising tariffs, the new service provider was allowed to raise them.[83]

In *SAUR*, the measure in question was taken after the post-crisis successful negotiation of contractual arrangements establishing a new tariff regime. The provincial authority failed to abide by the new contractual terms by delaying its approval of a tariff increase although it was fully aware that such a raise was correct and appropriate.[84] According to the tribunal, the gravity of the provincial authority's breach of the new tariff regime prevented it from accepting the host State's defence based on its regulatory power.[85]

In cases arising directly from measures taken to cope with the Argentinian economic crisis at the turn of the millennium, tribunals deferred from finding some regulations unlawful, but, at the same time, they did step in to identify some possibilities to reconcile public interests and investors' legitimate expectations.

In *Impregilo*, the concession contract concluded between a local operating company of which the investor is a major shareholder (42.58 per cent) and the provincial authority, which provided that:

> [t]he calculation of applicable tariffs [. . .] shall be based on the general principle that tariffs shall cover all operating expenses, maintenance expenses and service amortization and provide a reasonable return on concessionaire's investment subject to efficient management and operation by the Concessionaire and strict compliance with the applicable service quality and expansion goals.

The tribunal considered this provision to be 'an essential basis for the concession which would have to be upheld even in a changing economic climate'.[86] After the financial crisis, the provincial government introduced a 'social tariff for low-income residential owners'. Although such a new element evidently unfavourably affected the concessionaire, the provincial government appeared reluctant to renegotiate the concession

---

[82] *Azurix Corp*, n. 6 above, para. 376.
[83] Ibid., para. 375.
[84] *SAUR International*, n. 6 above, para. 404.
[85] Ibid., paras 402–405.
[86] *Impregilo*, n. 6 above, para. 324.

contract. The tribunal thus held that the host State, 'by failing to restore a reasonable equilibrium in the concession, aggravated its situation to such extent as to constitute a breach of its duty under the [Argentina-Italy] BIT to afford a fair and equitable treatment to Impregilo's investment'.[87] Needless to say, nothing prevents the host State from introducing such a social tariff. However, if it affects the concessionaire's financial situation so severely that the viability of the concessionaire's water-related service is put into question, the host State ought to take some measures to ensure the concessionaire's sustainable operation. This is what was not done in *Impregilo*.

Similarly, the *Aguas Argentinas/Suez* and *Aguas Provinciales de Santa Fe/Suez* tribunal held, with respect also to the host State's persistent refusal to revise the tariff in accordance with the concession contract, that:

> [t]here is no question that under the legal framework Argentina had the right to regulate [. . .] But AASA [the investors' local company] and the Claimants, as participants in any regulated industry, had the legitimate expectation that the Argentine authorities would exercise that regulatory authority and discretion within the rules of the detailed legal framework that Argentina had established for the Concession. But when faced with the crisis, Argentina refused to do this. It still refused once the crisis had abated.[88]

The tribunal characterised the acts of Argentina as 'an abuse of regulatory discretion'. It further stated:

> if Argentina's concern was to avoid an increase in tariffs during a time of crisis, it might have relieved AASA, at least temporarily, of investment commitments that were placing a crippling burden on the Concession so long as tariffs did not increase. If Argentina's concern was to protect the poor from increased tariffs, it might have allowed tariff increases for other consumers while applying a social tariff or a subsidy to the poor, a solution clearly permitted by the regulatory framework.[89]

The tribunal thus considered that the host State should have been able to exercise its right to regulate while respecting its obligations under the concession contract and the applicable investment treaties (Argentina-France BIT and Argentina-Spain BIT).

---

[87] Ibid., paras 328–331.
[88] *Suez*, Case No. ARB/03/19, n. 6 above, para. 237. *See also Suez*, Case No ARB/03/17, n. 6 above, para. 217;
[89] *Suez*, Case No. ARB/03/19, ibid., para. 235. *See also Suez* Case No ARB/03/17, ibid., para. 215.

CONCLUSION

From our analysis of the arbitral awards on urban water utilities, both optimistic and pessimistic implications can be drawn. Contrary to the often manifested perception that investment arbitration is inimical to the human right to water, a close examination of arbitral decisions dealing with water-related issues indicates that the host State's regulatory power has been recognised and that only an abusive use of such power is considered to be incompatible with relevant provisions (FET or indirect expropriation) of applicable investment treaties. This leads us to look at the function of investment arbitration optimistically as a mechanism that 'reduc[es] the space for unprincipled and arbitrary actions of the host state and thus contribute to good governance'.[90]

The importance of a healthy regulatory framework, or more broadly good water governance, is often emphasised in policy and economic analyses of past failed projects as well as in intensive exchanges of good practices.[91] In order to strengthen the regulation and governance, various policy guidelines have also been prepared by institutions with water-related expertise.[92]

The importance of public policy as the process of the realisation of the human right to water is attracting increasing attention.[93] The General Comment No. 15 and other documents prepared by international organisation or other water agencies thus emphasise the need of establishing the

---

[90] R Dolzer, 'The impact of international investment treaties on domestic administrative law', (2004) 37 *New York University Journal of International Law and Politics* 953.

[91] Notably, *UNDP, Human Development Report 2006*, n. 48 above, 100 (with the specific reference to the weak regulatory framework of the Buenos Aires concession, out of which *Aguas Argentinas/Suez*, *Azurix*, and *Impregilo* cases emerged). For the examples of scholarly works, *see* e.g. K Bakker, *Privatizing Water: Governance Failure and the World's Urban Water Crisis* (New York: Cornell University Press 2010); A Post, *Foreign And Domestic Investment in Argentina: The Politics of Privatized Infrastructure* (New York: Cambridge University Press 2014).

[92] The examples include the OECD Principles on Water Governance under the Water Governance Initiative (C/MIN(2015)12), and the Lisbon Charter: Guiding the Public Policy and Regulation of Drinking Water Supply, Sanitation and Wastewater Management Services (2015) prepared by the International Water Association.

[93] *See also* P Hunt and others, 'Implementation of economic, social and cultural rights', in S Sheeran and Sir N Rodley (eds), *Routledge Handbook of International Human Rights Law* (London: Routledge 2013) 545.

independent regulatory body.⁹⁴ Similarly, the 2007 UNHCHR Report stresses that 'a clear and efficient regulatory framework' should be introduced to achieve human rights goals.⁹⁵ In a Handbook prepared by the Special Rapporteur on the human right to water and sanitation, the expectations and challenges for regulatory bodies are expressed as follows:

> [a]s the regulatory body is responsible for setting and monitoring affordability standards and targets, this body should also set tariffs. However, regulatory bodies may face challenges from two directions. Because low service charges are frequently a vote-catcher, politicians may intrude on the decision-making process for tariffs, pushing the prices down to secure a better outcome in local elections. On the other hand, service providers may push for higher tariffs to secure better profits. In both cases, the regulatory body must have a legal mandate for independent tariff-setting.⁹⁶

This statement resonates with the findings of the arbitral tribunals examined in this chapter. The ongoing discussion with the water specialists at international settings seem to show convergence toward the consensual understanding of the regulatory framework that effectively assures and implements the human rights standards, including, but not limited to, the affordability standard and sustainable cost-recovery. Investment arbitral tribunals, for their part, function as a watchdog for disguised 'regulatory' measures. As research on the regulatory framework for good water governance goes on, future tribunals will have access to materials on which they can rely when assessing the conformity of the host State's acts with the applicable investment treaty. Since the concept of good

---

⁹⁴ Where water services (such as piped water networks, water tankers, access to rivers and wells) are operated or controlled by third parties, States parties must prevent them from compromising equal, affordable, and physical access to sufficient, safe and acceptable water. To prevent such abuses *an effective regulatory system must be established*, in conformity with the Covenant and this general comment, which includes *independent monitoring*, genuine public participation and imposition of penalties for non-compliance.

CESCR, 'General Comment No. 15', n. 35 above, para. 24 [emphasis added]. *See also* SMA Salman and S McInerney-Lankford, *The Human Right to Water: Legal and Policy Dimensions*, (World Bank 2004) 74; C Dubreuil, *Right to Water: From Concept to Implementation* (World Water Council 2006) 49; COHRE and others, *Manual on the Right to Water and Sanitation* (2007) 44.

⁹⁵ 'Report of the United Nations High Commissioner for Human Rights', n. 1 above, para. 53.

⁹⁶ Special Rapporteur on the Human Right to Water and Sanitation, *Realizing the Human Rights to Water and Sanitation: A Handbook, Financing* (Portugal 2014) 18. *See also* de Albuquerque, n. 35 above, para. 51.

governance tends to restrict the host State's discretion, it is not likely that the deference accorded by tribunals to the exercise of State powers would become greater.[97]

Nevertheless, the present study also leaves concern over investment arbitration. Admittedly, the investment arbitration mechanism has suffered from its 'legitimacy deficits'.[98] For instance, the fact that party-appointed arbitrators decide on the lawfulness of the host State's measures taken to protect and promote public interest has been a serious issue of concern.[99] The procedural innovation of the tribunals to accept *amicus* briefs is one of the potential means to ensure the quality of the reasoning of awards, thereby strengthening the credibility of arbitral decisions. The inclination of the tribunals in water-related cases to allow NGOs to submit briefs echoes the growing expectation for participation in water-related decision-making in general.[100] However, this procedure is not free from criticisms.[101] Such institutional questions of investment arbitration are not specific to water-related issues and need to be discussed in a larger context.

ACKNOWLEDGEMENT

This work was partly supported by JSPS KAKENHI Grant Number 15J09910.

---

[97] Our finding does not deny the possibility of future tribunals' reliance on the human rights instruments to justify a host State's measure as suggested by some commentators (Tanzi, in Treves, Seatzu and Trevisanut (eds), n. 11 above, 333–4), but it does not seem to us that such decisions will be frequent. *Cf* M Jacob and S W Schill, 'Going soft: Toward a new age of soft law in international investment law?' (2014) 8 *World Arbitration and Mediation Review* 1.

[98] E.g. S Franck, 'The legitimacy crisis in investment treaty arbitration: Privatizing public international law through inconsistent decisions', (2005) 73 *Fordham Law Review* 1521.

[99] *See* the recent proposal of the European Union to establish a permanent court of investment. European Commission, *Fact Sheet: Why the new EU proposal for an Investment Court System in TTIP is Beneficial to Both States and Investors* (Brussels 2015) <http://europa.eu/rapid/press-release_MEMO-15-6060_en.htm>.

[100] L Boisson de Chazournes, *Fresh Water in International Law* (Oxford: Oxford University Press, 2013) 164–7; M Cuq, *L'eau en droit international : Convergences et divergences dans les approches juridiques* (Bruxelles: Larcier, 2013) 114–22; O McIntyre, 'The human right to water as a creature of global administrative law', (2012) 37 *Water International* 654.

[101] E.g. Harrison, in Dupuy, Francioni and Petersmann (eds), n. 56 above, 396.

9. The provision and violation of water rights (the case of Pakistan) – a human rights based approach

Sikander Ahmed Shah*

1. INTRODUCTION

Water is a basic necessity of life. However, its value is determined differently in different parts of the world.[1] The scarcity of water causes its value to increase incrementally, a phenomenon witnessed in the developing world where in some localities its value is comparable to gold.[2] Furthermore, with the process of industrialization accelerating in the developing world, need for water will increase incrementally and its main use will not be limited to agriculture and domestic consumption.[3] Over one billion people globally do not have access to basic water supplies and half of the developing world's population suffers from disease due to the contaminated supply of water.[4] The international governance regimes are therefore faced with a difficult task; they have to categorize water in a manner that promotes the standard of living of the global citizenry most effectively.[5]

---

* The author is extremely grateful to Anoshay Fazal, a brilliant legal scholar, for providing him with endless and invaluable research and editorial assistance.
[1] Note, *What Price for the Priceless?: Implementing the Justiciability of the Right to Water* (2007) 120 HARV L REV 1067.
[2] Ibid.
[3] Lee-Yee Huang, *Not Just another Drop in the Human Rights Bucket: The Legal Significance of a Codified Human Right to Water* (2008) 20 FLA J INT'L L 353.
[4] Erik B Bluemel, *The Implications of Formulating a Human Right to Water* (2004) 31 ECOLOGY L Q 957.
[5] Ibid., 959.

1.1 The Theoretical Basis for Classifying Water as a Human Right

The quintessential question presented is why should water be classified as a human right and not viewed as an economic good or as an object of environmental protection. Those in favor of classifying water emancipation as a human right, either in the form of a civil-political or socio-economic right, argue that the human rights framework is the most effective way to provide access to adequate and healthy water. Because of the presence of an established legal framework through which emancipation is most pragmatically realizable, violations of the right are adequately ascertainable and hence state conduct can be most effectively monitored for implementation.

The problem with categorizing water as an economic good is that the inequitable distribution of water is tolerated.[6] There is the fear with this approach that access to water will be determined only by market forces and not by equity and need.[7] For instance, the economic-good-based approach can be used to promote the privatization of water[8] on the basis of the full cost recovery principle,[9] with the aim to improve the water supply system infrastructure.[10] This phenomenon will however lead to the non-provision of water to those who cannot afford it, such as that witnessed in the *Cochabamba* case.[11] Incidentally in the opinion of the leading expert on water law, the Committee on Economic, Social and

---

[6] Ibid.

[7] Ibid.

[8] For a detailed discussion about the cost and benefits for privatization of water for the promotion of human rights see, Malgosia Fitzmaurice, *'Symposium: Environmental Protection and Human Rights in the New Millennium: Perspectives,' Challenges, and Opportunities* (2007) 18 FORDHAM ENVTL L REV 537.

[9] "Full cost recovery means that the state or private water supplier should be able to recover the full costs of supplying water to all users." Fitzmaurice, ibid., 964. See Julien Chaisse and Chistian Bellak, *Navigating the Expanding Universe of Investment Treaties—Creation and Use of Critical Index*, 18 J. INT'L ECON. L. 79 (2015), for understanding the link between treaties that states entered into with the increase of the FDI.

[10] Bluemel, *supra* note 4 above, 965.

[11] See generally Julien Chaisse and Marine Polo, *Globalization of water privatization: Ramifications of investor-state disputes in the "blue gold" economy*, (2015) 38 BOSTON COLLEGE INTERNATIONAL AND COMPARATIVE LAW REVIEW 1; Erik J Woodhouse, Note, *The "Guerra del Agua" and the Cochabamba Concession: Social Risk and Foreign Direct Investment in Public Infrastructure* (2003) 39 STAN J INT'L L 295; Andrew Nickson and Claudia Vargas, *The Limitations of Water Regulation: The Failure of the Cochabamba Concession in Bolivia* (2002) 21 BULL OF LATIN AM RES 128.

Cultural Rights ("ESCR Committee"),[12] does not per se determine the privatization of water to be a violation of the human right to water.[13]

The problem of confronting water issues via the paradigm of environmental protection is that it focuses solely on conservation and protection, and solutions are derived from soft law principles and non-binding agreements and arrangements. This approach is also constrained by sovereignty and economic considerations,[14] with violations primarily subject to interstate negotiation, mediation and arbitration with the interest of non-state parties not adequately factored in.[15]

The positive of the human-right based approach is that it examines water-based rights also from an anthropocentric perspective and can more concretely identify state violations and create pressure on states to fulfill their obligations to provide and improve water infrastructure.[16] Furthermore, classifying water emancipation as a human right brings its enforceability to the grass-root level, where remedies can even be claimed by individuals in international forums historically not open to non-state based participation[17] and at the national and municipal levels where there are a number of adequate judicial remedies.[18]

Some experts are, however, critical of the human right-based approach to emancipation. They argue that the approach is simplistic; the malleability of human right language promotes double standards and can be used by developed states to retard the development of third world nations and as a consequence the classification of water as a human right impedes the realization of other hierarchically superior human rights.[19] Moreover,

---

[12] See UN Comm on Economic, Social, and Cultural Rights, The Right to the Highest Attainable Standard of Health: General Comment No. 14: art. 12 of the International Covenant on Economic, Social, and Cultural Rights, § 4, UN Doc E/C12/2000/4 (2000) 31.

[13] Fitzmaurice, *supra* note 8 above, 552.

[14] Lee-Yee Huang, *supra* note 3 above, 359.

[15] Lee-Yee Huang, ibid., 361. See also Julien Chaisse, *Assessing the Exposure of Asian States to Investment Claims,* 6 CONTEMP. ASIA ARB. J. 187, 201 (2013); Julien Chaisse, *Exploring the Confines of International Investment and Domestic Health Protections—Is a General Exceptions Clause a Forced Perspective?,* 39 AM. J. L. & MED. 332 (2013).

[16] Lee-Yee Huang, *supra* note 3 above, 359.

[17] First Optional Protocol to the International Covenant on Civil and Political Rights, art 1, Dec 16, 1966, 999 UNTS 302 [hereinafter First Optional Protocol]; American Convention on Human Rights, art 44, Nov 22, 1969, 1144 UNTS 144 [hereinafter AMCHR].

[18] Lee-Yee Huang, *supra* note 3 above, 359.

[19] Ibid., 367.

the approach does not account for political economy[20] that effectively dictates environmental policy. According to Smets, the problems with that approach are that "1. It creates international liability; 2. It prevents commoditization of water; 3. It implies free access to water; 4. It hinders liberalization or privatization of water utilities; 5. It creates obstacles to free trade; and 6. It facilitates legal harassment of water utilities or public authorities."[21]

In Smets' view water can be a commodity and a right concurrently. He supports his point by highlighting the successful privatization regime of water in the United Kingdom, where it is illegal to disconnect water, while even as an enforceable, fundamental and standalone human right in South Africa, guaranteed under the South African Constitution, water is frequently disconnected for segments of the population.[22]

1.2 The Human Right to Water under International Law

1.2.1 Historical overview

The right to water has not achieved the status of customary international law.[23] Moreover, foundational international treaties and declarations do not explicitly mention water as a fundamental human right. Proponents of the human right to water argue that because the right to water is so fundamental and apparent, it was unnecessary to explicitly enumerate the existence and protection of such a right in documents such as the Universal Declaration of Human Rights ("UDHR").[24] Under this view, the existence of the human right to water can be assumed and be substantiated by the fact that other lesser rights and goals are listed in major treaties and documents, whose realization is completely dependent on the provision of the right to water.[25]

---

[20] Ibid., 360.
[21] Henri Smets, *Economics of Water Services and the Right to Water* in FRESH WATER AND INTERNATIONAL ECONOMIC LAW (Brown-Weiss and others (eds) 2005) 177.
[22] Amy Hardberger, *Whose Job is It Anyway?: Governmental Obligations Created by the Human Right to Water* (2006) 41 TEX INT'L L J 533.
[23] Amy Hardberger, *Life, Liberty and the Pursuit of Water: Evaluating Water as Human Right and the Duties and Obligations it Creates* (2005) 4 NW U J INT'L HUM RTS 331.
[24] Article 25 of the UDHR is the most relevant provision on the basis of which the right to water can be implied. It states that "everyone has the right to a standard of living adequate for the health and well-being of himself and his family, including food. . .."
[25] Hardberger, *supra* note 23 above, 345.

In 1977, at the Mar Del Plata Conference in Argentina the idea of the human right to adequate quantity and quality of drinking water was explicitly introduced.[26] The first human rights treaty to explicitly recognize the right to water was the Convention for the Elimination of All Forms of Discrimination against Women ("CEDAW") in 1979.[27] Subsequently, the Convention on the Rights of the Child ("CRC") explicitly recognized the right of children to clean drinking water.[28] Unfortunately explicit reference to the right of water is limited to only these two thematic human rights treaties, which aim to provide protection for particular vulnerable groups in society.[29]

The United Nations Committee on Economic, Social and Cultural Rights adopted General Comment No.14 in 2000, linking the enumerated right of health under Art.12 of the International Covenant on Economic, Social and Cultural Rights ("ICESCR"), with the right to "access to safe and potable water."[30] The Committee further enumerated that state obligations included refraining from polluting water resources.[31]

The right to water was further established under General Comment 15.[32] In interpreting Art. 11 and 12 of the ICESCR, the Committee indicated that water was "one of the most fundamental conditions for survival"[33] and linked standard of living.[34] For there to be a realization of the right to water, the comments indicate that water emancipation can only be

---

[26] Report on the United Nations Water Conference, Mar del Plata, GA Res 32/158, UN GAOR, 107thPlenMtg, UN Doc E77IIA12 (1977).

[27] States Parties shall ensure to [rural] women the right. . . to enjoy adequate living conditions, particularly in relation to . . . water supply. See Convention on the Elimination of All Forms of Discrimination against Women, GA Res 34/180, at art. 14 (2)(h), U.N. GAOR, 34th Sess, Supp No 46, UN Doc A/34/46 (1979) [hereinafter CEDAW].

[28] Convention on the Rights of the Child, GA Res 44/25 annex, at art. 24(2)(c), UN GAOR, 44th Sess, Supp No 49, UN Doc A/44/49 (1989).

[29] Hardberger, *supra* note 23 above, 347.

[30] UN Comm on Economic, Social, and Cultural Rights, The Right to the Highest Attainable Standard of Health: General Comment No. 14: art 12 of the International Covenant on Economic, Social, and Cultural Rights, § 4, UN Doc E/C12/2000/4 (2000).

[31] Ibid., §§ 30, 34.

[32] See UN Comm on Economic, Social, and Cultural Rights, Substantive Issues Arising in the Implementation of the International Covenant on Economic, Social and Cultural Rights: General Comment No 15: The right to water: arts. 11 and 12 of the International Covenant on Economic, Social and Cultural Rights, UN Doc E/C12/2002/11 (Nov 26, 2002) 1 (stating that the "depletion and unequal distribution of water is exacerbating existing poverty") [hereinafter General Comment 15].

[33] Ibid., 3.

[34] Ibid., 1.

achieved when there is availability of water supply for continuous personal and domestic use. The water should be safe, of a quality free from hazardous contaminants and of an "acceptable colour, odour and taste" and it needs to be physically and economically accessible in a non-discriminatory fashion.[35]

Even though General Comments of the ICESCR committee are non-binding and of an advisory nature,[36] they are meant to elucidate and interpret existing rights, which would be binding under the ICESCR. However, the Committee lacks the power to expand existing rights or create new ones.[37] The General Comment is a form of notice to all the states parties to the ICESCR; its Committee expects information on steps taken towards the realization of enumerated rights, when states submit their general reports.[38] General Comment 15 is a detailed and comprehensive document and clearly recognizes water as a human right, however it does not mandate the enforceability of the right to water and deference to a state's response on account of limited resources, is respected.[39]

Recently numerous international conferences or declarations have either explicitly or implicitly recognized access of the right to water. Prominent ones include the Declaration on the Right of Development,[40] the 1992 Dublin Statement on Water and Sustainable Development, the 2000 Ministerial Declaration of the Second Water Conference and the 2005 Millennium Project, commissioned by the Secretary General of the United Nations, under which one goal "is to ensure that the proportion of people without access to safe drinking water and basic sanitation is halved by 2015."[41]

Importantly, the United Nations High Commissioner for Human Rights has stated that:

---

[35] Ibid., 5.
[36] Lee-Yee Huang, *supra* note 3 above, 357.
[37] Hardberger, *supra* note 23 above, 348.
[38] Stephen C McCaffrey, *Small Capacity and Big Responsibilities: Financial and Legal Implications of a Human Right to Water for Developing Countries* (2009) 21 GEO INT'L ENVTL L REV 679.
[39] UN Econ and Soc Council [ECOSOC], Comm on Econ, Soc, and Cultural Rights, Substantive Issues Arising in the Implementation of the International Covenant on Economic, Social and Cultural Rights, UN Doc E/C.12/2002/11 (2003) 41 [hereinafter ECOSOC], <www.unhchr.ch/tbs/doc.nsf/0/a5458d1d1bbd-713fc1256cc400389e94/$ FILE/G0340229.pdf>.
[40] Declaration on the Right of Development, GA Res 44/128 (1986), art. 8.
[41] Fitzmaurice, *supra* note 8 above, 547.

it is now time to consider access to safe drinking water and sanitation as a human right, defined as the right to equal and non-discriminatory access to a sufficient amount of safe drinking water for personal and domestic uses drinking, personal sanitation, washing of clothes, food preparation and personal and household hygiene to sustain life and health.[42]

In 2006, The United Nations Development Programme ("UNDP") recommended that states should make water a human right.[43] The right to water is also protected under International Humanitarian Law. Deprivation of water and protection of water sources are most at issue. The Hague Resolutions,[44] the Geneva Conventions and Customary International laws are often invoked.[45] For instance, under the 1949 Geneva Convention Relative to the Treatment of Prisoners of War, drinking water has to be provided adequately.[46] Art. 89 of the Fourth Geneva Convention pertaining to civilian protection, states that "sufficient drinking water shall be supplied to internees. . .."[47] Art. 54 of Protocol I of 1977 to the Geneva Convention (Protocol I), prohibits a state to "attack, destroy, remove or render useless. . .drinking water installations and supplies and irrigation works."[48]

Specialized International agreements, such as the Prohibition of Military

---

[42] UN General Assembly, Human Rights Council, Report of the United Nations High Commissioner for Human Rights on the Scope and Content of the Relevant Human Rights Obligations Related to Equitable Access to Safe Drinking Water and Sanitation Under International Human Rights Instruments, P 66, UN Doc A/HRC/6/3 (2007).

[43] UN Dev Programme, 'Beyond Scarcity: Power, Poverty and the Global Water Crisis' (2006) 4, <http://hdr.undp.org/ en/media/HDR06-complete.pdf>; but see Editorial, 'Clean Water Should be Recognized as a Human Right' *PLoS Med* (2009) 6, www.plosmedicine.org/article/info%3Adoi%2F10.1371%2Fjournal.pmed.1000102>. (In 2009 at the World Water Forum, among other nations the US, Canada and Russia all rejected classifying water as a human right).

[44] See Convention Respecting the Laws and Customs of War on Land art 23 (a), annexed to Convention [No IV] Respecting the Laws and Customs of War on Land, Oct 18, 1907, 37 Stat 2277. (Prohibiting use of poison and the conventions general scope 'allows its application to the purposeful contamination of water sources'). See also Hardberger, *supra* note 22 above, 552.

[45] Hardberger, ibid., 549.

[46] Geneva Convention Relative to the Treatment of Prisoners of War arts 20, 26, 29 and 46, Aug 12, 1949, 6 UST 3316, 74 UNTS 135, <www.ohchr.org/english/law/prisonerwar.htm>.

[47] Geneva Convention (IV) Relative to the Protection of Civilian Persons in Time of War art 89, Aug 12, 1949, 75 UNTS 287, 6 UST 3516 [hereinafter Fourth Geneva Convention].

[48] Protocol Additional to the Geneva Conventions of 12 August 1949, and Relating to the Protection of Victims of International Armed Conflicts, opened for

or Any Other Hostile Use of Environmental Modification Techniques ("ENMOD"), which relate to the protection of the environment under humanitarian law, include the right to protect water.[49] As regards this agreement, apart from environmental protection, four other provisions in the agreement are relevant to water security.[50]

The incorporation of legal protection for the right to water is also witnessed in multilateral agreements that relate to water issues, but are not human rights treaties. One example of such a treaty is the 1997 United Nations Convention on Non-Navigational Uses of International Watercourses.[51] Furthermore, numerous regional treaties have explicitly recognized not just the right to water but "healthy water," as a fundamental human right.[52]

1.2.2 The scope of the human right to water

The right to water does not mean that everyone is entitled to a limitless quantity of water for all needs and wants. The right is limited to access of water of sufficient quantity and quality, for fundamental uses relating to the adequate protection of a human life and health, for purposes of consumption, for instance in order to prevent dehydration, for hygiene and sanitation and "for cooking, cleaning and subsistence agriculture," it does not include the right to water "for commercial, industrial or large-scale agricultural or irrigation activities."[53]

1.2.3 Whether the right to water is a progressive or immediately realizable right

As the source of authority of General Comment 15 is derived from Arts 11 and 12 of the ICESCR, the realization of this right under the covenant is progressive in nature and states are under no obligation to

---

signature Dec 12, 1977, 1125 UNTS 3, art 54 [hereinafter Victims of International Armed Conflicts].

[49] Convention of the Prohibition of Military or Any Other Hostile Use of Environmental Modification Techniques, May 18, 1977, 31 UST 333, TIAS No 9614, [hereinafter ENMOD].

[50] (1) poison as a means of warfare; (2) destruction of enemy property; (3) attack on objects necessary for civilian survival; and (4) attacks on installations that contain dangerous forces.

Hardberger, *supra* note 22 above, 552.

[51] Fitzmaurice, *supra* note 8 above, 544.
[52] Ibid.
[53] Leticia K Nkonya, *Socioeconomic Rights: Empowerment for Global Justice: Realizing the Human Right to Water in Tanzania* (2010) 17 HUM RTS BR 25.

give immediate effect to this right.[54] This determination is affirmed by Art. 2(1) of the ICESCR.[55] This raises a conflict of sorts with the core obligations enumerated in General Comment No. 15, which are to be realized immediately[56] and states cannot justify non-compliance on the non-derogable core values, set out in paragraph 37 of General Comment No. 15.[57]

1.3 The Domestic Justiciability and National Implementation of the Human Right to Water

If one is to assume that water is a human right, then where does it stand in the hierarchy of rights? Is it an independent human right, or is it a subordinate right? And is it a means for achieving an explicitly established right, such as the right to life or health? It is also important to determine that as a right, whether primary or subordinate, is it subject to immediate realization? Or is it programmatic in nature, subject to only progressive realization and implementation based on state resources?

In examining this issue it is important to look at how the justiciability of the human right to water has been pursued in different jurisdictions. Two countries, India and South Africa, have determined justiciability of the right to water, but have done it in very different ways. The intrinsic nature of the right, its content and implementation, has all been viewed differently.

Under the Indian Constitution there is no enumerated justiciable right to water.[58] The right to water is derivative of the constitutional and fundamental right to life, a justiciable civil and political right. On the other hand, the Indian Constitution lists socio-economic and cultural rights under the Directive Principle of State Policy,[59] which are rights subject to progressive implementation and are non-justiciable under

---

[54] Fitzmaurice, *supra* note 8 above, 543.
[55] States parties to the ICESCR are obliged to only "take steps ... to the maximum of its available resources, with an eye towards achieving progressively the full realization of the rights recognized in the Covenant."
[56] General Comment No 15, *supra* note 32 above, 37.
[57] Ibid., 40.
[58] Note, *What Price for the Priceless?*, *supra* note 1 above, 1080.
[59] INDIA CONST., art. 38. The Directive Principles provide that: "the State shall strive to promote the welfare of the people by securing and protecting as effectively as it may a social order in which justice, social, economic and political, shall inform all the institutions of the national life."

Art. 37 of the Indian Constitution.[60] The Supreme Court of India has affirmed the justiciability of the right to water on numerous occasions.[61] Other South Asian states like Bangladesh and Pakistan have followed the Indian model.[62] In addition to access to water, the Indian Supreme Court has also held the pollution of water as a violation of the human right to water.[63]

---

[60] INDIA CONST., art. 37: "The provisions contained in this Part shall not be enforceable by any court, but the principles therein laid down are nevertheless fundamental in the governance of the country and it shall be the duty of the State to apply these principles in making laws."

[61] See *Attakoya Thangal v Union of India* (1990) 1 KLT 583:

the administrative agency cannot be permitted to function in such a manner as to make inroads, into the fundamental right under Art 21. The right to life is much more than the right to animal existence and its attributes are many fold, as life itself. A prioritisation of human needs and a new value system has been recognized in these areas. The right to sweet water, and the right to free air, are attributes of the right to life, for, these are the basic elements which sustain life itself.

See also *AP Pollution Control Bd II v Prof MV Nayudu* (2001) 2 SCC 62 (holding that the right of access to drinking water is fundamental to life, by creating a state duty under art. 21 to provide such access to its citizens); *Vellore Citizens' Welfare Forum v Union of India* (1996) 5 SCC 647 ("the constitutional and statutory provisions protect a person's right to fresh air, clean water and pollution-free environment, but the source of the right is the inalienable common law right of clean environment."). The Court has also articulated the idea that the right to life necessitates a right to a "healthy environment," making water pollution a justiciable issue. See, e.g., *Kumar v Bihar*, (1991) 1 SCC 598 (holding that the right to life "includes the right to enjoyment of pollution-free water and air for full enjoyment of life").

[62] Note, *What Price for the Priceless?*, *supra* note 1 above, 1079 n, 51. See also, Jona Razzaque, "Access to Environmental Justice: Role of the Judiciary in Bangladesh" (unpublished manuscript) 1, <www.eng-consult.com/BEN/papers/Paper-jona.PDF>.

[63] See *MC Mehta v Union of India* (1988) 1 SCC 471 (the government was ordered by the apex court to improve the sewage system and stop the throwing of burnt corpses into the river Ganges); See *Vellore Citizens' Welfare Forum v Union of India*, ibid., (tanneries were violating citizens' rights by emptying untreated waste into local drinking water supplies and agricultural areas); *MC Mehta v State of Orissa*, AIR 1992 Ori 225 (the court after finding out that the government knew beforehand that sewage was mixing with river water and causing waterborne diseases held the state was obligated to stop and prevent the pollution for the maintenance of wholesome water for consumption). See *AP Pollution Control Bd v Prof MV Nayudu*, ibid., 3, the court held that the right to access to drinking water is fundamental to life and that the state has a duty under art. 21 to provide clean drinking water to its citizens. In *MC Mehta v Union of India*, (2004) 3 SCR 128, 45, the apex Court recognized groundwater as a public asset with citizens having

Alternatively, the South African approach explicitly recognizes the right to water as an independent, justiciable and legally enforceable right under its Constitution. The right however is socio-economic in nature; it is therefore a positive right and unlike negative liberties, is understood not to be subject to immediate realization but to progressive implementation. Therefore, the South African Constitution recognizes the right to water,[64] but subjects this right to the state's ability to fulfill it in light of available resources.[65] The South African courts have held the non-provision of water to be unacceptable when there is a proven inability to pay for basic water service.[66] A number of African nations have followed the South African approach.[67]

As a whole the South African approach has been favored. From a development rights perspective, the Indian approach has been criticized for not being effective in securing affirmative rights and entitlement to water. This is because the vessel for protection, the right to life, is a negative right that protects individuals from interference; what is needed, however, is for states to be obligated to provide healthy water in adequate quantities to their populations. The Indian approach is focused more on respect and protection, rather than on the fulfillment of rights.

As the Indian approach safeguards the negative right of freedom from interference, water freedom as a derivative negative right, leads to a passive

---

the right to the use of air, water, and earth as protected under art. 21 of the Constitution. See also *MC Mehta v Kamal Nath*, (1997) 1 SCC 388.

[64] S AFR CONST (1996) §§ 27(1) "Everyone has the right to have access to...b. sufficient food and water...(2) The state must take reasonable legislative and other measures, within its available resources, to achieve the progressive realisation of each of these rights".

[65] Note, *What Price for the Priceless?*, *supra* note 1 above, 1083; S AFR CONST (1996) §§ 27(2).

[66] *Residents of Bon Vista Mansions v Southern Metropolitan Local Council* 2002 (6) BCLR 625 (W) (S Afr); See also *Lindiwe Mazibuko and Others v The City of Johannesburg and Others* 2008 High Court of South Africa (Witwatersrand Local Division) Case No 06/13865 3 (S Afr) [hereinafter Mazibuko]. (Prepayment water system was "unconstitutional and unlawful" and the city must "provide each applicant and other similarly placed residents of Phiri Township with . . . free basic water supply of 50 litres per person per day and . . . the option of a metered supply installed at the cost of the City of Johannesburg.") Mazibuko, 183. (City water policy also held to be discriminatory against women because of water cut off, women were the ones who were generally forced to travel long distances to obtain water). See Mazibuko, 159.

[67] See Constitutions of Gambia, Uganda, and Zambia. GAM CONST, art. 216(4); UGANDA CONST, Nat'l Objectives and Directive Principles of State Policy XIV; ZAMBIA CONST. (Constitution Act 1991) art. 112(d).

approach to water emancipation which is ineffective in dealing with the prevalent global water crisis being witnessed today.[68] There is also a limit to which the expansive reinterpretation of negative rights can lead to the provision and realization of positive rights through judicial activism.

The South African approach provided legislative protection to the right to water by enumerating it as a positive state-based obligation, with the role of the judiciary limited to determining whether the government is fulfilling its constitutional obligations, or violating the law. Unlike the Indian approach, which also gives rise to judicial unpredictability,[69] such vesting of authority in the judiciary makes its assessments primarily legal and hence judges are not forced to indulge in policymaking which is not within the ambit of their jurisdiction. This in turn also accords more sanctity to legal decisions and allows the judiciary to effectively monitor the government for compliance, without infringing upon the authority of other governmental organs.

Furthermore, courts are in a position to direct the government to allocate funds for water emancipation initiatives and then subsequently monitor such spending.[70] The judicial power to monitor the government to positively provide is not unfettered, but is constrained by the reasonableness test, as enumerated in the South African Constitution under Art.27(2).[71]

2. PAKISTAN AND THE HUMAN RIGHT TO WATER

2.1 Pakistan and its International Law Commitments on Water

Pakistan has ratified or acceded to all major human rights treaties including the CRC in 1990,[72] and the CEDAW in 1996.[73] It ratified the ICESCR in 2008 and recently the International Covenant on Civil and Political Rights ("ICCPR") and the Convention against Torture, Cruel, Inhuman and Degrading Treatment or Punishment ("CAT")

---

[68] See Note, *What Price for the Priceless?*, *supra* note 1 above, 1086.
[69] Ibid., 1088.
[70] Ibid., 1087.
[71] See generally *South Africa v Grootboom*, 2000 (11) BCLR 1169 (CC)(S Afr).
[72] Pakistan ratified the Convention on the Rights of the Child on December 12, 1990.
[73] Pakistan ratified the Convention on the Elimination of All forms of Discrimination Against Women on March 12, 1996.

in 2010.[74] All reservations and declarations made by Pakistan do not impact upon its obligation to protect the human right to water, under its international law commitments. The reservations primarily relate to Pakistan's commitments being subject to the injunctions of Islam, the dictates of the Constitution or not recognizing the jurisdiction of various adjudicative bodies like the ICJ.[75] Pakistan has also ratified all the Geneva Conventions,[76] but not its optional protocols.[77]

Hence the obligations concerning water emancipation under General Comment 15 as well as under treaty and customary law, are fully applicable on Pakistan.

2.2 The Constitutional and Legislative Protection to the Right of Water in Pakistan

There are numerous constitutional provisions that are relevant to the protection of water. Many of them relate to fundamental rights and are subject to immediate realization. The most relevant is the security of person under Art. 9.[78] Others include the inviolability of dignity of man under Art. 14,[79] the equality of citizens under Art. 25, complaints as to

---

[74] Pakistan ratified the International Covenant on Economic, Social and Cultural Rights on April 17, 2008; the International Covenant on Civil and Political Rights on June 28, 2010 and the United Nations Convention Against Torture on June 3, 2010.

[75] See for instance Pakistan's reservation to art. 6 which relates to the right to life. "The Islamic Republic of Pakistan declares that the provisions of Articles . . . 6. . . shall be so applied to the extent that they are not repugnant to the provisions of the constitution of Pakistan and the sharia laws."

[76] Convention (I) for the Amelioration of the Condition of the Wounded and Sick in Armed Forces in the Field. Geneva, 12 August 1949; Convention (II) for the Amelioration of the Condition of Wounded, Sick and Shipwrecked Members of Armed Forces at Sea. Geneva, 12 August 1949; Convention (III) relative to the Treatment of Prisoners of War. Geneva, 12 August 1949. Convention (IV) relative to the Protection of Civilian Persons in Time of War. Geneva, 12 August 1949. Pakistan ratified these Conventions on June 12, 1951.

[77] Protocol Additional to the Geneva Conventions of 12 August 1949, and relating to the Protection of Victims of International Armed Conflicts (Protocol I), 8 June 1977. Protocol Additional to the Geneva Conventions of 12 August 1949, and relating to the Protection of Victims of Non-International Armed Conflicts (Protocol II), 8 June 1977.

[78] PAKISTAN CONST., art. 9: "Security of person. – No person shall be deprived of life or liberty save in accordance with law."

[79] Ibid., art. 14: "Inviolability of dignity of man, etc – (1) The dignity of man and, subject to law, the privacy of home, shall be inviolable."

interference with water supplies under Art. 155[80] and Art. 184, relating to the original jurisdiction of Supreme Court.[81]

Under federal legislation, the relevant provisions relating to the human right to water, including the prevention of water pollution, include numerous provisions of the Environmental Protection Act 1997,[82] Art. 14 relating to the disposal of wastes and effluents and Art. 20 relating to drinking water, of the amended Factories Act of 1934.[83] Furthermore, there is criminal penalty under the Pakistan Penal Code for corrupting the water of any public spring or reservoir.[84] Other relevant legislation includes the

---

[80] Ibid., art. 155:

Complaints as to interference with water supplies –

(1) If the interests of a Province, the Federal Capital or the Federally Administered Tribal Areas, or any of the inhabitants thereof, in water from any natural source of supply [or reservoir] have been or are likely to be affected prejudicially by –

   (a) any executive act or legislation taken or passed or proposed to be taken or passed, or
   (b) the failure of any authority to exercise any of its powers with respect to the use and distribution or control of water from that source. . ..

[81] Ibid., art. 184: Original jurisdiction of Supreme Court.

(1) The Supreme Court shall, to the exclusion of every other Court, have original jurisdiction in any dispute between any two or more Governments. . ..

(3) Without prejudice to the provisions of Article 199, the Supreme Court shall, if it considers that a question of public importance with reference to the enforcement of any of the Fundamental Rights conferred by Chapter I of Part II is involved, have the power to make an order of the nature mentioned in the said Article.

[82] See, Pakistan Environmental Protection Act 1997, §§ xix–xxiv.

[83] See art. 14: "Disposal of wastes and effluents. – (1) Effective arrangements shall be made in every factory for the disposal of wastes and effluents due to the manufacturing process carried on therein." See also, art. 20:

Drinking Water. –

(1) In every factory effective arrangements shall be made to provide and maintain at suitable points conveniently situated for all workers employed therein a sufficient supply of whole-some drinking water.

(2) All such points shall be legibly marked "Drinking Water" in a language understood by the majority of the workers and no such point shall be situated within twenty feet of any washing place, urinal or latrine, unless a shorter distance is approved in writing by the Chief Inspector.

[84] Pakistan Penal Code Chapter XIV: Of Offences Affecting The Public Health, Safety, Convenience, Decency And Morals 277:

Fouling water of public spring or reservoir: Whoever voluntarily corrupts or fouls the water of any public spring or reservoir, so as to render it less fit for the

Pakistan Council of Research in Water Resources Act 2007, which set up the Pakistan Council of Research in Water Resources. The functions of this body are primarily research oriented with an aim to improve the technology for the advancement as well as the conservation of existing water resources. This body is also required to provide recommendations to the government, regarding the quality of water that needs to be maintained and how existing water sources may be utilized and conserved.[85]

Furthermore, various water- and sanitation-based policies and guidelines have been approved by the national government. Under the National Drinking Water Policy approved by the Federal Cabinet on 28 September 2009, the government recognized that access to clean drinking water is the basic human right of every citizen;[86] the government, through the formulation of this policy is committed to providing access to clean and safe, affordable drinking water in adequate quantities to the entire population. The policy also identified the current disparity and inaccessibility of water in Pakistan and acknowledged how this situation leads to various water- and sanitation-related diseases in the country. The policy defined drinking water as "water used for domestic purposes including drinking, cooking, hygiene and other domestic uses."[87] Safe drinking water is defined as water that complies with national drinking water quality standards. With respect to access and adequacy, the policy mandates that the water be accessible to both urban and rural areas at a distance of no more than 30 minutes and adequacy be between 45 and 120 liters per capita per day.[88] These parameters seem to have been established in conformance with similar standards outlined under General Comment No. 15.[89] Under S. 6.12, the policy sets

---

purpose for which it is ordinarily used, shall be punished with imprisonment of either description for a term which may extend to three months, or with fine which may extend to [one thousand five hundred rupees], or with both.

[85] See Section 4 – Functions of the Pakistan Council of Research in Water Resources Act 1 of 2007 – Pakistan Council of Research in Water Resources Act, 2007:

...b) Design, develop and evaluate water conservation technologies for irrigation, drinking and industrial water... [a]dvise the government and submit the policies recommendations regarding water quality, development, management, conservation and utilisation of water resources... publish scientific papers, reports and periodicals as well as to arrange seminars, workshops and conferences on water related issues;...

[86] See National Drinking Water Policy 2009, Foreword.
[87] Ibid., Foreword § 2.
[88] Ibid.
[89] See General Comment No 15, *supra* note 32 above, 5.

out that various forms of legislation are to be enacted to ensure implementation of these measures, including the Pakistan Safe Drinking Water Act.[90]

Furthermore, under the National Sanitation Policy of September 2006, guidelines are provided to the federal and provincial governments, federally administered territories and local governments to develop their policies regarding sanitation for improving the quality of life for citizens. It recognizes the alarming lack of sanitation facilities available to the citizens of Pakistan, save a few urban cities and where sewerage arrangements are almost non-existent. This of course has led to various health problems. The policy highlights that the needs of women and children, vulnerable groups that had previously been ignored, be kept in mind when implementing the policy guidelines.[91] The policy mandates the development of by laws by provincial governments, which need to be implemented by the Tehsil Municipal Administration (TMA). Furthermore, all levels of the government are required to create awareness, promote research and enable capacity building to address sanitation issues.[92]

Finally, the National Environment Policy 2005 seeks to provide a framework for addressing the various environmental issues facing Pakistan, particularly the pollution of fresh-water bodies, air pollution, lack of waste management, etc. Amongst its objectives, it recognizes the need to meet international obligations effectively, in line with national objectives.[93] In listing its sectoral and cross-sectoral guidelines, the policy recognizes both the need for water supply and management and the concerns regarding health and environment. In addressing water supply and management, the policy lists a number of guidelines for the government to ensure sustainable access to water supply that is safe to use.[94]

Provincial legislation such as the Baluchistan Ground Water Rights Administration Ordinance, 1978 Ordinance IX of 1978, established regulatory and supervisory functions for the Provincial Water Board[95] and a Water Committee to overlook the implementation of the policies of the Water Board. It also set up and laid out the functions for the Baluchistan Water and Sanitation Agency (B-WASA), requiring it to plan,

---

[90] See National Drinking Water Policy, *supra* note 86 above, Foreword, § 6.12: "... (i) Pakistan Safe Drinking Water Act will be enacted to ensure compliance with the National Drinking Water Quality Standards. ..."

[91] See ibid., Foreword § 4(iii); 6(m).

[92] See generally ibid., Foreword § 6 – Policy Measures.

[93] See National Environment Policy, § 2.2(d).

[94] Ibid., § 2.2(d), Section 3.1. Water Supply and Management.

[95] See Baluchistan Ground Water Rights Administration Ordinance, 1978 ordinance Ix of 1978. § 3 Establishment and functions of Provincial Water Board.

construct and maintain water supplies in addition to providing sanitation to Municipal Corporation and the Quetta Development Authority.[96]

However, these duties have devolved over the years to municipal authorities. For instance, under the Punjab Local Government Ordinance (2001),[97] a number of provincial functions including water management and sanitation have been entrusted to the Tehsil Municipal Administration.[98] The functions of the Tehsil Municipal Administration and Union Administration concerning water management are diverse and include the development of water resources, regulating sanitation services and disposal,[99] water supply and its maintenance,[100] as well as the preservation of public resources of drinking water, such as wells, ponds, etc. Serving the Tehsil Municipal Administration is the Union Nazim, who has been entrusted with corresponding duties, including the prevention of health hazards and breach of watercourses that fall within his area of jurisdiction.[101] Lastly, the Village Council is required to adhere to the requirements of the ordinance and prevent the contamination of water, and develop and improve water supply sources.[102]

---

[96] See, Functions of WASA in Baluchistan as established under the Baluchistan Ground Water Rights Administration Ordinance, 1978 — ordinance IX of 1978: –

... Plan, design, construct, operate and maintain water supply, sewerage and sanitation system within the service area of the Water and Sanitation Authority to be established under Section 3 of this Ordinance... Monitor and control water resources in the Area, both surface and underground and issue licenses for abstraction of water from such resources in the Area in accordance with regulations made by the authority....

[97] Punjab Local Government Ordinance 2001.
[98] See ibid., § 52.
[99] Ibid., § 54: – Functions and Powers of the Tehsil Municipal Administration – Punjab Local Government Ordinance (2001) –

...(h) provide, manage, operate, maintain and improve the municipal infrastructure and services, including –

(i) water supply and control and development of water sources, other than systems maintained by the Union and Village Councils

...

(iv) sanitation and solid waste collection and sanitary disposal of solid, liquid, industrial and hospital wastes....

[100] Ibid., s. 54-A – Functions and Powers of the Tehsil Municipal Administration.
[101] Ibid., s. 80 – Functions of Union Nazim.
[102] Ibid., s. 96: – Functions of Village Council and Neighbourhood Council – "...(a) develop and improve water supply sources; ... (d) take measures to prevent contamination of water...."

The City District Government[103] and the Tehsil Municipal Administration[104] are also responsible for the enforcement of punishment for offences, as determined by the court, relating to the contamination or pollution of water, failure on the part of industries to dispose of hazardous waste, or offenses relating to the provision of contaminated water for human consumption.[105] Other forms of offenses such as failure to stop leakage of drain pipes, the obstruction of water pipes, etc. have been made punishable by the issuance of tickets rather than through the court and are the responsibility of the Tehsil Town Officer.[106]

It is therefore clear that with the devolution of power from the provinces to the municipal and district governments,[107] municipal services[108] including water supply, access and sanitation have now become the responsibility

---

[103] See ibid., Fourth Schedule – Part B:

...8.Discharging any dangerous chemical, inflammable, hazardous or offensive article in any drain, or sewer, public water course. . .
10. Supplying or marketing drinking water for human consumption in any form, from any source which is contaminated or suspected to be dangerous to public health. . ..

[104] See ibid., Fourth Schedule – Part D.
[105] Ibid.
[106] Ibid., Eighth Schedule 9:

Obstructing or tampering with any main pipe, meter or any apparatus or appliance for the supply of water or sewerage system. Fine: Rs. 1000/. . ..
28.Failure by the owner or occupier of any land or building to clean, repair, cover, fill up or drain off any private well, tank or other source of water supply, which is declared under this Ordinance to be injurious to health or offensive to the neighbourhood. Fine: Rs. 1000/.

[107] See ibid., art. 52:

Entrustment of certain decentralised offices to Tehsil Municipal Administration.– Provided further that Water and Sanitation Agencies coming under the control of District Government under sub-section (3) of section 182 functioning in a tehsil shall further be decentralized to the concerned Tehsil Municipal Administration: Provided also that Water and Sanitation Agency or similar agencies functioning in a City District and coming under the control of City District under sub-section (3) of section 182 may further be decentralised to the City District Administration or, according to requirements of service delivery, may be decentralised to towns in a city district.

[108] The SBNP Local Government Ordinance 2001 s. 2(xxii): "'municipal services' include, but not limited to intra-city or intra or inter-town or tehsil network of water supply, sanitation, conservancy, removal and disposal of sullage, refuse, garbage, sewer or storm water, solid or liquid waste, drainage, public toilets. . .."

of the local government, especially the Tehsil Officer (Infrastructure and Services) and the Union Administration.[109]

2.3 Seminal Judgments on Water Emancipation in Pakistan

The judicial treatment accorded to the right to water in Pakistan, emulates the position of the Indian Judiciary on the matter. The incorporation and significance of the most fundamental non economic right, the right to life, in the written Constitution of both Pakistan and India is a result of following US jurisprudence.[110] In three seminal Superior Court judgments, denial of the water right has been viewed as a violation of the constitutional right to life.[111] Other constitutional provisions expressly invoked include, the inviolability of dignity of man[112] and Art. 184(3), which deals with the original jurisdiction of the Supreme Court.[113]

In *Shehla Zia and others v WAPDA*,[114] a petition was filed over the possible health concerns as a result of the building of a grid station in a residential area. It was argued that the electromagnetic field generated by the presence of high voltage transmission lines, posed a serious health hazard to residents. The respondent raised the objection that the facts of

---

[109] See Punjab Local Government Ordinance, art. 53. "Structure of the Tehsil Municipal Administration. – . . . (3)(ii) Tehsil Officer (Infrastructure and Services) who shall be responsible for water, sewerage, drainage, sanitation. . ..."; art. 76: "Functions of the Union Administration – (j) to provide and maintain public sources of drinking water, including wells, water pumps, tanks, ponds and other works for the supply of water. . ..."; art. 94: "Water supply – (1) The concerned local government shall provide or cause to be provided to its local area a supply of wholesome water sufficient for public and private purposes"; art. 95: "Private source of water supply. – (1) All private sources of water supply within the local area of a concerned local government shall be subject to control, regulation and inspection by the local government."

[110] See *Shehla Zia and Others v WAPDA* (PLD 1994 SC 693) para. 14 Saleem Akhtar J, acknowledging the impact of US jurisprudence concerning the right to life on south Asian case law and citing the seminal US case, *Munn v Illinois*, (94 US 113 (1877)), ("By the term 'life,'. . .something more is meant than mere animal existence. The inhibition against its deprivation extends to all those limbs and faculties by which life is enjoyed. . . . [t]he deprivation not only of life, but of whatever God has given to everyone with life for its growth and enjoyment, is prohibited . . ..)."; see also *Griswold v Connecticut* 381 US 479 (1965) ("specific guarantees in the Bill of Rights have penumbras, formed by emanations from those guarantees that help give them life and substance.")

[111] PAKISTAN CONST., art. 9: see *supra* note 78 above.
[112] Ibid., art. 14: see *supra* note 79 above.
[113] Ibid., art. 184: see *supra* note 81 above.
[114] See *Shehla Zia*, *supra* note 110 above.

the case do not justify intervention under Art. 184 of the Constitution. The respondent argued that the grid station and the transmission line were being constructed after a proper study had been conducted, taking into consideration the related risks, economic considerations and the requirements of a particular area. The Court in interpreting Art. 9 of the Constitution stated that the right to life included all such amenities and facilities, which a person born in a free country is entitled to enjoy with dignity, both legally and constitutionally; and a person is entitled to the protection of law from being exposed to the hazards of electromagnetic fields, or any other such hazards which may be the result of the installation and construction of any grid station, factory, power station or such like installations. The court then directed the respondents to re-evaluate their scheme.[115]

The utilization of the right to life principle for promoting human right in Pakistan through judicial activism has been established; however, the *Shehla Zia* case has been the principle vehicle through which the courts have exercised their powers in relation to this matter. An important example of this approach can be seen in the *Benazir Bhutto* judgment,[116] which followed the *Shehla* case and examined the right to life by stating that:

> it is a sacred right, which cannot be violated, discriminated or abused by any authority. The word "life" is very significant as it covers all facets of human existence...Life includes all such amenities and facilities which a person born in a free country is entitled to enjoy with dignity, legally and constitutionally.[117]

Furthermore, the court derived support for its position from US jurisprudence by stating that "constitutional law in America provides an extensive and wide meaning to the word 'life' which includes all such rights which are necessary and essential for leading a free, proper, comfortable and clean life." The court finally held that, "any action taken which may create hazards of life will be encroaching upon the personal rights of a citizen to enjoy the life according to law."[118]

In *Mrs. Anjum Irfan v Lahore Development Authority through Director*

---

[115] See *Shehla Zia*, ibid., para. 15: "Any action taken which may create hazards of life will be encroaching upon the personal rights of a citizen to enjoy the life according to law...In our view the word life constitutionally is so wide that the danger and encroachment complaint of would impinge fundamental right of a citizen."
[116] See *Benazir Bhutto v President of Pakistan* (PLD 1998 Supreme Court 388).
[117] Ibid.
[118] Ibid.

General and others,[119] the petitioner had submitted that an alarming amount of untreated water was being drained out into the River Ravi. As a consequence, pollution in the waters had reached hazardous levels and was resulting in the spread of diseases including jaundice and typhoid. The court, considering s. 11 of the Pakistan Penal Code ("PPC"),[120] determined that the provisions relating to the prohibition of pollution of water, do not require any particular *mens rea*; and that corruption or fouling of water of any public spring or reservoir, so as to render it unfit for the purpose for which it is ordinarily used, is punishable under s. 277, PPC. Considering the problem of pollution, the court also referred to the *Shehla Zia* case,[121] and following this judgment took an expansive view as to the meaning to the fundamental right to life, read in conjunction with Art. 14 of the Constitution.[122] The court directed the respondents to implement the relevant provisions of the Pakistan Environment Protection Act 1997 both in letter and spirit, and frame all the necessary rules and regulations.

In *General Secretary, West Pakistan Salt Miners Labour Union (CBA) Khewra, Jhelum v. The Director, Industries and Mineral Development, Punjab, Lahore*,[123] the petitioners filed a claim for access to clean and unpolluted drinking water. As a result of leasing and excessive mining in the area, the water catchment area was reduced and the mining operations posed a serious danger to the already scarce source of water in the mountainous area.

The court invoked both Arts 9 and 14 of the Constitution, to hold that in hilly areas, where access to water is scarce, difficult or limited, the right to have water free from pollution and contamination is a right to life. This, however, does not mean that persons residing in other parts of the country where water is available in abundance do not possess such a right. The court added that the right to have unpolluted water is the right of every person, wherever he lives. The court in its judgment directed the miners to shift within four months to a different location, so as to avoid pollution of the waters.

There are a number of other cases that indirectly deal with the right

---

[119] *Mrs Anjum Irfan v Lahore Development Authority through Director-General and Others* (PLD 2002 Lahore 555).
[120] Pakistan Penal Code (Act XLV of 1860) §. 11: "The word 'person' includes any Company or Association, or body of persons, whether incorporated or not."
[121] See *Shehla Zia, supra* note 110 above.
[122] PAKISTAN CONST. art. 14: see *supra* note 79 above.
[123] *General Secretary, West Pakistan Salt Miners Labour Union (CBA) Khewra, Jhelum v The Director, Industries and Mineral Development, Punjab, Lahore* (1994 SCMR 2061).

of water, including those that relate to the contamination of coastal areas with nuclear waste[124] and where waste from leather factories is contaminating water and is amounting to nuisance.[125] However, an in-depth discussion of such case law would be outside the ambit of this chapter.

It is apparent from the view of relevant judgments in Pakistan concerning water rights, that the approach utilized by the country's judiciary is similar to its Indian counterpart.[126] Even leading academic jurists in Pakistan have openly advocated positive measures for the progressive implementation of the right to life.[127] The efficacy of this approach in dealing with water emancipation is, however, highly questionable since the implementation of such a progressive interpretation, has not led to any tangible measures in providing access to healthy water or in addressing public health concerns with regard to the general population.

2.4 An Assessment of the Status of Water Rights in Pakistan

There are two recent expositions of the current status and functioning of water rights in Pakistan. The first is the commencement of the CDWA project. In light of the UN Millennium Development Goals ("MDG"),[128] Pakistani authorities as well as international stakeholders such as the World Health Organization ("WHO") and the United Nations Children's Fund ("UNICEF") have made development efforts in Pakistan. The federal government in pursuance of the MDG in 2004 initiated a project known as the Clean Drinking Water for All ("CDWA"), which aimed at installing about 7000 water purification plants all over Pakistan. About Rs.22 billion was allocated for the entire project, out of which Rs.13 billion

---

[124] *Ch, Riaz Ahmad Yazdani v The Federation of Pakistan and 8 others* (1990 CLC 1406).
[125] *Abdul Latif v Additional Sessions Judge*, Sahiwal (2001 CLC 1139).
[126] *See* Julien Chaisse et al., *Deconstructing Service and Investment Negotiating Stance: A Case Study of India at WTO GATS and Investment Fora*, 14 J. WORLD INV. & TRADE 44 (2013).
[127] See generally Justice (R) Fazal Karim, [2006] *Judicial Review of Public Actions,* Pakistan Law House: Karachi.
[128] See generally, UN Summit on the Millennium Development Goals 20–22 (2010) <www.un.org/ millenniumgoals/> (concluded with the adoption of a global action plan to achieve the eight anti-poverty goals by their 2015 target date and the announcement of major new commitments for women's and children's health and other initiatives against poverty, hunger and disease.) UN Millennium Development Goals < http://www.un.org/millenniumgoals/ >

was allocated to the Province of Punjab. It is important to note that of the 7000 plants that were to be installed, so far only a total of 2000 plants have been installed throughout Pakistan. And of a total of 3494 plants planned for Punjab, only 300 have been installed.

However, despite the number of stakeholders and expectations about this project, it now stands abandoned. There are a number of problems that have contributed to this failure. First, there were problems regarding the political ownership of the project. For instance, in Punjab, the project contractors were prequalified and financed by the Pakistan Peoples Party ("PPP") led establishment/federal government, whereas the government of Punjab was only entrusted to take care of the execution phase. Although the Punjab government signed the project contract, they always had an issue with the structure sketched by the federal government for its implementation. Secondly, there were multifaceted problems with the financial arrangements between the federal government and the Punjab government for the implementation of the project. The total cost of the project was Rs.22 billion and Punjab was allocated Rs. 13 billion.

However, the federal government only transferred Rs.4 billion and the remaining amount was not forthcoming, the argument being that transfers were to be made on the basis of progress. The Punjab government voiced its concern at not being given the whole amount and was dissatisfied with the arrangement.

Bad governance is another reason for the government's inability to deliver. The local government and community development department of the Punjab government formed a Provincial Project Management Unit ("PPMU"), which was manned by technical experts (environmental engineers, chemical engineers and material engineers). This initiative failed because a bureaucrat was put in charge of the project, a technocrat or an individual having technical expertise in project management would have been a better candidate. Furthermore, funding problems led to contractual failures.[129]

Projects like CDWA strive to achieve a temporary solution to provide limited access to clean drinking water which is also not portable i.e. supply of clean drinkable water in taps. Clean water is only accessible to those people who can directly access the water from the plant area. To make matters worse, water purification plants have a durability of only ten years. Such projects do not envisage a permanent

---

[129] Out of the Rs 4 billion received by the Punjab government, Rs 1.8 billions was given to contractors. However, the remaining Rs 2.8 billion was never paid to the contractors in violation of their contractual rights.

solution to the core problems concerning clean water availability and accessibility for the general population. A permanent solution for portable access is not possible until and unless the faulty underground water distribution network is rebuilt and properly maintained by the relevant authorities.

An example of water and sewerage distribution mismanagement can be viewed in the case of the Sozo Water Park in Lahore, where the allowance of inlets of sewage of the Park, which include three housing societies, eight squatter settlements and slums, into the Lahore Canal have given rise to serious public health concerns.[130] The polluted water is certain to give birth to waterborne diseases and infection. Therefore, the Secretary of the Environmental Protection Department ("EPD") issued notices to the concerned authorities[131] for a meeting to discuss steps necessary to shift these 12 inlets away from the canal.[132] Following the meeting held on May 5, 2011, solutions outlined included the blocking of the sewerage inlets and effectively policing for enforcement, laying down sewage pipelines in the area that empty into the Ravi river and installing dustbins along the canal in order to facilitate people disposing of solid waste. The Director of the Laboratories at the EPD expressed the view that the district government should plan and devise a long-term solution instead of building a sewer trunk that would end up discharging wastes into River Ravi, which is already affecting the quality of irrigational waters and ultimately the crops in southern Punjab.

2.5 Food Security and the Human Right to Water

"Food security [is] a situation that exists when all people, at all times, have physical, social and economic access to sufficient, safe and nutritious food that meets their dietary needs and food preferences for an active and

---

[130] Data obtained through the Environment Protection Department ("EPD").

[131] These notices were sent to the Managing Director of WASA, the Lahore Waste Management Company, the District Coordination Officer of Lahore, District Officer of the EPD, the Director General of the Lahore Development Authority, and the Secretary of the Irrigation Department. The TMO's of Allama Iqbal Town, Wagha Town and Aziz Bhatti Town were also informed

[132] The EPD is a regulatory authority and according to s. 16 of Pakistan Environment Protection Act 1997, the EPD issues an Environment Protection Order ("EPO") to the concerned authority allowing 30 days for preventive measures. In case of no compliance, the EPD forms a tribunal, which examines reports and samples before taking a decision with which the concerned authority has to comply.

healthy life".[133] Food security is heavily reliant on water resources, and agriculture is the largest consumer of freshwater.[134]

Statistics show that irrigation for agriculture consumes almost 98 percent of the fresh water resources in Pakistan.[135] With increasing demand for water amid climate change and increasing population, food security has become a progressively serious global threat. Food demand in Pakistan is estimated to increase by 40 percent based on the projected increase in the country's population by 2025.[136] The risk associated with food insecurity is apparent with under-nourishment exponentially on the rise in Pakistan as indicated by a recent Food and Agriculture Organization ("FAO") report.[137]

In Pakistan, the right to food is constitutionally protected. Under Art. 38 relating to the promotion of social and economic well-being of the people, the state is responsible for the provision of basic necessities of life, such as food.[138] Furthermore, Pakistan is required to improve and develop its agricultural industry including modes of food production under the ICESCR.[139]

---

[133] Food and Agriculture Organization, *The State of Food Insecurity in the World 2001* (Rome 2002) 10.

[134] Ibid., 12.

[135] Paper 667, 'Meeting Future Food Demands Of Pakistan Under Scarce Water Situations' (70th Annual Session Proceedings, Pakistan Engineering Congress 2007) 238.

[136] Alam, M, and Bhutta, M N, 'Availability of water in Pakistan during 21st century' *Proceedings of the International Conference on Evapotranspiration and Irrigation Scheduling* (San Antonio, Texas, USA 1996).

[137] Food and Agriculture Organization, *supra* note 133, 9.

[138] PAKISTAN CONST., art. 38:

Promotion of social and economic well-being of the people. – The State shall – ...(d) provide basic necessities of life, such as food, ...for all such citizens, irrespective of sex, caste, creed or race, as are permanently or temporarily unable to earn their livelihood on account of infirmity, sickness or unemployment; ....

[139] General Comment 15, *supra* note 32, art. 11(2):

The States Parties ... shall take, individually and through international co-operation, the measures, including specific programmes, which are needed:
(a) To improve methods of production, conservation and distribution of food by making full use of technical and scientific knowledge, by disseminating knowledge of the principles of nutrition and by developing or reforming agrarian systems in such a way as to achieve the most efficient development and utilization of natural resources;
(b) Taking into account the problems of both food-importing and food-exporting countries, to ensure an equitable distribution of world food supplies in relation to need.

Previously, Pakistan lacked a legislative framework which focally addressed food security concerns. There, however, was legislation, which though poorly enforced, could have been utilized for achieving minimum levels of food safety. Examples of such legislation include the Pure Food Ordinance 1960[140] and the Cantonment Pure Food Act 1966,[141] which regulate standards for the provision of pure and unadulterated food materials for consumption, but do not address food security.

In Punjab, however, the provision of food security may now be possible in light of the recently approved Punjab Food Safety and Standards Act, 2011.[142] This Act not only defines standards on food quality but also seeks to establish the Punjab Food Authority, which is being created with the responsibility to ensure the availability of food safe for human consumption. Most importantly, water is included within the definition of food under this legislation.[143]

In relation to irrigation, the relevant regulatory authority is the Indus River System Authority ("IRSA") which was set up in 1992 in order to implement the Water Appointment Accord as agreed between the provinces. Although relatively effective in distributing water to the provinces in the past, the authority has recently indicated that it expects acute water shortages for irrigation. In this regard, the Chairman of IRSA has highlighted the urgent need to institute various measures for the purpose of achieving adequate water storage and management in order to mitigate against this developing crisis.[144] The chief executive of the South Asian Conservation Agriculture Network has also raised concerns about the depleting water resources and the water and economic losses faced because of the current irrigation network. He has suggested ways for reforming the water management system for better collection and

---

[140] See Pure Food Ordinance 1960 – Section 2.
[141] See The Cantonment Pure Food Act 1966 – Explanation I.
[142] "Cabinet meeting approves Food Authority Act", *The Express Tribune* <http://tribune.com.pk/story/132658/ cabinet-meeting-approves-food-authority-act/> accessed on 13/05/2011.
[143] See Punjab Food Safety And Standards Act 2011, s. 2(f):

Food means any article used as food or drink for human consumption other than drugs, and includes–...Water in any form, including ice, intended for human consumption or for use in the composition or preparation of food: Provided that the Government may declare, by notification in the Official Gazette, any other article as food for the purposes of this Act.

[144] Pakistan to face 25 MAF water shortage: IRSA – Pak Observer <http://pakobserver.net/201103/10/ detailnews.asp?id=80250> accessed on 13/05/2011.

storage of water resources.¹⁴⁵ It is therefore imperative that access to water for irrigation be improved. 90 percent of the agricultural production of Pakistan comes from irrigated lands and the country is increasingly becoming water deficient.

Another problem that threatens food security is the lack of planning and coordination between the different regulatory bodies responsible for the development of food and water policy. As food production is wholly dependent on the provision of water, deprivation of water rights to landowners combined with governmental policy that subsidizes or promotes water intensive crops is bound to lead to crop failures and will subsequently result in food shortages. Furthermore, the inequitable distribution of water between landowners on unwarranted grounds negatively impacts agrarian production. For activities such as diverting water channels to influential landholders has resulted in water shortages and hindrances in cultivating crops.¹⁴⁶

3. CONCLUSION

Legally speaking, Pakistan has numerous international and domestic law obligations to its people for the provision of adequate supply of clean uncontaminated water for a diverse number of purposes – some of these uses constitute fundamental human rights. In supporting these obligations, the superior judiciary of Pakistan has examined the right to water and has held its deprivation to be a violation of the fundamental right to life, guaranteed under the Constitution of Pakistan. Yet, water emancipation is on the decline in Pakistan; and in practice, the overwhelming majority of the population is deprived of this essential human right or resource.

The reasons why the state continues to fail miserably to meet its water-based obligations are multifaceted. The main reasons include corrupt and incompetent governmental functionaries, lack of accountability and transparency of water-based regulatory authorities, systematic organizational deficiencies within the regulatory frame-work and no substantive coordination between the relevant departments. Then there are insufficient resources, both monetary and non-monetary, including

---

¹⁴⁵ Water Management Practices in Pakistan Issues and Options for Productivity Enhancement-Round table Discussion on Agriculture and Water in Pakistan (2011) <http://siteresources.worldbank.org/PAKISTANEXTN/Resources/ WMPracticesinPakistan.pdf> accessed on 13/05/2011.

¹⁴⁶ Shortage of water at tail-end may hit cotton crop: SCA- Dawn News, <www.dawn.com/2011/04/16/ shortage-of-water-at-tail-end-may-hit-cotton-crop-sca.html> accessed on 14/05/2011.

human capital and technical expertise.[147] Other reasons include distrust and discord between the federal and provincial governments, unwarranted intrusion from international donor agencies including the International Monetary Fund ("IMF") and the World Bank, lack of public discourse and debate, the presence of centralized water policy formation without engaging with and incorporating input from vested stakeholders, including grass root organizations, civic society groups and national and international NGOs and lastly the desire from the government at all levels to achieve short-term solutions rather than instituting measures that strive to achieve a long term resolution of the problem.

Unfortunately, all these deficiencies point to the fact that it will be extremely hard, if not impossible, for Pakistan to meet the targets set out by the Millennium Goals relative to the human right to water.

---

[147] For instance under the CDWA project, Pakistan sought to import rather than develop the technology itself. This decision is strong evidence of Pakistan's lack of expertise in developing and maintaining a sustainable water supply.

10. The human right to clean water and sanitation – a perspective from Nigeria

Cosmas Emeziem*

INTRODUCTION

Water is one of humankind's earliest sources of survival, sustenance and inspiration. It has inspired poets since the beginning of time.[1] It has fed nations and also caused sorrows where it is insufficient. As soon as a human being is born, she must be washed and made clean. This washing may spell doom or death if it is done with unwholesome water. So from the cradle to the grave the need for wholesome water for life is an undying need. Equally, as human societies and populations enlarge, the demand for wholesome water enlarges exponentially – it heightens the need for safe water. Over time, poverty and other factors like global warming, drought and pollution have further put pressure on existing sources of safe and sufficient water for living across the globe. This has led to a near emergency situation of non-availability of safe and sustainable drinking water for many. Nigeria as part of the human community is affected by this common problem.[2] To understand this, this Introduction looks at the make-up of Nigeria and its water peculiarities.

The Federal Republic of Nigeria is situated in tropical West Africa. It is a country with diverse ethnic nationalities and about 175 million inhabitants. Its landmass is put at 933,770 square kilometers. It operates a federal system of government with constitutional powers devolving between the central government in Abuja and 36 states of the Federation.

---

* This chapter was developed as a paper in the International Environmental Law Seminar Spring 2015, guided by Professor Keith S. Porter. Any error attributable to it is mine.

[1] 'Like a deer that yearns for running streams...' (the Bible, Psalm 42).

[2] The *Guardian Newspaper* of Nigeria (21/03/2015) reported water shortage and fear of epidemics for the city of Calabar in Southern Nigeria.

The government structure also includes 774 local government units across the country.³ Nigeria has warm weather and good but irregular precipitation which tapers off to near-desert conditions in the northernmost parts.

Nigeria has broad ecological zones ranging from the mangrove swamp forest in the Niger Delta to the rain forest belt, the lush green grasslands and then the Sahel belt in the North. Besides its vast oil mineral wealth,⁴ it is also well drained by a network of rivers and streams. Its geographic base is not prone to tragic earth movements because it is not situated along the known tectonic fault-lines.⁵ Thus, at a glance one sees the providential privileges of Nigeria.

According to the Food and Agricultural Organization:

> The hydrology of Nigeria is dominated by two great river systems, the Niger-Benue and the Chad systems. With the exception of a few rivers that empty directly into the Atlantic Ocean (Cross River ... Osun ... Imo, and a few others), all other flowing waters ultimately find their way into the Chad Basin or down the lower Niger to the sea...
>
> The two river systems (...) are separated by a primary watershed extending north-east and north-west from the Bauchi Plateau which is the main source of their principal tributaries...
>
> The rivers flowing into Lake Chad emanate both from the central highland and from the high plateau and converge to form the Yobe River just before flowing into Lake Chad. Some rivers flowing into the lake originate from the Cameroon Mountains. Only a small part of Lake Chad lies within Nigeria. Along the Nigerian border, the lake is little more than a vast swamp...
>
> Within Nigeria the River Niger is fed by rivers flowing into it from all directions with headwaters originating from the central plateau in the north, from the Yoruba highlands in the south, from Benin Republic to the west and from the eastern highlands. A significant flow from outside Nigeria comes from the watersheds stretching westwards right up to the Fouta Djallon Mountains of Guinea...
>
> Of the other rivers flowing to the Atlantic, the Cross River is fed by many tributaries originating in the Cameroon Mountains. It flows east and then turns southwards and empties into the Atlantic Ocean with limited delta formation. Both the Ogun and Osun rivers are fed by rivers originating from the Yoruba highlands. They flow slowly from north to south into the Lagos lagoons before discharging through creeks and swamps into the Atlantic Ocean.⁶

---

³ See the Constitution of the Federal Republic of Nigeria 1999, ss 1, 2, and 3.

⁴ Nigeria has an estimated 159 trillion cubic feet (Tcf) of proven natural gas reserves. <www.nnpcgroup.com/NNPCBusiness/BusinessInformation/OilGasinNigeria/DevelopmentoftheIndustry.aspx> accessed 05/02/2016. She also has 37billion barrels of proven crude oil reserves <www.eia.gov/countries/country-data.cfm?fips=ni > accessed 05/02/2016.

⁵ Nigeria is not prone to earthquakes.

⁶ See < www.fao.org/docrep/005/t1230e/t1230e02.htm > accessed 05/02/2016.

In its hydrology Nigeria has natural aquifers with different yielding capacities. The notable aquifers are:

1. The Sokoto Basin Zone comprises (of) sedimentary rocks in northwest Nigeria. Its yields range from below 1.0 to 5.0 litre per second (L/s).
2. The Chad Basin Zone comprises sedimentary rocks. There are three distinct aquifer zones thereat; Upper, Middle, and Lower. Borehole water yields are about 1.2 to 1.6 L/s from the upper unconfined aquifer and 1.5 to 2.1 L/s from the Middle aquifer.
3. The Middle Niger Basin Zone comprises sandstone aquifers yielding between 0.7 and 5.0 L/s and the Alluvium in the Niger Valley yielding between 7.5 and 37.0 L/s.
4. The Benue Basin Zone is the least exploited basin in Nigeria extending from the Cameroon border to the Niger-Benue confluence. The sandstone aquifers in the area yield between 1.0 and 8.0 L/s.
5. The South-western Zone comprises sedimentary rocks bounded in the south by the coastal Alluvium and in the north by the Basement Complex.
6. The South-Central Zone is made up of Cretaceous and Tertiary sediments centered on the Niger Delta. Its water yields are from 3.0 to 7.0 L/s.
7. The South-eastern Zone comprises Cretaceous sediments in the Anambra and Cross River basins. Water borehole numbers are low due to abundant surface water resources.
8. The Basement Complex comprises over 60 percent of the country's area. It consists of low permeability rocks and groundwater occurs in the weathered mantle and fracture zones with yields of between 1.0 and 2.0 L/s.[7]

Many Nigerians depend on these open water sources for their daily water needs. The huge population of Nigeria – 175 million and at an annual growth rate of about 2.5 percent – and its water resources provides good grounds for a careful review of the right to clean water and sanitation. Significantly, the success or otherwise in recognizing, respecting and fulfilling the human right to clean water and sanitation in Nigeria can have a positive effect in the region. This chapter explores the extent of the realization of the human right to water and sanitation in Nigeria. It is argued and recommended, inter alia, that a social justice approach involving critical

---

[7] Jim Kundell (ed.), *Water Profile of Nigeria* <www.eoearth.org/view/article/156977/ > accessed 13/02/2016.

social engagement, citizenship participation, legislative intervention and corporate social responsibility are possible measures that could solve the problem.

This chapter aims at analyzing the practical ramifications of the human right to clean water and sanitation in the context of a large developing country, Nigeria. The chapter begins by an overview of the human right to clean water and sanitation. Then, the chapter gives the options for a full realization of the human right to clean water and sanitation in Nigeria. Finally, the notion and role of social justice are explored to explain the role they play in the specific context of a developing country.

I. HUMAN RIGHT TO CLEAN WATER AND SANITATION: ORIGIN AND DEFINITION OF BASIC CONCEPTS

The human right to clean water and sanitation has gained currency over time. It is however, an ancient question of humankind. This can be seen in biblical stories and cultural anecdotes worldwide. In modern times and amid an expanding human population, it has become more critical.[8] Thus from Sao Paulo[9] to Sudan, Tibet to Turkana, water for living has become an irrepressible discourse. In the African world view, it is perceived as a wrong to deny anybody drinking water. Hence natural water sources like springs where the people get water for sanitation and drinking were communally shared and it was immaterial that it was located in any individual's or particular community's property. However, modernization and pollution compromised this value; but there is a reawakened consciousness globally that it is immoral for humanity to allow a huge number of persons to go without water for drinking and sanitation.

---

[8] See generally J Chaisse and M Polo, 'Globalization of water privatization: Ramifications of investor-state disputes in the "blue gold" economy', (2015) 38 *Boston College International and Comparative Law Review* 1; Erik J Woodhouse, Note, 'The "Guerra del Agua" and the Cochabamba Concession: Social Risk and Foreign Direct Investment in Public Infrastructure' (2003) 39 *Stanford Journal of International Law* 295; Andrew Nickson and Claudia Vargas, 'The limitations of water regulation: The failure of the Cochabamba Concession in Bolivia' (2002) 21 *Bulletin of Latin American Research* 128.

[9] *The Guardian Newspaper* of London had on May 21, 2014 reported critical water shortage in Sao Paulo Brazil. < www.theguardian.com/sustainable-business/sao-paulo-water-shortage-world- cup > See also the Flint Michigan water crises reported February 13, 2016 <www.theguardian.com/us-news/2016/feb/13/flint-water-crisis-governors-aides-knew-of-issues-within-weeks-records-suggest >

Deducible from this is a perception shared by many scholars and international bodies that access to safe, or wholesome water for daily living and sanitation has attained the level of a human right. The rationale being that water is not just another commodity on the market shelf. It is rather an essential need for life. It is indispensable and other aspects of living are dependent on it. Take away water and life will shrink away and die. Thus without clean water for drinking and sanitation, other rights like life, dignity and liberty become futile dreams.

A recent World Health Organization (WHO) publication on 'health through clean drinking water and basic sanitation' reaffirms the above assertions. In the said publication; it is stated that:

> About *2.6 billion people* . . . lack even a simple 'improved' latrine and *1.1 billion* people has no access to any type of improved drinking source of water. As a direct consequence, *1.6 million* people die every year from diarrheal diseases . . . attributable to lack of access to safe drinking water and basic sanitation and 90% of these are children under 5, mostly in developing countries. Also *160 million* people are infected with schistosomiasis causing tens of thousands of deaths yearly; 500 million people are at risk of trachoma from which *146 million* are threatened by blindness and 6 million are visually impaired.[10]

As regards Nigeria:

> Data from the Water Supply and Sanitation Baseline Survey (WSSBS), gathered in 2007, reported a national access figure of 54.3 percent for water supply and 65.6 percent for improved sanitation. That survey also found that 18.8 percent of the population resorts to open defecation. . .[11]

According to the JMP, access to improved water supply in Nigeria nationally was 47 percent in 1990. By 2008 the percentage of the population with access had increased to 58 percent (86 million), spread across 75 percent of the urban population and 42 percent of the rural population.

The 2008 access data implies that as many as 63 million Nigerians have no access to improved water supply. In respect of sanitation, the JMP reports that 37 percent of the total population had access to improved sanitation facilities in 1990. In addition, . . .26 percent used shared facilities. . . By 2008, 32 percent of the total population had access to improved sanitation, indicating a fall in the percentage of people served, even though the absolute number of people with access increased by 11.5 million.[12]

---

[10] See <www.who.int/water_sanitation_health/mdg1/en/ >.
[11] UNICEF, *Progress on Drinking Water and Sanitation* (2014) < www.theguardian.com/sustainable-business/sao-paulo-water-shortage-world- cup.
[12] A Joint study by the World Bank, and UNICEF, 'Water Supply and Sanitation in Nigeria: Turning Finance into Services for 2015 and beyond' also

This perception – that water is a human right – reached a crescendo in July 2010 when the United Nations General Assembly (UNGA) adopted resolution 64/292, recognizing access to wholesome water and sanitation as a human right.[13] The UNGA in that resolution articulated the array of prior declarations and statements evidencing the wide appreciation of access to clean water as a human right. It called on states parties and international organizations to ensure the provision of financial resources, and all other means of upgrading efforts aimed at providing safe, clean, accessible and affordable drinking water and sanitation for all.

In the same vein the United Nations Committee on Economic, Social and Cultural Rights had prior to this time in November 2002 issued its General Comment No. 15 stating:

> The human right to water entitles everyone to sufficient, safe, acceptable, physically accessible and affordable water for personal and domestic uses. An adequate amount of safe water is necessary to prevent death from dehydration, to reduce the risk of water related diseases and to provide for consumption, cooking, personal and domestic hygienic requirements.[14]

Indeed, the UN pegs the minimum standard for safe, accessible and affordable water at between 50 to 100 liters per person per day. The water should be free of pathogens, chemicals or other contaminants which may be injurious to health. It must be physically accessible – usually within 30 minutes reach of the homestead.[15]

Illustrative of the convergence of opinion between scholars and global policy makers on the right to water is the view expressed recently that 'water is an essential and indispensable element of life. It is a finite

---

observed the wide disparities in access among Nigeria's 36 states (Lagos state was 81 percent in 2007, for Sokoto state it was 13 percent).

[13] It is argued that the resolution is a restatement of international law with origins earlier than the year 2010.

[14] UN Committee on Economic, Social and Cultural Rights (CESCR), General Comment No. 15: The Right to Water (Arts. 11 and 12 of the Covenant) (2003) E/C12/2002/11, <www.refworld.org/docid/4538838d11.html> accessed 14/02/2016. The UNGA also recalled among others its resolutions 58/217 adopted 23 December 2003 proclaiming the international decade of action 'water for life' 2005–2015; 61/192 adopted 20 December 2006, proclaiming 2008 the international year of sanitation and the Mar del Planta action plan of 1997 adopted by the UN Water Conference and the Rio declaration of 3–14 June 1992.

[15] Indeed, 100 liters of water per person, per day is insufficient yet many do not have access to it.

resource without any alternative and upon which there is total dependence for survival.'[16]

Despite this shared opinion on water and sanitation, there is a huge gap in terms of the reality of access to clean water and sanitation around the world.[17] How can we ensure the actualization of this human right? How do we transcend from slogans to practical programs and policies? How do we reduce this gap and give a dignified existence to the masses who today are yearning for drinking water? It is only a proactive answer to these questions that can give meaning to the right to water and sanitation.

Like all other environmental concerns of humankind, the actualization of this right demands action at the local level. Thus while having an eye on the global scene there is also a need to act locally. This explains our present attempt to explore the human right to clean water and sanitation using Nigeria as the prism. We shall also look at possible collaborative grounds with other countries in the African Region.[18] This is essential because water issues are rarely limited to one nation. Indeed, the major aquifers in Nigeria have international or cross-boundary extensions. The international nature of water issues is also seen in the challenges that arise from an attempt, for instance, to dam a river.[19] This chapter is mainly aimed at ascertaining the extent of recognition, enforcement and availability of the human right to water and sanitation in Nigeria.

To give us a handle on Nigeria and the issue of human right to water and sanitation we shall proceed to give at least working definitions of the basic concepts which we shall encounter in this discourse.

Human Rights: It is ordinarily difficult to articulate an uncontested definition of human rights. That notwithstanding:

> Human rights are rights inherent to all human beings, whatever our nationality, place of residence, sex, national or ethnic origin, colour, religion, language, or any other status. We are all equally entitled to our human rights without discrimination. These rights are all interrelated, interdependent and indivisible.

---

[16] Salman M A Salman, 'The human right to water and sanitation; is the obligation deliverable' (2014) <http://dx- doi.org/10.1080/02508060.215.986616 > accessed 12/02/2016.

[17] See < www.who.int/water_sanitation_health/mdg1/en/ >.

[18] The Abuja Declaration by African Ministerial Conference on Water (AMCOW), April 29–30, 2002 seeks to 'strengthen intergovernmental co-operation in order to halt and reverse the water crisis and sanitation problems in Africa' <www.africanwater.org/Documents/amcow_declaration.pdf> accessed 13/02/2016.

[19] Gabčíkovo-Nagymaros Project (Hungary/Slovakia).

Universal human rights are often expressed and guaranteed by law, in the forms of treaties, customary international law, general principles and other sources of international law. International human rights law lays down obligations of Governments to act in certain ways or to refrain from certain acts, in order to promote and protect human rights and fundamental freedoms of individuals or groups.[20]

There is also a shared sense amongst scholars that human rights are difficult to define, as can be seen in some works.[21] However, all human rights are perceived as inseparable and interdependent whether they are civil and political rights, or social and cultural rights. Thus the enhancement or protection of one right facilitates the advancement of the others and vice versa. It suffices however to note that the essential attribute of human right is that it is a right inherent in a human being simply because she is human. Her age, nature, race or circumstances of birth cannot diminish her claim to it. They inhere to the human person and there is an obligation incumbent on society to recognize, respect and fulfil them.

Social Justice is seen as:

> one or more equitable resolutions sought on behalf of individuals and communities who are disenfranchised, underrepresented, or otherwise excluded from meaningful participation in legal, economic, cultural, and social structures with the ultimate goal of removing barriers to participation and effecting social change.[22]

According to the United Nations – *Social Justice in an Open World*:[23]

> Social justice may be broadly understood as the fair and compassionate distribution of the fruits of economic growth…currently, maximizing growth appears to be the primary objective, but it is also essential to ensure that growth is sustainable, that the integrity of the natural environment is respected, that the use of non-renewable resources is rationalized, and that future generations are able to enjoy a beautiful and hospitable earth. The conception of social justice must integrate these dimensions, starting with the right of all human beings to benefit from a safe and pleasant environment…

---

[20] < www.ohchr.org/EN/Issues/Pages/WhatareHumanRights.aspx >.

[21] Joseph Raz, 'Human Rights Without Foundations' *Oxford Legal Studies Research Paper No 14/2007* (2007) <http://ssrn.com/abstract=999874 or http://dx.doi.org/10.2139/ssrn.999874 > accessed 10/02/2016.

[22] Bryan Garner, *Black's Law Dictionary* (10th edn, Thomson Reuters 2014) 996.

[23] *Social Justice in an Open World: The Role of the United Nations* <www.un.org/en/events/socialjusticeday/> accessed 12/02/2016.

Clean Water can be described as water that is good enough for drinking and sanitation. This means that the water is not contaminated by microbes, pathogens, chemicals or other substances that are inimical to health. This type of water may also be described as wholesome water. The World Health Organization has also provided what is referred to as drinking water quality standard, which is aimed at setting drinking water quality standards around the world.[24]

Sanitation generally refers to the provision of facilities and services for the safe disposal of human urine and faeces. Inadequate sanitation is a major cause of disease worldwide and improving sanitation is known to have a significant beneficial impact on health both in households and across communities. The word 'sanitation' also refers to the maintenance of hygienic conditions, through services such as garbage collection and wastewater disposal.[25]

Access to safe water is measured by the proportion of the population with access to adequate amount of safe drinking water located within a convenient distance from the user's dwelling.[26]

From these definitions the nature of the human right to water and sanitation and their fulfilment considering the existing global statistics about want of access to them is apparent.

II. LEGAL FRAMEWORK ON WATER AND SANITATION IN NIGERIA

Nigeria as a sovereign nation has many legal instruments relevant to water and sanitation. These instruments are both international and domestic. For ease of appreciation, we shall group the documents broadly into two categories – international and domestic.

---

[24] <www.who.int/entity/water_sanitation_health/dwq/guidelines/en/-> accessed 05/02/2016.
[25] <www.who.int/topics/sanitation/en/ > accessed 13/02/2016.
[26] <www.un.org/esa/population/pubsarchive/chart/12.pdf > accessed 10/02/2016.

a. **International Legal Documents Acceded to by Nigeria**

Nigeria is a state party to many treaties and conventions on water and sanitation thereby assuming obligations under them.[27] Some of these documents are:

i. The African Charter on Human and Peoples' Rights (adopted 27 June 1981, entered into force 21 October, 1986) (Ratification and Enforcement Act Cap A8 Laws of the Federation of Nigeria 2004);
ii. The International Covenant on Civil and Political Rights (ICCPR) (adopted 16 December 1966, entered into force 23 March 1976) ratified by Nigeria in 1993;
iii The International Convention on Elimination of all forms of Racial Discrimination (adopted 21 December 1965 entered into force 4 January 1969) ratified by Nigeria in 1967;
iv The International Covenant on Economic, Social and Cultural Rights (ICESCR) (adopted 16 December 1966, entered into force 3 January, 1976) ratified by Nigeria in 1993;
v Convention on the Elimination of all Forms of Discrimination Against Women (CEDAW) (adopted 18 December 1979, entered into force 3 September 1981) Ratified by Nigeria in 1985;
vi The Universal Declaration of Human Rights (UDHR) (adopted 10 December 1948) Nigeria joined the UN on October 7, 1960 without any reservation to the UDHR.

b. **Domestic Legislations and Other Statutory Instruments**

i. The Constitution of the Federal Republic of Nigeria 1999;
ii. The National Environmental Standards and Regulations Enforcement Agency (Establishment) Act, 2007;
iii. The River Basins Development Act 1987;[28]
iv. The Water Resources Act 1993.

---

[27] The ICCPR in para. 1 of its preamble recognizes the inherent dignity and the equal and inalienable rights of all members of the human family.
[28] The Act created 12 River Basin Development Authorities with the mandate to undertake comprehensive development of all water resources for multi-purpose use.

c. **Comments and General Review of These Instruments as they Relate to the Human Right to Clean Water and Sanitation**[29]

Nigeria is a full member of the African Union and has since 1986 ratified the African Charter on Human and Peoples' Rights. Article 16 of the African Charter recognizes the right of every individual to enjoy 'the best and attainable state of physical and mental health.' Equally Article 24 provides that 'all peoples shall have the right to a general satisfactory environment favorable to their development.' The African Charter has a direct enforceability and Nigeria made no reservation to any provision(s) therein before ratification.

The Supreme Court of Nigeria has also interpreted the African Charter and held that it is a statute with international flavor. Therefore, where a conflict arises between it and another statute, its provisions will prevail over those of that other statute for the reason that 'it is presumed that the legislature does not intend to breach an international obligation. To this extent...the Charter possesses "a greater vigor and strength" than any other domestic statute.'[30]

Instructively, that the Supreme Court also laid out a clear standard for the construction of the provisions of the African Charter when it stated that:

> the spirit of a Convention or a treaty demands that the interpretation and application of its provision should meet international and civilized legal concepts. That means those concepts which are widely acceptable and at the same time of clear certainty in application. Thus, the courts will not construe a statute so as to bring it into conflict with international law.[31]

The ICESCR, in tenor much like the African Charter, provides that: '...States Parties to the present Covenant recognize the right of everyone to an adequate standard of living for himself and his family, including adequate food, clothing and housing, and to the continuous improvement of living conditions.'[32]

Further in Article 12 it recognized the essence of giving everyone the highest attainable standard of physical and mental health which includes environmental hygiene. The Covenant states thus:

---

[29] We shall use a few of these instruments to highlight of their significance to the human right to water and sanitation.
[30] *Abacha v Fawehinmi* (2000) 6NWLR (Pt 660) 228.
[31] Ibid.
[32] ICESCR art. 11.

> ...States Parties to the present Covenant recognize the right of everyone to the enjoyment of the highest attainable standard of physical and mental health. The steps to be taken by the States Parties to the present Covenant to achieve the full realization of this right shall include those necessary for:
> (a) The provision for the reduction of the stillbirth-rate and of infant mortality and for the healthy development of the child;
> (b) The improvement of all aspects of environmental and industrial hygiene ...

Nigeria ratified this Covenant in 1993. Although we could not find any judicial construction of this provision in the Nigerian courts in the course of this research, it is submitted that going by the interpretation of the African Charter as we saw in *Abacha v Fawehinmi*[33] it is a basis of obligation and redress before the Nigerian courts. Equally, the courts by the common law doctrine of precedent are bound to follow this progressive interpretation established by the apex court.

Other international human rights instruments acceded to by Nigeria have similar provisions and hence the obligations of Nigeria to its citizens and the international community via them is established. It is also clear that they are sufficient to elicit actions and sanctions against relevant agencies within the country in matters of human right to water and sanitation. This is so because the obligation to recognize, respect and fulfil these human rights subsists. Equally all human rights are imprescriptible, indivisible and unalienable.[34]

Domestically, the Constitution of the Federal Republic of Nigeria 1999 is the foundation of other laws in the country and enjoys supremacy over them.[35] The provisions of Chapter 2 of the Constitution[36] can be called the social charter of the country. The chapter emphasizes social justice, freedom, welfare and the provision of basic needs like food and shelter as the fundamental objectives and directive principles of state policy. This is a clear assimilation of the principles visible in the international obligations assumed by Nigeria which are subsisting.[37]

---

[33] *Abacha v Fawehinmi* (2000) 6NWLR (Pt 660) 228.
[34] The Vienna Declaration on Program of Action on Human Rights 1993.
[35] Constitution of Nigeria 1999, s. 1 provides for its supremacy.
[36] Ibid., ss 13–24.
[37] Section 14(1): 'the Federal Republic of Nigeria shall be a state based on the principles of democracy and social justice.'

III. TOWARDS A FULL REALIZATION OF THE HUMAN RIGHT TO CLEAN WATER AND SANITATION IN NIGERIA

The above review of the existing international and domestic laws of Nigeria, shows a number of laws upon which the human right to water and sanitation in Nigeria is anchored. Our concern, however, is how to make these laws realize their professed objectives. This concern is made crucial by the dichotomy existing between civil and political rights in Nigerian jurisprudence. There is not yet any domestic legislation expressly acknowledging the human right to water despite the clarity of the international obligations. The question is: how do we make for meaningful recognition, respect and fulfilment of this human right?

a. Judicial Enforcement of Access to Clean Water and Sanitation in Nigeria

Judicial enforcement of socioeconomic rights is a difficult challenge worldwide. In Nigeria water and sanitation are perceived as socioeconomic and cultural rights. They are seen as welfare issues and thus limited in terms of judicial enforcement. Many countries still have this perception of socioeconomic and cultural rights and even those who have enacted them as part of their constitutions have also developed a limiting interpretative jurisprudence about them. Such phrases as 'core standards', 'gradual realization' and 'evolving capacity' are seen in most judicial interpretations. In other places you see judicial passivity towards enforcement of the socioeconomic and cultural rights of which water and sanitation are core aspects. The contest between business rights and bilateral investment treaties has also become a stumbling block to it.[38] Hence the startling statistics on the number of persons globally without access to adequate water and sanitation. There is therefore no significant social transformation.[39]

---

[38] Julien Chaisse and Christian Bellak, 'Navigating the expanding universe of investment treaties – creation and use of critical index', (2015) 18 *Journal of International and Economic Law* 79. On the rise of investment claims, see Julien Chaisse, 'Assessing the exposure of Asian states to investment claims', (2013) 6 *Contemporary Asia Arbitration Journal* 187, 201.

[39] David Landau, 'The reality of social rights enforcement' (2011) 53 *Harvard International Law Journal* < http://ssrn.com/abstract> accessed 12/02/2016.

Illustrative of this is the South African case of *Mazibuko v City of Johannesburg*:[40]

> The soul-searing facts of the case is that at about 2 am on March 27, 2005 one Phiri resident named Vasimuzi Paki was roused from sleep by frantic shouts by a tenant who was trying to put out a fire outbreak in one of the apartments in the neighbourhood. Other neighbours also gathered and assisted in the first crucial minutes to put out the fire. They were using water from the prepaid water meter supplied by the Johannesburg water Company which was recently installed in all the neighbourhood. The prepaid meter was meant to control the quantity of water use by households and this was capped at 6 kilolitres per household per month. . . While the residents battled to put out the fire the water went off automatically. They tried unsuccessfully to reach the police. They were therefore forced to scoop up ditch water in a frantic effort to stop the fire. After battling for about an hour they finally put out the fire but not before the shack had burnt down. It was only after Paki's tenant returned in the morning from her night shift that everyone discovered the horror that her two small children had been sleeping in the shack. They both died in the fire.[41]

The issue became notorious and was tested in the Constitutional Court of South Africa. Two points of decision in the case were:

1. What is the legality of prepaid water meters in a predominantly poor black neighborhood by which water supply automatically stops after a household had used six kiloliters of water a month without notice or opportunity of hearing?
2. How reasonable is this water measurement which does not take into consideration the number of persons per household and the special needs of households having sick persons like HIV/AIDS patients?

The Court held the prepaid meter unlawful but contrary to the international standard of 50 liters per person per day as the minimum daily water need of individuals the court approved 42 liters. The danger in this is that it limits the rights and sometimes becomes a policy cap in the hands of state officials.

In Nigeria the right to clean water has not received any serious judicial consideration. This is not surprising because by the provisions of

---

[40] *Mazibuko and Others v City of Johannesburg and Others* (CCT 39/09) [2009] ZACC 28; 2010 (3) BCLR 239 (CC); 2010 (4) SA 1 (CC) (8 October 2009) <www.saflii.org/za/cases/ZACC/2009/28.html>.

[41] Patrick Bond and Jackie Dugard, 'The case of Johannesburg Water: what really happened at the prepaid parish pump' (2008) 12 *Law Democracy & Development* 1.

section 6(6) of the Nigerian Constitution socioeconomic and cultural rights under Chapter 2 of the Constitution are not justiciable.[42] This has left socioeconomic and cultural rights, of which the right to water and sanitation forms part, as mere platitudes.[43] This is wrong because the Maastricht guidelines[44] on enforcement of socioeconomic rights views all human rights as one hence they are inseparable and should be recognized, respected and fulfilled as a unit. Comparatively the South African interpretation with all its limitations is more progressive because it recognizes the right to water as a human right.

In a more proactive interpretation of fundamental objectives and directive principles of state policy, in the case of *Minerva Mills Ltd.. v Union of India*,[45] it was held thus:

> to a large majority of people who are living in subhuman existence in conditions of abject poverty and for whom life is one long unbroken story of want and destitution, notions of individual freedom and liberty though representing some of the cherished values of free society would sound as empty words bandied about in the drawing rooms of the rich and well to do and the only solution for making these rights meaningful to them is to make the material conditions and usher in a new socio-economic order where socio economic justice will inform all institutions in public life so that the pre-conditions for fundamental liberties for all may be secured.

In the case of *SERAC v. Nigeria*[46] the African Court of Human Rights relying on Articles 16 and 24 of the African Charter found Nigeria liable for compromising the right to a healthy environment and means of sustenance like food and drinking water. This is the judicial situation in

---

[42] This means that they cannot be grounds for judicial actions seeking enforcement.

[43] Dakas CJ Dakas, 'Judicial reform of the legal framework for human rights litigation in Nigeria: Novelties and Perplexities' <www.nials-nigeria.org/journals/Dakas-Judicial%20Reform-Legal%20Framework%20of%20Human%20Rights%20Litigation.pdf>.

[44] On the tenth anniversary of the Limburg Principles on the Implementation of the ICESCR, a group of experts met in Maastricht from 22–26 January 1997 (the Netherlands) to elaborate on the Limburg Principles regarding the nature and scope of violations of these rights and appropriate responses and remedies.

[45] AIR 1980 SC 1789. PN Bhagwati's view depicts progressive interpretation of socioeconomic and cultural rights.

[46] *Social and Economic Rights Action Centre (SERAC) v Nigeria* (2001) AHRLR 60 (ACHPR 2001). This was a case of pollution of water and farmlands in the Niger delta and thereby the destruction of the source of drinking water. <www.chr.up.ac.za/index.php/browse-by-subject/410-nigeria-social-and-economic-rights-action-centre-serac-and- another-v-ni > accessed 13/02/2016.

Nigeria which calls for urgent attention. It is expensive to seek judicial enforcement and most Nigerians who do not have access to adequate water for daily living cannot afford it.

b. Challenges to the Full Realization of the Human Right to Clean Water and Sanitation in Nigeria

There are many challenges bedeviling the recognition, respect and fulfilment of the right to water and sanitation in Nigeria.

First, the legal infrastructure. The domestic laws are not yet clearly defined as to the recognition, respect and fulfilment of the right to clean water and sanitation. Equally, there are many areas of uncoordinated legislations between the states and the Federal Government. There is an observable duplication of agencies across the country leading at times to some kind of confusion and bureaucratic redundancy.

Second, the issue of urban planning and development. There are different urban planning legislations across the states of the federation and compliance with urban planning and infrastructure development standards is quite low. This leads to a kind of uncontrolled urbanization without adequate water supply and sanitation infrastructure. This also creates general poor living conditions.

Third, the inadequate energy supply. To attain a sufficient water supply for sanitation and living energy must be adequate. According to a recent publication, the National Planning Commission of the Federal Republic of Nigeria reports that Nigeria needs at least 35,000 megawatts of electricity to drive its economy. However only about 3,500 megawatts of electricity is being produced.[47] This is inadequate to drive the economy and provide proper water and sanitation system to about 175 million people.

Fourth, corruption and a lack of political will also constitute obstacles and it is hoped that there will be a change soon in the spirit of the new leadership.

Fifth, the lack of a coordinated execution of a clear water and sanitation master plan appears to be the most difficult obstacle. Each of the 36 state governments has its own environmental and sanitation agency.[48] Sometimes sanitation is treated as if it is a distinct issue from clean water thus unwittingly downgrading the connection between them.

---

[47] <www.vanguardngr.com/2011/10/electricity-nigeria-needs-35000-megawatts-by-2020/>.

[48] For instance, there is the Lagos State Waste Management Authority, Enugu State Waste Management Authority, etc.

IV. SOCIAL JUSTICE TO THE RESCUE

There is a weak structure of recognition, respect and fulfilment of the human right to water and sanitation in Nigeria. This is self-evident, because the rights are not recognized as justiciable in courts by the Nigerian Constitution. It is therefore imperative that we devise other efficient means of making a timely intervention in order to rescue many citizens who are exposed to diseases because of inadequate water and sanitation. It is submitted that the tool for this intervention is social justice. Social justice is a conscious attempt to use *all means available to society* to reach out to the basic needs and rights of citizens especially with respect to such essentials as water, food, sanitation and shelter. This is imperative because strict adherence to the letter of the law may not timeously meet the pain of those citizens living under undignified conditions. It demands conscious intervention and the removal of obstacles like tyranny and poverty which robs the citizens of the 'capability' to live a meaningful and dignified life.[49] To this effect we have identified some major areas of action informed by the social justice motivation through which we should approach the problem of human right to clean water and sanitation. They are:

a. critical social engagement and active citizenship participation;
b. education and poverty eradication;
c. legislative intervention;
d. private sector engagement/collaboration; and
e. corporate social responsibility.

Critical social engagement entails that society as a whole must see the human right to water and sanitation as a common need of humanity. At present, there appears to be a limited appreciation of the problem and it is often seen as a problem of the poor. There is thus a need for a reorientation to create a massive awareness and genuine disposition towards tackling this problem.

Because governments alone cannot solve the problem of human right to water and sanitation there is a need for active citizenship participation at several layers of society. This can be through a careful utilization of the existing water and sanitation infrastructure. It can equally take the shape of private sector initiated funding of the provision of clean water and sanitation in society. One crucial point is also making sure that citizens

---

[49] Amartya Sen, *Development as Freedom* (Oxford University Press 1999). This Nobel Prize winner rightly sees a nexus between development and freedoms.

learn how not to pollute existing water sources. This will help in preserving what is available while searching for ways of expanding accessibility, sustainability and affordability of water and sanitation.

Education has a way of giving human beings an elevated appreciation of their rights. It also increases competence, because the educated learn to develop their own capacities to solve their basic life needs. Also education adds to the dignity and the value which citizens put on their lives. It reduces the disappearing number of the middle class, inequality and creates room for social class mobility. A capable middle class can then rise up to challenge the unacceptable situation thereby improving the well-being of other citizens. Poverty and lack of development are directly linked to education. An uneducated child develops no capacity and instead of becoming a contributing member of society becomes a perpetual dependent. Poverty diminishes human flourishing. To attain sustainable access to water and sanitation we must tackle poverty. Hence we must bring a social justice perspective to development. In a recent publication Amartya Sen had this to say:

> development...can be seen as a process of expanding the real freedoms that people enjoy. Focusing on human freedoms contrasts with narrow views of development such as identifying development with the growth of gross national product or with technological advance, or social modernization. Growth of GNP or of individual incomes can of course be very important...development requires the removal of the major sources of un-freedom, poverty as well as tyranny, poor economic opportunities as well as systematic social deprivation, neglect of public facilities as well as intolerance or repressive states ... sometimes the lack of substantive freedoms relates directly to poverty which robs people of freedom to satisfying hunger, or to achieve sufficient nutrition, or ... to be adequately clothed or sheltered, or to enjoy clean water and sanitary facilities.[50]

Poverty is thus a major obstacle and a reduction in poverty will undoubtedly expand access to clean water and sanitation.

Law is also a tool of social engineering, hence the legislature must enact laws that are social justice oriented. By law a given percentage of annual appropriation bills can be reserved for water and sanitation projects. Equally legislation can make the non-justiciable provisions of the Constitution relating to socioeconomic and cultural rights justiciable or amenable to judicial enforcement. Some kind of consumption tax could also be placed on the upper class so as to unlock the needed funds for water and sanitation project developments. This is the essence of

---

[50] Ibid., 3.

legislative interventions. Bilateral treaties on trade and investment should be made to recognize the essential nature of water and sanitation.

Private sector engagement and collaboration is also a way of finding sustainable answers to this problem of non-recognition, respect and fulfilment of the human right to water and sanitation. States can combine with other states and organizations in order to finance or share technical support for the purpose of realizing the human right to water and sanitation. In Nigeria there is the Lake Chad Basin Commission – a regional attempt to find solution to the problem but this appears not to have been effectually pursued. Equally the World Bank through the International Finance Corporation is very active in collaborating and supporting states which intend to pursue the human right to water and sanitation sustainably. Because of the heavy capital outlay involved in water and sanitation infrastructure, it is important that this type of regional and public/private sector collaboration is pursued. Indeed, investment in water and sanitation infrastructure can help in solving the power generation problem and release the needed energy for both industrial and private use.

Corporate social responsibility as a matter of duty is another reasonable means of ensuring access to clean water and sanitation. It is important that corporations operating in any community see that they owe a duty of care to the community and the environment wherein they do their business. Sustainable practices, which do not contaminate underground aquifers or surface water sources, should be pursued. Equally some percentage of their net profits should be set aside for social justice interventions in their host community. This can actually form a basis of legislative enactments to ensure that it is a basic requirement of corporate practice.

CONCLUSION

Access to clean water and sanitation are fundamental human rights. There are international, regional and domestic legal frameworks acknowledging this. There is also a convergence of opinion among scholars and international global policy formulating bodies that these are human rights. Indeed, both the Maastricht guidelines and Hamburg principles before it acknowledged that all human rights are intertwined. To ignore one is to ignore the other. They form a bundle of rights and to disregard the right to water and sanitation is also to desiccate the right to life.

Despite this understanding, there is a huge gap between the proclaimed rights and the actual respect and fulfilment of these rights. This has led to an inexplicable situation whereby a large human population are forced to live in inhuman and degrading conditions for want of clean water and sanitation. This has given rise to diseases and other inhuman dissonance which we can no longer ignore. Nigeria has a fair share of this disproportionate impact of lack of access to adequate sanitation and water.

In Nigeria the letter of the law has not been translated into actual enjoyment of these rights for many. There is great difficulty in the enforcement of the right to clean water and sanitation due to legislative, judicial and other institutional obstacles. We have attempted to proffer possible solutions to this problem by way of critical social engagement, collaboration between governments and the private sector, education and massive enlightenment, legislative intervention and corporate social responsibility amongst others. These have to be permeated by a social justice perception.

In reality, the boundaries of this chapter do not permit an exhaustive discussion of the fullest dimensions of this human right. However, we will be satisfied if we have commenced a conscious discussion of the means of finding lasting solutions to it.

Humankind must act now to give access to the many who lead diminished lives because of lack of access to clean water and sanitation. It is our duty to do so as it serves the best interest of everyone else we pay dearly in terms of diseases, child and maternal mortality and undue strain to existing infrastructure. It is immoral to think that the essential aspect of life is to accumulate wealth and grow our GNP without a commensurate elevation of the lives of members of the human family. Indeed 'ill fares the land to hastening ills a prey/ where wealth accumulate and men decay.'[51]

---

[51] Oliver Goldsmith, 'The Deserted Village' <www.poetryfoundation.org/poem/173557>.

11. Troubled waters: impact of the private sector in implementing the right to water
Preetha Mahadevan

INTRODUCTION

The United Nations Development Program's Human Development Report of 2006[1] indicates that as many as 1.1 billion people worldwide do not have access to safe drinking water. Since its acceptance as a recognized basic human right, the obligation to create of a framework for providing clean and safe drinking water to all human beings across the globe has fallen upon the respective government agencies. Yet, multinational corporations such as Veolia Environnement, Suez Environnement, Pepsi, Nestlé and Coca-Cola are among the major private sector participants in global water supply systems. This trend is often seen in countries where the government chooses to engage private corporations because of its inability to provide for the infrastructure and technical expertise required in building a public water supply system. Other instances where private participation is seen is where there is an obligation to employ private players in fulfillment of terms accepted in agreements with institutions providing loans for funding such projects. Another important reason for large-scale private participation is because water as a resource is the primary output product or is a prime resource in the supply chain of these industries.

This chapter attempts first, to trace the evolution of the Right in the international law regime, in brief. Secondly, the chapter attempts to ascertain the cost associated with the human right to water. Then, the chapter explores the measures undertaken by international organizations in enforcing this right. The chapter then examines methods such as privatization, public-private partnerships, concession

---

[1] 'Human Development Report 2006' (United Nations Development Program 2006) <http://hdr.undp.org/sites/default/files/reports/267/hdr06-complete.pdf> accessed 20 February 2016.

agreements and international investment agreements with the support of international organizations such as the World Bank, and the outcome of such efforts. Finally, the chapter explores how corporate entities – the private sector involved in construction of facilities and providing water utilities to the public on behalf of the government and the entities involved in packaging/bottling of water for commercial sales embrace the 'right to water', and the efforts in developing clear standards and metrics for what companies should/must do; ways to evaluate their compliance with these standards and the consequences of failure to comply.

INTERNATIONAL LAW RECOGNIZING 'RIGHT TO SAFE WATER'

The status of the 'right to safe water' as basic human right in international law continued to be ambiguous for most part of the twentieth century. Although the UN Water Conference in the 1977 Mar del Plata[2] recognized that all peoples have a right to access drinking water, the explicit recognition of the 'right to safe drinking water and sanitation' by the United Nations was made only in 2010[3] and immediately affirmed by the United Nations Human Rights Council.[4]

Other international instruments like the Convention on the Elimination of all forms of Discrimination Against Women (CEDAW)[5] in 1979; the Child Rights Committee[6] in 1989 have also provided for explicit recognition of the right to water and sanitation. The Committee assessing the International Covenant on Economic, Social and Cultural Rights (ICESCR) in 2002 issued a General Comment No. 15 wherein it impliedly extended the meaning of Art. 11(1) of the ICESCR to include the 'right to water and sanitation' within its ambit. The Convention on the Rights of Persons with Disabilities (CRPD) in Art. 28 included within the ambit of 'adequate standard of living and social protection', the right to persons with disabilities to access to clean water services.

Regional Instruments such as the African Charter on the Rights and

---

[2] See generally United Nations Water Conference: Summary And Main Documents (1978).
[3] UN General Assembly Resolution 64/292 (3 August 2010).
[4] UN Commission on Human Rights Res A/HRC/RES/15/9.
[5] Art. 14 of the UN Committee for the Elimination of all forms of Discrimination Against Women (12 December 1979).
[6] Art. 24 of the Convention on the Rights of the Child (20 November 1989).

Welfare of the Child[7] provided that the member states shall undertake to ensure the full implementation of the provision of adequate nutrition and safe drinking water. Similar provisions have also been included in the Protocol to the African Charter on Human and Peoples' Rights on the Rights of Women in Africa.[8] The European Charter on Water Resources[9] in Art. 5 recognized that everyone has the right to a sufficient quantity of water for his or her basic needs.

It is imperative to note that even prior to this explicit recognition, many nation states have affirmed this right – some by way of codification in their Constitution[10, 11] and some by judicial interpretation of the 'right to life'. (The Supreme Court of India in the case of *Narmada Bachao Andolan*[12] held that water is the basic need for the survival of the human beings and is a part of right to life and human rights as enshrined in Art. 21 of the Constitution of India.)

DOES RIGHT TO WATER INCLUDE RIGHT TO 'FREE ACCESS TO WATER'?

> Free or subsidized delivery of a public service like water leads to abuse of the resource
>
> Former World Bank President James Wolfensohn
> (April 12, 2000 after the Cochabamba struggle ended)

There has been a looming question as to whether there should be a cost associated in the supply of water when it is recognized as a basic human right.

The Dublin Statement on Water and Sustainable Development (Dublin Principles) adopted on January 31, 1992 in Dublin, Ireland recognized that an economic value can be affixed to water in all its uses and therefore that water should be recognized as an economic good. The Dublin Statement further affirmed that access to clean water should be provided at an affordable price and that failure to recognize the economic value of water in the past has led to the resource being wasted. It also acknowl-

---

[7] Art. 14 of the African Charter on the Rights and Welfare of the Child (July 1990) CAB/LEG/24.9/49.
[8] Art. 15 of the Protocol to the African Charter on Human and Peoples' Rights on the Rights of Women in Africa (13 September 2000) CAB/LEG/66.6.
[9] See the European Charter on Water Resources 2001 (17 October 2001) CO-DBP (2001) 8 [CO-P/documents/codbp2001/08c].
[10] See generally The Constitution of the Republic of South Africa, 1996.
[11] See generally The Constitution of Kenya, 2010.
[12] *Narmada Bachao Andolan v. Union of India* A.I.R. 2000 S.C. 375.

edged that recognizing water as an economic good supports 'achieving efficient and equitable use' which helps in turn in 'encouraging conservation and protection of water resources'.

This conflict can be discussed in two aspects: the *full cost recovery paradigm* which suggests that 'water and sanitation access for all people is not possible without recovery of all costs from the user'[13] and the *human rights paradigm* which suggests that 'water and sanitation access for all people is not possible without recognition of water as an inalienable human right'.[14]

According to the cost recovery paradigm full cost pricing and recovery will lead to a stable system that in turn leads to financially, operationally and environmentally sustainable operations. Not contrarily, the human rights paradigm recognizes that 'services must be affordable, not free, but that no one should be denied access for inability to pay'[15] which leads to a policy that water should be accessible to all people, even those in extreme poverty who are unable to pay for the utility.[16] This is probably the acceptable midpoint in creating a balance between the two paradigms in a step towards sustainable development.

With this as the standard for pricing, it is relevant to explore the impact the involvement of the private sector has had on the tariffing of water as a utility. The Cochabamba struggle of 2000 paints an unpleasant picture of how unsupervised private sector actions in the pricing of water can lead to concerns relating to protection of human rights. With respect to increasing the water network and accessibility, issues were seen in the Public Private Partnership (PPP) adopted by Guyana where raising the tariff upon the introduction of the PPP did not help in increasing the access to piped water as even the increased tariff was barely enough to cover the costs incurred by the water utility.[17]

The Report of the independent expert on the issue of human rights obligations related to access to safe drinking water and sanitation, Ms. Catarina de Albuquerque of the United Nations General Assembly

---

[13] P. Mader, 'Water Paradigms: Full Cost Recovery versus Human Rights' Working Paper (2011), Human Rights to Water & Sanitation Program, Harvard Kennedy School, Cambridge, MA, USA.
[14] Ibid.
[15] Sharmila L. Murthy, 'The Human Right(s) to Water and Sanitation: History, Meaning, and Controversy Over Privatization', (2013) 31 *Berkeley J. Int'l. L.* 89.
[16] See for example, Water Industry Act 1991 (UK), s. 61(A).
[17] Philippe Marin, *Public-Private Partnerships for Urban Water Utilities* (World Bank Publications 2009).

suggests that non-state actors cannot unilaterally determine the tariff structure and that the state should be involved in the process to ensure that the services are affordable even to the poorest and consider offering flexible payment plans such as 'phased connections charges, payment in installments and grace periods'.[18]

THE UNITED NATIONS AND OTHER INTERNATIONAL AGENCIES: RECENT INITIATIVES

The United Nations in 2003 by way of a Resolution[19] proclaimed the decade from 2005 to 2015 as the International Decade for Action, 'Water for Life'. In this endeavour, the UN-Water, the Secretariat of UN-Water, a United Nations interagency coordination mechanism for all freshwater related issues was created. The UNICEF and the WHO Joint Monitoring Programme for Water Supply and Sanitation operates under the umbrella of UN-Water. Other initiatives of the UN included the Programme on Capacity Development in Bonn and the Programme on Advocacy and Communication in Zaragoza, Spain. Several other initiatives of the UNDP, the World Bank, the UNICEF, UN-Habitat, UNEP and multi-stakeholder organizations such as Water Supply and Sanitation Collaborative Council are noteworthy.[20]

The UN Human Rights Council in 2008 established the mandate of (prior to 2010 – the Independent Expert) the Special Rapporteur on the human right to safe drinking water and sanitation to examine these crucial issues and provide recommendations to governments, to the UN and other stakeholders. The Special Rapporteur, Ms. de Albuquerque has periodically reviewed the right avidly with respect to the measures undertaken by the state actors, the non-state actors or the private sector and their obligations and responsibilities, operation of services, accountability and decision-making.

The UN General Assembly in 2000 adopted a resolution – 'United Nations Millennium Declaration'[21] wherein it was resolved to achieve by the year 2015, the eight-part plan popularly known as the *Millennium*

---

[18] UNGA, 'Report of the independent expert on the issue of human rights obligations related to access to safe drinking water and sanitation, Catarina de Albuquerque' (1 July 2009) UN Doc. A/HRC/12/24.
[19] A/RES/58/217, 23rd December 2003.
[20] A/65/297, 16th August 2010.
[21] A/RES/55/2, 18th September 2000.

Development Goals ('MDG'). The goals included the eradication of poverty and extreme hunger; achieving universal primary education; promoting gender equality and empowering women, reducing child mortality; improving maternal health; combatting HIV/AIDS, malaria and other diseases; ensuring environmental sustainability; and forming a global partnership for development.

Goal 7 of the MDG deals with ensuring environment sustainability and Target 7C specifically deals with a goal to halve the proportion of the population without sustainable access to safe drinking water and basic sanitation by the year 2015. The UN in an attempt to achieve this goal established global partnerships such as 'Sanitation and Water for All' which partners with 90 low- and middle-income developing countries, donor partners, UN and other multi-lateral agencies, development banks, civil societies, research and learning partners and sector partners. The 2013 Fact Sheet released by the UN reports that 'more than 2 billion people have gained access to *improved* drinking water sources since 1990, exceeding the MDG target'.[22]

The report of the Special Rapporteur of the Secretary-General, John Ruggie[23] on the issue of human rights and transnational corporations and other business enterprises was an attempt to map the international standards of responsibility and accountability for corporate acts.

With the endorsement of the UN Human Rights Council in 2011, the UN Guiding Principles[24] for Business and Human Rights became the initial corporate and human rights responsibility initiative to be endorsed by the UN. The Principles outline three major pillars defining how the governments and business should implement the framework:

- the state duty to protect human rights;
- the corporate responsibility to respect human rights;
- access to remedy for victims of business-related abuses.

---

[22] See generally UN Millennium Development Goals Fact Sheet (2013) <www.un.org/millenniumgoals/pdf/Goal_7_fs.pdf> accessed 20/02/2016.

[23] UNHRC, 'Report of the Special Representative of the Secretary-General on the issue of human rights and transnational corporations and other business enterprises, John Ruggie' (7 April 2008) UN Doc. A/HRC/8/5.

[24] UNHRC. 'Report of the Special Representative of the Secretary-General on the Issue of Human Rights and Transnational Corporations and Other Business Enterprises,' Guiding Principles on Business and Human Rights: Implementing the United Nations 'Protect, Respect and Remedy' Framework, (21 March 2011) UN Doc. A/HRC/17/31.

CORPORATE INVOLVEMENT AND IMPACT ON THE HUMAN RIGHT TO WATER

Water plays a colossal role in the functioning of most aspects of human activity. Empirical data suggests that the largest extractor and consumer of water, sector wise is the agricultural sector, followed by domestic use and then by the industrial sector.[25] Even though the industrial sector is not the largest extractor and consumer of water, economic activities are also heavily reliant of water for their sustenance. Corporations tryst with water as a resource can be attributed to the multitude of business interests. Businesses regularly consume water as a direct input in their business activity – for instance, in the various manufacturing industries and as an indirect input while relying on the agricultural sector for raw materials. Besides this, they are also involved in water utilities – drinking water supply and irrigation supply for agriculture. Moreover, corporations have also invested in the wastewater treatment sector and other such sectors that have a direct impact on the environment.

For the purpose of this chapter, the entities are divided as businesses that use water for 'industrial use' and as businesses that are involved in the 'water utilities'. This separation is essential to identify the dissimilar interests that these have towards the human right to water and the impact they leave in recognizing and respecting the right.

A. Industrial Use

The industries that are the largest consumers of water as a raw material or an input directly or indirectly in their supply chain are the agricultural industries, the extractive industry and the beverage companies. The availability of water directly has an impact on the location where such industries are located, costs involved in the production that has a direct impact on the market value of the final product.

The 'water footprint' of these industries indicates the water used directly and indirectly by a consumer using the product. For instance, a study[26] on the water footprint in the soft drink industry of a sugar-containing

---

[25] 'Vital Water Graphics – An Overview of the State of the World's Fresh and Marine Waters' (2008) 2nd Edition UNEP, Nairobi, Kenya (2008), <www.unep.org/dewa/vitalwater/article43.html> accessed 20 February 2016.

[26] A.E. Ercin, M.M. Aldaya and A.Y. Hoekstra, 'Corporate Water Footprint Accounting and Impact Assessment: The Case of the Water Footprint of a Sugar-containing Carbonated Beverage' (2011) 25 *Water Resour Manage* 721–41 < http://www.waterfootprint.org/Reports/Ercin-et-al-2011-CorporateWater Footprint-Softdrink.pdf> accessed 20 February 2016.

carbonated beverage is computed by taking into account the operational water footprint (this includes the water consumed during the process of production and not returned to the source of withdrawal, water polluted as a result of such process) and the supply-chain water footprint (this relates to the inputs used in the production and includes the water footprint of each of the other ingredients used in the manufacture product apart from water, and the water footprint of the products used in the post production such as bottling, cap, labelling, etc.). The study concludes that a 1.5 litre sugar-containing carbonated beverage uses between 169 litres and 309 litres of water to produce, depending upon several variables. This study and similar water footprint analyses of other industries reveals that the corporate interests of industries play a critical role in the availability of water leading to a direct impact on the human right to water.

The study by the JP Morgan Securities Inc.[27] prepared in collaboration with the World Resources Institute, based on a water footprint study indicates that the largest corporate consumers of water are the food and beverage industry as depicted in Figure 11.1 below.

In Kerala, India, following reports of contamination of the ground water with toxic chemicals, criminal claims are pending against Coca-Cola where damages of about $48 million are demanded. Operations in the Dasna plant, Uttar Pradesh were suspended in 2016 following communication received from the State Pollution Control Board.[28]

In Mehdiganj, India where a bottling plant of Hindustan Coca-Cola is located, the authorities of the State Pollution Control Board of Uttar Pradesh, India cancelled the Water Consent and Air Consent Order, restraining the company from functioning and also cancelled its approval for capacity enhancement, after finding violations of the terms of its license.[29]

The company has also faced rejection from the State Government for permission to set up a plant in Perundurai in Tamil Nadu.[30]

---

[27] 'Watching water: A Guide to Evaluating Corporate Risks in a Thirsty World' (Global Equity Research, JP Morgan Securities Inc., 31 March 2008) <http://pdf.wri.org/jpmorgan_watching_water.pdf > accessed 20 February 2016.

[28] Statement of the Hindustan Coca Cola Beverages Pvt. Ltd. dated 19th August 2016 available at http://www.coca-colaindia.com/statement-hindustan-coca-cola-beverages-pvt-ltd-4/ (last accessed on 31st August 2016).

[29] Instruction of the UP State Pollution Control Board dated 06.06.2014 available at http://www.indiaresource.org/documents/Mehdiganj2014/CancelNOC6-6-14.pdf (last accessed on 31st August 2016).

[30] Letter dated 20th April 2015 issued by the State Industries Promotion

billion liters

Company	Water Used (bn liters)	Ratio, liters of water per kg or liter of end product
Coca-Cola	288	2.4
Nestlé	155	4.1
Unilever	66	3.3
Kraft	54	6.0
Danone	51	2.8
Total	613	

Source: Company reports, company environmental/sustainability reports.

Figure 11.1 Food and beverage water consumption metrics

Even greater than the apprehension caused by the soft drink beverage industry are the concerns arising from the bottled water industry. These industries extract groundwater from springs and upon bottling, the price of the product is often determined based on its origin. There are often allegations that water is mined and extracted by companies in places where there is a negligible or no public water supply system, resulting in the position that, while the local community is unable to source water to meet its demand, the ground water continues to be depleted by the bottling companies. Nestlé's Pakistan plant was censured for not providing for drinking water to the local community in Bhati Dilwan village in the Sheikhupura region. Similarly, while half the population in Fiji do not have access to clean drinking water, it continues to bottle about half a million units every day for export to the US. This chapter does not explore the issues of the environmental damage caused by the use of unrecycled plastic bottles generating tremendous landfill.

The only possible deterrent effect that seems to have worked in response to the depletion of ground water caused by such corporations seem to be the impending threat of water scarcity that is predicted for the near future. The above-mentioned report also maps the sector-wide impact that water scarcity is expected to have on industries as below shown in Figure 11.2 below.

Acknowledging this as a potential concern, the management of the

---

Corporation of Tamil Nadu, available at http://www.indiaresource.org/documents/Perundurai/SIPCOTletterApr202015.pdf (last accessed on 31st August 2016).

Sector	Principal Impacts
Food and Beverages	Manufacturing disruptions, higher commodity costs, higher power costs, loss of access to sources of bottled water
Manufacturing	Production disruptions, problems with discharge of liquid wastes
Semiconductor Manufacturing	Production disruptions, higher costs for water purification, limits on expansion
Power Generation	Plant shutdowns due to lack of cooling water, high costs to purchase substitute power
Insurance	Positive impact due to demand for new coverages; costs from fire and drought claims
Extractive Industries	Potential restrictions on drilling, mining, use of slurry transport, and waste discharge

Source: World Resources Institute.

Figure 11.2 Water scarcity sector-wise impact on industries

Coca-Cola Company in their annual statement to their shareholders in 2006 and in 2009 declared that:

> **Water scarcity and poor quality could negatively impact the Coca-Cola system's production costs and capacity**
> Water is the main ingredient in substantially all of our products. It is also a limited resource in many parts of the world, facing unprecedented challenges from overexploitation, increasing pollution, poor management and climate change. As demand for water continues to increase around the world, and as water becomes scarcer and the quality of available water deteriorates, *our system may incur increasing production costs or face capacity constraints which could adversely affect our profitability or net operating revenues in the long run.*[31]
> [emphasis added]

Recognizing the effect that water scarcity is expected to have globally, corporations have adopted sustainability measures to help minimize the impact. Since the administration of a public water supply system rests with the governments, and corporations work with a profit motive, the

---

[31] Coca-Cola Company 10-K 2010, <http://www.wikinvest.com/stock/Coca-Cola_Company_(KO)/Filing/10-K/2010/F46738191> accessed 20 February 2016.

obligation is obviously upon state actors to develop strong regulations in all aspects including imposing a duty upon corporations who are involved in water mining to provide for the water supply system in the community. It is also the duty of state actors to regulate the usage of the groundwater and to control the hazardous effects on the environment.

Living in a time where corporate entities are a bigger economic enterprise than many countries, regulating the functioning of private actors would be impossible without a multi-stakeholder approach with efforts from all sectors – state actors, non-state actors, corporate entities, civic organizations and other interested institutions.

B. Water Utilities

The late 1980s and early 1990s saw a surge in the number of governments attempting to bring about reforms in the public utilities sphere. In the water and sanitation sector, large populations lacked access to piped water and many other regions received poor quality services because a plethora of issues plagued the system.[32] Rampant corruption, weak administrative and regulatory compliance led to deteriorating and substandard infrastructures coupled with unprecedented growth in population in the urban areas and few financial resources.

There was therefore an urgent need to revamp the entire system to step up the operations of the public water supply system. Since there was a need for capital investment, many of these initiatives were undertaken with the support of international financial organizations such as the World Bank and the IMF.[33] The inclusion of the private sector emerged at this time as a solution to the impediments faced by the developing nations; by a partnership with the public sector, the governments were able to access capital and the technical expertise required to build infrastructural capacities.

Over the years, different strategies have evolved for including private sector operators in the water supply system: (1) the Public-Private-Partnership (PPP) system for improved services with continual monitoring by the government; (2) the outright sale of the public water supply

---

[32] See generally J Chaisse and M Polo, 'Globalization of water privatization: Ramifications of investor-state disputes in the "blue gold" economy', (2015) 38 *Boston College International and Comparative Law Review* 1.

[33] See Julien Chaisse and Christian Bellak, 'Navigating the Expanding Universe of Investment Treaties – Creation and Use of Critical Index,' (2015) 18 *Journal of International Economic Law* 79 for understanding the link between treaties that states entered into with the increase of the FDI.

system – assets and the operation to the public sector; (3) concession agreements with the private sector providing the same model of the public water supply system for a specified period of time (lease system); and (4) an optimized public supply system with capital investment from the private sector.

Opponents of the privatization move noted the change merely as a result of financially strapped states with no inability to fund their own reforms, being forced to privatize in some form their water supply system to secure funding from financial institutions (a report of the International Consortium of Investigative Journalists[34] shows that one-third of the loans provided by the World Bank for water supply require the recipient state to privatize its water operations in some form before the funds are awarded) and the surrender to the desire of rich transnational corporations to milk benefits from this sector which was seen a cash cow.[35] (According to Rubenstein, the private sector is considered more profitable on a return on equity basis than other regulated industries.)

A graphical representation[36] of the global stakeholders in the private water supply system sector based on their revenue is provided in Figure 11.3.

Of the listed corporate players, it is estimated that the top three entities alone deliver water to 300 million consumers in more than 100 countries around the world. 'Their growth is exponential' argue Barlow and Clarke quoting figures showing aggressive growth when compared to a decade ago.[37] International Institutions defending privatization are quick to point to the successful operations of the private sector in providing access to clean water supply in regions such as Ghana, Mozambique, Senegal and urban Uganda.[38] With such a sizeable presence of the private sector in

---

[34] 'Promoting privatization', (International Consortium of Investigative Journalists, 3 February 2003) <http://www.icij.org/projects/waterbarons/promoting-privatization> accessed 20 February 2016.

[35] Edwin S. Rubenstein, 'The Untapped Potential of Water Privatization' 2000, A Hudson Institute Report For American Water Works, Inc. <www.esr-research.com/Theprivatewaterindustry.htm> accessed 20 February 2016.

[36] 'World's 10 Largest Water Companies: Who they are, what they do and how much revenue they collected' (18 January 2012) <http://www.etcgroup.org/content/worlds-10-largest-water-companies > accessed 20 February 2016.

[37] Maude Barlow and Tony Clarke, 'Water Privatization' (Polaris Institute, January 2004) <www.globalpolicy.org/component/content/article/209/43398.html> accessed 20 February 2016.

[38] 'Sanitation and Water Supply: Improving Services for the Poor (International Development Association, The World Bank) <http://siteresources.worldbank.org/IDA/Resources/IDA-Sanitation-WaterSupply.pdf> accessed 20 February 2016.

World's Largest Water Companies

Company (Headquarters)	What they do	Revenue 2009 (US$ million)
1. Veolia Environment (France)	Water supply and mgmt., waste mgmt., energy and transport services	45,519
2. Suez Environment (France)	Water supply, wastewater treatment, waste management	17,623
3. ITT Corporation (USA)	Water supply, treat wastewater, supply pumps etc. for handling toxic water	10,900
4. United Utilities (UK) (FY ending 3/31/2010)	Water supply and sewage treatment	5,894
5. Severn Trent (UK) (FY ending 3/31/2010)	Water supply and sewage	2,547
5. Thames Water (UK) (FY ending 3/31/2010)	Water supply and wastewater treatment	2,400
7. American Water Works Company (USA)	Water supply and wastewater mgmt	2,441
8. GE Water (USA)	Water treatment, wastewater treatment	2,500
9. Kurita Water Industries (Japan) (FY ending 3/31/2010)	Water/wastewater treatment/reclamation, soil and groundwater remediation	1,926
10. Nalco Company (USA)	Water treatment (Water-related revenue only)	1,628

Sources: Polaris Institute; Global Water Intelligence; ETC Group.

Figure 11.3 Global stakeholders in the private water supply system sector

the water utilities sector, it is important that initiatives are adopted as a multi-stakeholder approach for the achievement of the ultimate goal of implementing the right to access to clean and safe water supply.

International investment regime
Keeping in mind the position that the private sector has in the supply of water utilities, it is obvious that various large foreign investors have been interested in investing in this sector. The international investment regime is governed by Free Trade Agreements ('FTA') and by Bi-lateral Investment Treaties ('BIT'). Generally, corporations bid to operate water services in specified areas of a country and the successful bidder is awarded a 'concession agreement' under varying conditions.

Governments of countries, which are unable to meet the demand in water services either due to lack of technical expertise or because of poor infrastructure and management of the facilities, often seek foreign investor participation in the sector. In hindsight, there is another feature that can be linked to failures of the privatizations in the 1980s and 1990s in various regions across the world, which is that many of the concession agreements were entered into by developing nations under pressure from international financial organizations. At the time of entering into these agreements, many of the nations were struggling with weak bargaining power along with a very poor regulatory regime and rampant corruption. In this situation, by hastily adopting the regulatory templates of industrialized countries (for example, in Jamaica, the creation of a U.S.-style Public Utility Commission without the constitutional protections and well-developed rules of administrative due process prevalent in the United States led to regulatory instability that culminated in the nationalization of telecommunications in 1975), these developing nations ignored completely their macroeconomic conditions and all of these agreements were either renegotiated or terminated early. The World Bank in its Policy Research Report now acknowledges this causation.[39]

The failure of a concession agreement by way of termination meant that the investor corporation could approach the International Centre for Settlement of Investment Disputes ('ICSID') or similar dispute resolution forum applicable under the respective BITs seeking damages of expropriation under various grounds.[40]

---

[39] Ioannis N. Kessides, 'Reforming Infrastructure, Privatization, Regulation and Competition' (World Bank Policy Research Report, 2004) <http://elibrary.worldbank.org/doi/pdf/10.1596/0-8213-5070-6> accessed 20 February 2016.
[40] See Julien Chaisse, 'Assessing the Exposure of Asian States to Investment Claims,' 6 *Contemporary. Asia Arbitration Journal* 187, 201 (2013).

The most infamous of all was the case of the Aguas Del Tunari in the Republic of Bolivia[41] popularly known as the 'Cochabamba water war' case. Bolivia under pressure from the World Bank initiated the process of privatizing its water supply system in Cochabamba, in the late 1990s and following a failed tender process, entered into a concession agreement with a subsidiary of the Bechtel Corporation of the USA known as 'Aguas Del Tunari'. Immediately following the agreement, the corporation unilaterally hiked the water rates by about 35 per cent which sparked widespread violent protests in Cochabamba leading to the death of one civilian and injuries to hundreds of others. Following the protests, Aguas Del Tunari withdrew from Bolivia and sued under the BIT[42] before the ICSID Tribunal seeking damages. It is noteworthy that Bolivia in 2007 denounced the ICSID Convention on various grounds.

Although Del Tunari retracted the case, it played an important role in shaping the amendment to the ICSID Rules to allow for the participation of interested third parties as *amicus curiae* before the ICSID Tribunal where issues of human rights and public policy are affected. The amendment to the ICSID Arbitration Rule 37(2) on 10 April 2006 provides for written *amicus curiae* submissions, and new Rule 32(2) deals with the attendance of non-parties at hearings. In the case of *Biwater Gauff (Tanzania) Ltd v. Tanzania*, relating again to concession agreements for the supply of water, the ICSID Tribunal acceded to the *amicus curiae* petition and expressed the view that: 'it (the Tribunal) may benefit from the written submission by the Petitioners, and that allowing for making of such submission by these entities in these proceedings in an important element of overall discharge of the Arbitral Tribunal's mandate, and in securing wider confidence in the arbitral process itself'.[43]

In *Azurix Corporation v. Argentina*,[44] the investor state, Argentina did not provide the promised infrastructure capacity before the start of the corporation's function, which led to a algae bloom and poor quality in supply of water; the users defaulted on the payment to the corporation. The corporation then filed for an investment arbitration before the ICSID Tribunal. The tribunal though did not decide on the human rights aspect of the case on the ground that '[t]he matter has not been fully argued and the Tribunal fails to understand the incompatibility in the specifics of the

---

[41] *Aguas del Tunari, S.A. v. Republic of Bolivia*, ICSID Case No. ARB/02/3.
[42] Bolivia, Plurinational State of – Netherlands BIT (1992).
[43] *Biwater Gauff (Tanzania) Ltd v. Tanzania*, ICSID Case No. ARB/05/22, para 60.
[44] *Azurix Corporation v. Argentina*, ICSID Case No. ARB/01/12, para 261.

instant case'.⁴⁵ This case gives rise to a very important issue as to whether an international investment dispute resolution forum such as ICSID should have the jurisdiction over an issue affecting a basic human right such as water. The amendment to the ICSID Rules provides for some leeway for the human rights aspect of a dispute to be incorporated into the petition but rejected the request for oral hearings. However, since there exist specific forums for addressing human rights violations parties can now indulge in forum shopping. (The Quechuan Indian Nation declared that they would begin de novo proceedings before the Inter-American Court of Human Rights if the decision of the investment arbitral body were to be in favour of the investor.)

Acknowledging this issue, while not upholding its jurisdiction over investment arbitrations where the issue affects human rights, the ICSID Tribunal in the case of *Suez v. Argentina*⁴⁶ rejected the argument of Argentina and other *amicus* briefs and held that the human right to water does not trump a host state's investment treaty obligations.⁴⁷ The tribunal holding that the rights under the two treaties were not mutually exclusive explained that Argentina was 'subject to both international obligations, i.e. human rights and treaty obligation and must respect both of them equally'.⁴⁸

Defeating the multinationals cannot by any means be considered a success against privatization since in most of these cases, the investor state has had to pay, as expropriation, large amounts of money which could otherwise have been utilized in infrastructure and capacity building.⁴⁹ One highlight that can be taken from the string of termination of concession agreements is that the human right to water is established strongly at grassroots' level and it is noticeable that whenever an oppressive regime tries to underplay that right, the beneficiaries have repeatedly risen against it. However, it is important to also realize that this does not spell out any success in the attempt to make the water accessible to all societies across the world in fulfilment of the human right to water.

---

⁴⁵ *Ibid.*
⁴⁶ *Suez, Sociedad General de Aguas de Barcelona SA and Vivendi Universal SA v. Argentine Republic*, ICSID Case No. ARB/03/19 Decision on liability (20 July 2010).
⁴⁷ Heather L. Bray, 'ICSID and the Right to Water: An Ingredient in the Stone Soup', (2013) 29(2) *ICSID Review* 474.
⁴⁸ *Suez, Sociedad General de Aguas*, n. 46 above, 262.
⁴⁹ See generally Chaisse and Polo, n. 32 above. See also Julien Chaisse, 'Exploring the Confines of International Investment and Domestic Health Protections – Is a General Exceptions Clause a Forced Perspective?,' (2013) 39 *American Journal of Law & Medicine* 332.

Supporters for water privatization often argue that only multinational corporations with access to large amounts of capital will be able to afford the costs involved in creating an extensive network so that a water supply is accessible to the extremely poor and to those in remote locations who have been historically underserved. Recognizing the fact that corporations work with a profit motive and that associating a cost to water is crucial to conserving the resource, harmonizing the interests of the corporate sector and the human right to water is therefore significant.

Multi-stakeholder initiatives to develop standards and metrics for corporate entities impacting the right to water

Upon the recognition of the right as a basic human right by the UN in 2010, the focus turned to different institutions and sectors to follow suit and recognize and implement its provision and protection into their operational framework. Since such recognition is not binding on any institution, be it the government of a member state or a multinational corporation, acknowledgment can be characterized primarily as a voluntary measure. However, since the adoption of the Ruggie Framework, it is widely accepted that there is a responsibility on the private sector to ensure that their actions do not infringe upon the realization of human rights.

Activism was seen at a basic foundation level though shareholder initiatives in companies known to have an effect on the right (e.g. in 2010, shareholder resolutions were brought about in companies such as ExxonMobil, Intel etc. and later in Pepsi), by demanding that corporations adopt water policies articulating respect and commitment to the human right to water. While the corporate sector in some instances took up to the task of upholding this right on its own volition, for others in the sector, corporate social responsibility ('CSR') programmes, mandated by legislation, were influential in some respects for adopting policies encompassing the recognition and enforcement of the right to water. This apart, another important stimulus for companies to recognize the right was community concerns about water use and access which often led to large-scale protests and campaigns that frequently steered towards litigation. This was a major concern to multinational companies as the functioning of their operations could be stalled or even closed down if they were unable to comply with the requirements and interests of the local communities.

With such concerns imminent, the responsibility of the corporate sector to recognize and enforce the right was pursued even before the UN Resolution in 2010. Multi-stakeholder initiatives such as *2030 Water Resources Group* and the *CEO Water Mandate* were established under the auspices of the UN and the World Bank as a step towards directing the corporate sector in recognizing the right. As these multi-stakeholder initiatives predates the UN Resolution, food and beverage industry giants and the water utilities suppliers have welcomed the measure and affirmed their continued participation in implementing the right globally.

Some companies adopt formal policy goals that define their perspective on the human right to water and lay down their commitment to the cause. Other companies, however, follow a rights-based approach with no formal commitment setting forth their corporate policy towards the right. Both methods are known to have its pros and cons, and while corporates who adopt the policy approach argue that the policies raise awareness and create accountability, the ones in favour of the rights-based approach believe that it could be fully executed and a robust actionable commitment can be maintained. They also argue that adoption of a formal corporate policy could distract from meaningful actions and create unrealistic expectations from the companies.[50]

Multi-stakeholder initiatives According to the CEO Water Mandate White Paper published in 2010 by the Global Compact:

> [m]ulti-stakeholder processes create safe spaces for constructive dialogue between stakeholders with diverse views and competing interests. Where groups have previously been suspicious or hostile towards each other, they also build trust and mutual respect. In this way, they create an environment in which stakeholders are able to freely share experiences and knowledge and generate new ideas for resolving problems and reforming policies and practices. They also have the potential to improve governance and accountability systems by demonstrating that, if decision makers engage with all stakeholders when developing new policies, these are more likely to succeed and attract broad ownership.[51]

One of the earliest missions in involving multi-stakeholders in respect of the right to water was the Water Dialogues, which traced its origins to the 2001 Bonn Freshwater Conference where the German representative expressed support for a stakeholder dialogue to review issues surrounding private sector participation. The mission set to examine the issues of 'whether and how' private sector participation could contribute

---

[50] 'The CEO Water Mandate White Paper, The Human Right to Water: Emerging Corporate Practice and Stakeholder Expectations' (UN Global Compact – The Pacific Institute, November 2010) <www.pacinst.org/wp-content/uploads/sites/21/2013/02/ceo_water_mandate_human_right_to_water7.pdf> accessed 20 February 2016.

[51] Hillary Coulby, 'A Guide to Multistakeholder Work: Lessons from the Water Dialogues' (May 2009) <www.waterdialogues.org/downloads/new/Guide-to-Multistakeholder.pdf> accessed 20 February 2016.

to the affordable and sustainable water supply services in achieving the Millennium Development Goals.

As collaborative effort of the UN, The Swedish Government and several committed companies and specialized organizations in the Leaders Summit of 2007 formed the *CEO Water Mandate* as a public-private initiative to develop strategies and solutions to tackle the emerging global water crisis. In 2011, the mandate was amended to include the UN Resolution recognizing the human right to water. The mandate recognized six areas of operation to focus on an attempt to bring about a comprehensive approach to water management, recognizing the role of business in realizing the human right to water.

2030 Water resources group In 2008, another multi stakeholder initiative known as the '2030 Water Resources Group' (WRG) was formed to contribute new insights into the critical issue of water resource scarcity. This initiative included many corporate members, the World Bank led by IFC and a consortium of business partners, including The Barilla Group, The Coca-Cola Company, Nestlé, New Holland Agriculture, SABMiller, Standard Chartered and Syngenta International. Veolia Environnement joined the group for the second stage. 2030 WRG subscribes to the CEO Water Mandate as guiding principles for private sector's engagement in water resources policy.[52]

Evaluating compliance and consequences of non-compliance The UN Global Compact makes it abundantly clear that the CEO Water Mandate initiative is not 'designed, nor does it have the mandate, or resources, to monitor or measure participants' performance'[53] and is not a compliance-based initiative. However, in an attempt to ensure that the integrity of the initiative is safeguarded at all times, the mandate has developed measures on the recommendation of the Global Compact Advisory Council on issues relating to *failure to communicate progress* and *allegations of systematic or egregious abuses*.

The Coulby report[54] on multi-stakeholder work considers issues of conflict that arise between the members to a multi-stakeholder initiative. It suggests the use of mediation as a technique to resolve conflicts in such groups. In addition to the inbuilt compliance mechanisms in these

---

[52] See generally <www.2030wrg.org/about/ceo-mandate/> accessed 20 February 2016.
[53] See generally <http://ceowatermandate.org/files/Integrity_Measures_Note_EN.pdf> accessed 20 February 2016.
[54] Coulby, n. 51 above.

multi-stake-holder initiatives, there are certain other standards that are additional guidance mechanisms such as GEMI Water Tool, WBSCD Global Water Tool, Global Reporting Initiative Water Protocol, UNESCO World Water Assessment Programme, UNEP Stakeholder Engagement Manual, Accountability Stakeholder Engagement Standard, etc.

These measures seem rather inadequate to ensure compliance within a multi-stakeholder framework initiative. None of the guidelines and the frameworks themselves prescribe any specific method to handle non-compliance and obviously do not impose any punitive or other kind of damages in case of non-compliance. Ousting a member from a multi-stakeholder imitative for failure to comply would only inflict minimal reputational damage that would not serve the purpose of the initiative in the larger picture.

CONCLUSION

At present, there exists no international agreement that makes recognizing the human right to water legally binding. Taking into consideration the fact that the water is both an essential good as well as an economic commodity, there has been a need for institutions across all spheres of human activity to engage in fulfilling the aspiration of making the human right to water practicable. Transnational corporations, simply because of their global presence and economic dominance, play a major role in implementing this right. In this endeavour, as established in the previous sections, corporations have been involved in a substantial manner in respecting, recognizing and in implementing access to clean and safe drinking water for all.

These multi-stakeholder initiatives cannot however be considered as a panacea for solving all the issues that evolve out the interaction between the human right to water and corporate activities. There are several aspects of concern in relation to the use and supply of water that this system is unable to remedy. Taking into account all the apprehensions that exist in the system, and in conclusion, the following suggestions may be taken into consideration with respect to the multi-stakeholder initiatives and consequences of non-compliance with the metrics and standards developed by such initiatives:

- Inclusion of state actors in multi-stakeholder initiatives in order to bring about effective compliance standards through state mechanisms. Including the state players may help in imposing punitive damages and also in criminal prosecution in the state judiciary for viola-

tions that are heinous. This way, the compliance mechanism can be effectively outsourced to the state actors.
- Creating a national or regional level division of each of the multi-stakeholder initiatives for effective decentralization and monitoring the functioning of the programme and for the effective compliance of the companies that are a party to it.
- Grievance redressal mechanisms such as an *ad hoc* tribunal including members from all spheres of operations could be established to address internal conflicts that arise within the multi-stakeholder initiatives. Although non-binding, this will help in setting standards for future deliberations and functioning of the initiatives.

12. Sanitation rights, public law litigation and inequality: a case study from Brazil
Ana Paula de Barcellos

INTRODUCTION

Public law litigation has been used strategically throughout the world to advance human rights. However, I contend that when it comes to health rights, such litigation has been less strategic and has focused primarily on access to pharmaceuticals and medical procedures in hospitals.[1] Important as this approach may be for plaintiffs and for the right to health from an individual perspective, there is a risk that it may weaken health systems as a whole by concentrating health resources into pharmaceuticals and hospitals. More than half of the Brazilian Unified Health Care System (SUS) budget in 2010 was spent on pharmaceuticals and hospital procedures. SUS is the Brazilian national health system, a unified, public, and tax-funded health system, which is in charge of providing health care in a universal basis and free of charge. SUS spending on pharmaceuticals has increased every year since 1998; between 2003 and 2007, expenditure on medicines 'for exceptional use,'—and usually expensive—increased 252 percent.[2]

Health litigation cases have also been increasing in Brazil. In 2002, there was only one health-related purchase made as a result of a judicial

---

[1] See A Yamin and S Gloppen (eds), *Litigating Health Rights. Can Courts Bring More Justice to Health?* (Harvard University Press 2011), 43, 76, 106.

[2] C Carias and others, 'Medicamentos de dispensação excepcional: histórico e gastos do Ministério da Saúde do Brasil (Exceptional circumstance drug dispensing: history and expenditures of the Brazilian Ministry of Health)' (2011) 45 Revista de Saúde Pública 2; F Vieira, 'Gasto do Ministério da Saúde com medicamentos: tendência dos programas de 2002 a 2007 (Ministry of Health's spending on drugs: program trends from 2002 to 2007)' (2009) 43 Revista de Saúde Pública (http://www.scielo.br/pdf/rsp/2009nahead/534.pdf), 674.

decision. In 2011, there were 8,549 purchases, mostly of pharmaceuticals.³ There is an estimated 90 percent success rate for individual lawsuits that request medicines and medical treatment in Brazilian lower courts.⁴ In 2014, the Brazilian Federal Court of Accounts reported that litigation was contributing to the concentration of public health expenditure on pharmaceuticals and medical procedures over other priorities.⁵ It can also be argued that successful plaintiffs benefit and receive more from the health system than those who are unwilling or unable to go through the courts, which promotes inequality in a tax-funded and universal health system like the SUS. Furthermore, evidence suggests that plaintiffs are not from the most disadvantaged sectors in the population. I discuss this more fully in the Results and Discussion sections.

Sanitation is a challenging issue in Brazil. Brazilian law requires the government (usually municipalities) to provide universal sanitation services, with funding assistance from the federal government. Despite this, sewage collection takes place in only 55.15 percent of municipalities and sewage treatment occurs in just 28.52 percent of municipalities. In 2008, 2,495 municipalities did not have sewage collection systems and 3,977 did not have any sewage treatment system. Moreover, even in the cities with sanitation services, the services did not reach the whole population and were less available to the poorest communities.⁶ Brazilian federal expenditure on sanitation makes up a consistently low percentage of GDP.⁷ In 2009, for instance, the Brazilian Ministry of Health spent 45.93

---

³ Advocacia Geral da União/Consultoria Jurídica/Ministério da Saúde/Brasil (Federal Public Attorney's Office/Legal Department/Ministry of Health/Brazil), *Intervenção judicial na saúde pública. Panorama no âmbito da Justiça Federal e apontamentos na seara das Justiças estaduais (Court's intervention on public health. Overview from Federal Courts and notes on States Courts)* (2012) http://portalsaude.saude.gov.br/portalsaude/arquivos/Panorama.pdf accessed 2009.

⁴ F Hoffman and F Bentes, 'Accountability for Economic and Social Rights in Brazil' in Gauri and Brinks (eds), *Courting Social Justice: Judicial Enforcement of Social and Economic Rights in the Developing World* (Cambridge University Press 2008), 100.

⁵ Tribunal de Contas da União (Brazilian Federal Court of Accounts), *Relatório Sistêmico de Fiscalização da Saúde* (Health System Report), 2014.

⁶ Instituto Brasileiro de Geografia e Estatística - IBGE/Brasil (Brazilian Institute of Geography and Statistics/Brazil), *Atlas do saneamento (Sanitation atlas)* (Rio de Janeiro, RJ: IBGE 2011) and Instituto Brasileiro de Geografia e Estatística - IBGE/Brasil (Brazilian Institute of Geography and Statistics/Brazil), *Pesquisa nacional de saneamento básico (National survey for basic sanitation)* (Rio de Janeiro, RJ: IBGE 2008).

⁷ See <http://tabnet.datasus.gov.br/tabdata/livroidb/2ed/CapituloE.pdf>. For data on sanitation expenses, see 258.

238   *Charting the water regulatory future*

percent of its budget in specialized hospital care and only 2.13 percent on sanitation initiatives.[8]

International human rights law has regarded access to sanitation as part of the human right to health at least since the 1989 Convention of the Rights of the Child. In 2010, UN Resolution 64/292 recognized that access to water and sanitation is a human right essential to the realization of all human rights.[9] The Brazilian Constitution expressly provides for the right to health, and domestic law describes sanitation as a determinant of health. This shows coherence between health care and sanitation policies.[10]

The right to health therefore has an enforceable dimension, and health rights' litigation can and should seek to make underlying determinants of health available and accessible to all people in Brazil. Health rights litigation should also promote public policies that guarantee clean water and the collection and treatment of sewage.[11]

However, compared with an order to provide a plaintiff with a medicine or a clinical procedure, public health litigation can be more complex within judicial systems. The implementation of a court order to provide permanent clean water or a sanitation system in a municipality requires technical expertise, planning, and budgeting. This raises practical and philosophical questions regarding the court's capacity and role in dealing with such complexities, including whether judicial intervention in public health policy is appropriate in democracies.

Drawing on international literature, and documentation from the courts

---

[8] IPEA (2012) Políticas sociais: acompanhamento e análise. Brasília Instituto de Pesquisa Econômica Aplicada, Diretoria de Estudos e Políticas Sociais, 79, 105.

[9] For a summary of international law on water and sanitation, see <www.un.org/waterforlifedecade/pdf/human_right_to_water_and_sanitation_milestones.pdf>. See also COHRE, WaterAid, SDC and UN-HABITAT, *Sanitation: A human rights imperative* (Geneva 2008), 17. See also J Chaisse and M Polo, 'Globalization of water privatization: Ramifications of investor-state disputes in the "blue gold" economy', (2015) 38 *Boston College International and Comparative Law Review* 1.

[10] See 1988 Brazilian Constitution, arts 6, 196 and 220; and Law 8080/90.

[11] A Barcellos, *A eficácia jurídica dos princípios constitucionais. O princípio da dignidade da pessoa humana* (*The Legal Effect of the Constitutional Principles. The Principle of Human Dignity*) (3rd ed, Rio de Janeiro, RJ: Renovar Book Publisher, 2011), 351, and A Barcellos, 'O direito a prestações de saúde: Complexidades, mínimo existencial e o valor das abordagens coletiva e abstratas' ('The right health care services: complexities, minimum standards and the value of collective and abstract approaches'), in S Guerra and L Emerique (eds) *Perspectivas Constitucionais Contemporâneas* (*Contemporary Constitutional Perspective*) (Rio de Janeiro, RJ: Lumen Iuris Book Publisher 2010), 230.

in Brazil, this chapter makes two main claims. First, public law litigation can help foster public health policies by intervening in the political process in which public health priorities are set, and then monitoring policy implementation through public health services. Second, public law litigation can and should be used strategically to target initiatives that reach the most disenfranchised communities in the municipalities. This is consistent with a rights-based approach to health: It begins with the development of plans to ensure people's rights to health are fulfilled, and then strategically meets the rights of the most vulnerable and disadvantaged people first.

THE BRAZILIAN LEGAL SYSTEM AND SANITATION

Brazil, population 190,732,694 (2010), is a federal state comprising 26 states and the Federal District, which contains the country's capital. There are appellate courts of all 26 states, the District Capital Superior Court, and five federal regional courts. The databases of these 32 courts were researched for this chapter as explained below.

The Brazilian legal system allows public civil action and popular action, which are both class actions to protect the public interest and defend diffuse and collective rights. Associations, public prosecutors, and other public institutions can file public civil actions, and any citizen can file a popular action. In a popular action, the plaintiff is said to be acting on behalf of society, and the court decisions in these cases are expected to benefit society.

Public prosecutors in Brazil are public officials in charge of protecting the public interest, mostly through litigation, which makes them the main public law litigator. The Brazilian Constitution provides safeguards to guarantee their independence, so much so that it is commonplace for them to bring actions against the government, both federal and state, which also pays them. The activity of other public interest litigators (for example, NGOs and neighborhood associations) is still incipient in relation to sanitation rights.

Brazilian law has defined sanitation as a compulsory public service since 1978 (Law 6528/78).[12] In 1988, the new Brazilian Constitution mandated all levels of government to improve sanitation conditions (Article 23, IX). Municipalities are in charge of delivering the services on a universal basis,

---

[12] Rezende and Heller, *O saneamento no Brasil. Políticas e interfaces* (*Sanitation in Brazil. Public policies and Interfaces*) (Editora UFMG 2002).

although not necessarily free of charge for customers (Article 30, V); in addition, states may be involved with service delivery (Article 25, paragraph 3).[13] Many municipalities have agreements with state-owned water and sanitation companies for them to provide the services.[14] The federal government is responsible for funding the system and establishing national guidelines for sanitation (Article 21, XX) under the 2007 Federal Law 11445. This law includes clean water, sewage collection, treatment and adequate discharge, waste collection and adequate disposal, as well as water management, under the term 'sanitation.' Municipalities and states are legally required to have a sanitation plan and to execute it. Federal Decree 7217/2010 determined that municipalities and states would not receive federal funds for sanitation after 2014 if their sanitation plans had not been finalized, but Federal Decree 8211/2014 extended this to December 31, 2015.

Despite the legislative requirement for sanitation services to be provided, public funding has remained inadequate to meet the obligation. It is worth mentioning that Brazilian law also requires minimum investment in the public health system by all three levels of government (federal, state and municipal), but expenses on sanitations services cannot be included under this item of expenditure. Federal Law 11445 provides for technical support from federal government and agencies to cities and states in the preparation of their sanitation plans, so the failure to complete plans cannot be attributed to a lack of technical support. Rather, governments are not prioritizing sanitation policies and services, which in turn has led to the use of courts to seek provision of the services.

Brazilian law does not describe sanitation services directly as a right, but this has not prevented the courts from considering the human rights duty imposed upon the government as enforceable. Court decisions frequently refer to a right to sanitation services as a social and economic right, relying on an understanding of health rights (Articles 6 and 196), environmental rights (Article 225) and, in some cases, housing rights (Article 6) explicitly inscribed in the Brazilian Constitution. Brazilian courts refer to social and economic rights that are enshrined in the Constitution, usually framed as domestic obligations, not international ones.[15]

---

[13] Brazilian Supreme Court, Adin 2340, Adin 2077 and Adin 1842.
[14] S Rezende and L Heller, *O saneamento no Brasil. Políticas e interfaces* (*Sanitation in Brazil. Public policies and interfaces*) (Editora UFMG 2002).
[15] Yamin and Gloppen, n. 1 above, 3; A Klein, 'Judging as Nudging: New Governance Approaches for the Enforcement of Constitutional Social and Economic Rights' (2007/2008) 39 *Columbia Human Rights Law Review* 351.

METHODOLOGY

A review was undertaken of all the decisions made available in the online databases of the 32 Brazilian courts mentioned above from January 2003 to March 2013. The keywords in the initial search were 'sewage' (*esgoto*), 'public civil action' (*acao civil publica*) and 'popular action' (*acao popular*), producing 5,512 results. These court decisions were further refined to select decisions that had adjudicated requests for the provision of sanitation services—collection and/or treatment of sewage—made against government, public agencies, or publicly controlled companies responsible for providing them.

Cases were excluded if they requested only damages (torts) against the government, since such cases would have extremely indirect impact on the delivery of public services. Although judgments mandating damages can sometimes create incentives for change in a market environment, it was decided that, for the purposes of this research, this incentive mechanism is too dilute to measure when applied to the government. In Brazil, the government generally pays damages several years after the initial judicial decision, often under a different administration than the one responsible for the judicial action in the first place. Additionally, data are not available to compare the costs of providing sanitation services versus the damages the government was ordered to pay, making it impossible to determine whether it would have been more cost effective for the government to provide the services in the first instance.[16]

No differentiation was made between judicial decisions resulting from appeals against preliminary injunctions or final decisions. The decisions were not necessarily final (*res judicata*), because further appeals to the Superior Court of Justice and to the Brazilian Supreme Court can take several years before a decision is reached.

The Superior Court and Supreme Court websites were also searched using the same keywords for the same ten-year period. No judgment was found that overruled a court decision to grant a plaintiff collective goods

---

[16] See generally W Landes and R Posner, *The Economic Structure of Tort Law* (Harvard University Press 1987); GT Schwartz, 'Reality in the Economic Analysis of Tort Law: Does Tort Law Really Deter?' (1994/1995) 42 *University of California Law Review* 377; DJ Levinson, 'Making Government Pay: Markets, Politics, and the Allocation of Constitutional Costs' (2000) 67 *University of Chicago Law Review* 345; and ME Gilles, 'In Defense of Making Government Pay: The Deterrent Effect of Constitutional Tort Remedies' (2000/2001) 35 *Georgia Law Review* 845.

requests for sanitation services. Moreover, some decisions were found in which the Brazilian Supreme Court praised judicial interventions in public policies.[17]

There were some limitations in the databases; for example, the Court of Appeals for the state of Bahia had no decisions prior to 2012, and the Court of Appeals for the state of Mato Grosso do Sul had no decisions after July 2012. The Brazilian Federal Agency in charge of local sanitation projects (FUNASA) estimates large, complex sanitation plants can take up to 20 years to be planned, designed, and built before they start functioning. The ten-year time frame of the research may not have captured the enforcement of the decisions on schemes of this size.[18]

RESULTS

The review identified 258 cases that were relevant to this study. Courts granted the plaintiffs' requests, to some extent, in 76 percent of these cases. In four out of five cases where plaintiff's petitions were partially granted (79.69 percent), this meant setting a deadline (with an associated fine for non-compliance) for defendants to implement the sanitation service, or presenting a plan describing how it would be implemented over time. In those cases, therefore, the courts did not get involved in technical issues about how the service would be delivered. The original deadlines, however, were not met in any of the cases: courts usually accept the government's requests for postponements following much communication and negotiation on the issue.

The records show that 4 percent of the courts' decisions granting plaintiffs' requests were fully implemented. The timing of the review meant that there were cases where implementation was still under way. In some decisions it was shown that, in the context of the lawsuits, municipalities developed agreements with federal agencies to obtain resources and technical support to provide sanitation services to the population, thus

---

[17] *Ministério Público do Estado de São Paulo v Município de Sorocaba* (Brazilian Supreme Court, RE 254764/SP, August 24, 2010); and *Companhia Estadual de Águas e Esgotos – CEDAE v Ministério Público Federal* (Brazilian Supreme Court, RE 417408 AgR / RJ, March 20, 2012).

[18] See Ministério da Saúde and Fundação Nacional de Saúde, 'Termo de Referência para elaboração de planos municipais de saneamento básico. Procedimentos relativos ao convênio de cooperação técnica e financeira da Fundação Nacional de Saúde – FUNASA/MS, Brasília' [2012] 44.

impacting local public policy and implementation.[19] Examples of the impact of court decisions were identified. The implementation of a court order rendered in 2008 was reported, in 2010, to have resulted in the water filtration plan being fixed and in water being made available to everyone in Parecis, a municipality of almost 5,000 people in the northern Brazilian state of Rondonia.[20] In 2011, the Federal Regional Court for the 4th Circuit ordered the water and sanitation company of the state of Santa Catarina and other defendants to prepare and implement a plan to provide sewage collection and treatment for Barra do Sul (8,500 people), a city in the southern state of Santa Catarina. In 2014, the plan had already been presented and implementation begun.[21]

Public prosecutors filed 87 percent of the 258 cases. Of the court decisions examined, 47 percent dealt with requests for sanitation services for communities (community cases), 7 percent involved lack of sanitary infrastructure in public buildings (for instance, public hospitals and prisons), and 46 percent adjudicated claims that could be described as environmental cases, mainly dealing with pollution of water from the discharge of untreated sewage. The line between community cases and environmental cases can be indistinct; for example, cases involving public buildings can directly impact nearby communities and pollute community water sources.

All court decisions examined for this chapter related to areas in 177 municipalities. Even if these 177 municipalities belong to the 2,495 that completely lack both sewage collection and treatment systems, this would mean that litigation has reached, so far, only 7 percent of municipalities in need of sanitation systems.

Most lawsuits affected areas within cities with the same or higher Human Development Index (HDI—2000) than the regional average. There is more updated data on HDI (2010) but I decided to use the 2000 numbers as this may describe more accurately the reality of the cities when the lawsuits were filed. In the country's North Region, 70 percent of lawsuits were filed in cities with the same or higher HDIs than the regional

---

[19] Some examples are the Cities of Propriá and Moita Bonita (state of Sergipe), Barra Velha and Penha (state of Santa Catarina) and Pilar and São Miguel de Taipu (state of Paraíba).

[20] *Ministério Público do Estado de Rondonia v Companhia de Águas e Esgotos de Rondônia – CAERD* (TJRO, Appeal n. 1100324-60.2008.8.22.0018, December 9, 2010).

[21] *Ministério Público Federal v Municipio de Barra do Sul, Cia Catarinense de Águas e Saneamento – CASAN, IBAMA and others* (TRF4, Appeal n. 0002755-71.2003.404.7201/SC, June 1, 2011).

average, as were 61 percent of lawsuits in Northeast Region, 71.43 percent in Center-West Region, 69.50 percent in South Region, and 50 percent in Southeast Region.[22] The Southeast and South Regions, followed by Center-West, are the more affluent areas, while the Northeast and North Regions are the poorest ones (see Table 12.1).[23]

DISCUSSION

The Courts' Role in Improving Sanitation

The first claim of this chapter is that public law litigation can contribute to public health policies, sanitation in particular. Public policy development is complex, its implementation even more so, and neither can occur exclusively in the courtroom. However, this chapter argues that the courts can and should play an important role in public health policy and implementation.

The numbers of court cases addressing sanitation issues in Brazil is evidence that the courts are helping to improve access to sanitation services. In 2011, the Chief Justice of the Brazilian Superior Court of Justice commented on a case where a court had denied a request to suspend construction of a sewage treatment system: 'In a country where there are no sewage systems because it is an invisible service that, therefore, does not pay with votes, we cannot lose the opportunity of avoiding damage to public health and environment.'[24]

Court orders are not always carried out as directed, and therefore, it may be suggested they have limited ability to engender social change

---

[22] Programa das Nações Unidas para o Desenvolvimento – PNUD and Instituto de Pesquisa Econômica Aplicada – IPEA (United Nations Program for Development and Institute of Applied Economics), *Atlas do Desenvolvimento Humano no Brasil 2013* (*Brazilian Human Development Atlas 2013*) <www.atlasbrasil.org.br/2013/>.

[23] Data for the table: Instituto Brasileiro de Geografia e Estatística - IBGE/ Brasil (Brazilian Institute of Geography and Statistics/Brazil), *Síntese dos indicadores* (*Summary of social indicators*) (Rio de Janeiro, RJ: IBGE 2010), 35, 231; and Instituto Brasileiro de Geografia e Estatística - IBGE/Brasil (Brazilian Institute of Geography and Statistics/Brazil), *Contas regionais do Brasil* (*Regional accounts for Brazil*) (Rio de Janeiro, RJ: IBGE 2010), 17, 18.

[24] *SETEP construções SA v Companhia Catarinense de Águas e Saneamento – CASAN* (Brazilian Superior Court of Justice, AgReg na SS 2418, March 16, 2011).

Table 12.1 Regional inequality in Brazil: some indicators

	Brazil	North	Northeast	Center-West	Southeast	South
Population (2011)	195,243,000	16,499,000	54,226,000	14,576,000	82 067,000	27,875,000
Life expectancy (2009)	73.1	72.2	70.4	74.3	74.6	75.2
Percentage of national GDP (2010)	–	5.3%	13.5%	9.3%	55.4%	16.5%
GDP per capita (R$) (2010)	19,766.33	12,701.05	9,561,41	24,952.88	25,987.86	22,722.62
Child mortality (per 1,000 live births - estimates for 1996)[a]	60.7	–	96.4	41.1	36.7	35.2
Maternal mortality rate (per 100,000 live births - estimates for capitals/2002)[b]	54	60.5	67.7	49.4	47.9	41.9

Sources:

a   <ibge.gov.br/home/estatistica/populacao/condicaodevida/indicadoresminimos/tabela1.shtm>.
b   C Luizaga, S Gotlieb, M Jorge and R Laurenti, 'Maternal deaths: reviewing the adjustment factor for official data' (2010) 19/1 Epidemiologia e serviços de saúde, http://www.producao.usp.br/bitstream/handle/BDPI/13503/art_LUIZAGA_Mortes_2010. pdf?sequence=1&isAllowed=y, 11, 12.

and shape public policies.²⁵ It is true that processes leading to social change are frequently long term and complex, with multiple factors, some unpredictable, at play.²⁶ But evidence is now available showing concrete results from the examined court decisions.

Resorting to litigation does not exclude other means of bringing about social change, and can indeed promote alternative action.²⁷ A favorable court decision can launch social change processes, and media and social mobilization, for instance, are important to the implementation of judicial decisions.²⁸ Thus, courts, politics, and other social means should be used to promote access to sanitation services. Considering the need for sanitation services in Brazil and many other places, it would be unjustifiable not to employ a useful resource (litigation), along with other traditional means of social mobilization.

Brazilian courts have demonstrated a commitment to the promotion of social and economic rights.²⁹ Just as courts have been amenable to private goods' requests for social and economic rights—that is, goods that will be consumed solely by the plaintiff—this research has found that courts are also amenable to collective goods requests, as public policies, notwithstanding their complexity.

Health systems around the world, even in more developed countries,

---

²⁵ H Hershkoff, *Public Interest Litigation: Selected Issues and Examples* (2005) <http://siteresources.worldbank.org/INTLAWJUSTINST/Resources/PublicInterestLitigation%5B1%5D.pdf>, 11; D Horowitz, *The Courts and Social Policy* (Washington, DC: The Brookings Institution 1977), 22, 255; and G Rosenberg, *The Hollow Hope: Can Courts Bring About Social Change* (2nd ed, Chicago: The University of Chicago Press 2008), 157, 247.

²⁶ See note 1 for real impacts of litigation. I agree with Jacobson and Soliman's conclusion about the issue: 'We should not underestimate the ability of litigation to captivate public attention and force an issue onto the policy agenda. At the same time, we must be careful not to overestimate the ability of litigation to result in desirable policy changes': P Jacobson and S Soliman, 'Litigation as Public Health Policy. Theory or Reality' (2002) 30 *Journal of Law, Medicine and Ethics* 224.

²⁷ T Birkland, 'Agenda Setting in Public Policy' in F Fischer, G Miller, and M Sidney (eds), *Handbook of Public Policy Analysis. Theory, Politics and Methods* (Boca Raton, London, New York: CRC Press 2007), 63; and J Kingdon, *Agendas, Alternatives and Public Policies* (2nd ed, New York: Harper Collins 2010), 45, 197.

²⁸ S Cummings and D Rhode, 'Public Interest Litigation: Insights from Theory and Practice' (2009) XXXVI *Fordham Urban Law Journal* 603.

²⁹ For a critical perspective on courts' activism, see R Sieder, L Schjolden and A Angell (eds), *The Judicialization of Politics in Latin America* (New York, NY: Palgrave Macmillan 2009), 231; and R Gargarella, P Domingo and T Roux (eds), *Courts and Social Transformation in New Democracies. An Institutional Voice for the Poor?* (Aldershot/Burlington: Ashgate 2006), 13, 35, 61, 185.

have been struggling with priority-setting processes.[30] Just as decisions about public health systems are shaped in part by ethical, technical, and political criteria, so too are decisions about sanitation policies and services. The benefits of sanitation systems may take some years to be realized, whereas decisions to fund medicines can reap benefits immediately, which means sanitation projects may not be prioritized by usual electoral and political incentives.[31] And, as mentioned before, sanitation services are not evenly distributed among the population: the poor suffer more without sanitation services than do the better off.[32]

The Brazilian Government has been legally obligated to provide sanitation services for decades. It was democratically determined that sanitation services should be available to everyone. Despite this, basic sanitation services are still not provided to 68 percent of the population.[33] This research has found no evidence that a court decision results in sanitation services being provided any faster than they would have been without the judgment, and judicial intervention has limited itself to the negotiation of deadlines requiring that sanitation services are prioritized by the Executive branch. Therefore, asking the courts to support sanitation rights can be seen as another mechanism to support democratically determined policies, and should not be seen as a means of bypassing the political branches of government.

A Range of Technical and Democratic Options in Adjudication

One could argue that courts need to demonstrate a range of technical competence and democratic legitimacy to adjudicate requests that ask for

---

[30] I Williams, S Robinson and H. Dickinson, *Rationing in Health Care. The Theory and Practice of Priority Setting* (Bristol, UK: The Policy Press 2012), 85; N Daniels, *Just Health. Meeting Health Needs Fairly* (New York, NY: Cambridge University Press 2008), 313; J Butler, *The Ethics of Health Care Rationing. Principles and Practices* (New York, NY: Cassell 1999), 13; and M Malek (ed.), *Setting Priorities in Health Care* (New York, NY: John Wiley & Sons 1994), 7.

[31] For an analysis of Brazilian public expenditure on health, see Instituto de Pesquisa Econômica Aplicada – IPEA/Brasil (National Institute of Applied Economics/Brazil), *Políticas sociais: acompanhamento e análise 20 (Welfare public policies: monitoring and analysis 20)* (Brasília: IPEA 2012) 105.

[32] Instituto Brasileiro de Geografia e Estatística - IBGE/Brasil (Brazilian Institute of Geography and Statistics/Brazil), *Atlas do saneamento (Sanitation atlas)* (Rio de Janeiro, RJ: IBGE 2011), 11, 46.

[33] The number is only for sewage collection. L Heller, 'Relação entre saúde e saneamento na perspectiva do desenvolvimento' ('*Relationship between health and environmental sanitation in view of the development*') (1998) 3 Ciência e Saúde Coletiva 73.

implementation of public policies. Courts are likely to be ill equipped to decide, for example, which kind of sewage collection system is best suited to a specific area. However, courts may simply ask defendants to prepare and present a plan of action to provide sanitation services, or establish reasonable deadlines for the services to be delivered, in which case the court itself does not require specialist knowledge. Courts may also employ a newer public law litigation model, sometimes called experimentalism, which asks defendants to propose how they will comply with a broad order. The court may then need to negotiate and monitor the defendant's subsequent performance.[34] This approach has been used, for instance, in the structural reform cases of schools, prisons, and mental health facilities in the US.[35]

The results reveal that almost 80 percent of all the court decisions granting plaintiffs' requests did not provide directions on technical or service delivery issues, deferring to the Executive branch to make those decisions, as happens in other areas of public law litigation. The issue more frequently discussed in the cases is that of timing. Defendants do not deny they must provide the services, but they do not want to commit to specific deadlines. To decide when to implement a public policy is, of course, a political decision, as it requires the prioritization of resources. At the same time, though, if public officials do not implement a public policy set by law for years or decades, what does the rule of law mean? And what should courts do?

The Brazilian courts' answer to these questions is to negotiate deadlines and impose fines for non-compliance. This process was used in 79.69 percent of the cases in which the plaintiff's request was granted to some extent. Courts deferred universally to government requests for extensions, and, in fact, all the originally imposed deadlines were unmet. The conclusion from this research is that courts are aware of their technical

---

[34] E Neff, 'From Equal Protection to the Right to Health: Social and Economic Rights, Public Law Litigation, and how an Old Framework Informs a New Generation of Advocacy' (2010) 43 *Columbia Journal of Law and Social Problems* 151.

[35] A Chayes, 'The Role of the Judge in Public Law Litigation' (1975/1976) 89 *Harvard Law Review* 1281; O Fiss, 'The Supreme Court 1978 Term. Foreword: The forms of Justice' (1979) 93 *Harvard Law Review*, 18; O Fiss, 'The Social and Political Foundations of Adjudication' (1982) 6 *Law and Human Behavior* 121–8, 121, 122; C Sabel and W Simon, 'Destabilization Rights: How Public Law Litigation Succeeds (2004) 117 *Harvard Law Review* 1015; P Bergallo, 'Justice and Experimentalism: The Judiciary's Remedial Function in Public Interest Litigation in Argentina (2005) <digitalcommons.law.yale.edu/yls_sela/44/>, 1022, 1029, 1034.

limits, and they adjudicate requests for sanitation services in ways that do not go beyond their competence and legitimacy.

Public Law Litigation on Sanitation in Brazil: The Poor and the Worse-off

Lawsuits demanding pharmaceuticals and medical procedures for individuals have been criticized for promoting inequality because they concentrate resources on a small number of plaintiffs, and because plaintiffs usually do not come from the more disenfranchised groups within the population.[36] Critics argue that plaintiffs in these lawsuits get more from the health system than the rest of the population, and that the worse-off are disproportionately carrying the costs.[37]

In 2010, Brazil's federal expenditure on pharmaceuticals was US$24.29 per person. In the same year, the Brazilian Federal Government and eight Brazilian states combined spent US$2,074.86 in drugs per lawsuit in 240,980 lawsuits.[38] In 2006, the Brazilian state of São Paulo spent US$32,400,000 to comply with court orders to deliver pharmaceuticals for 3,600 plaintiffs in the city of São Paulo, producing a cost of US$9,000 per plaintiff. In the same year, São Paulo spent US$1,100 per person to fund its program for special and high-cost drugs, improving the lives of 380,000 people. There is also evidence that most plaintiffs in the state of São Paulo, for example, were represented by private lawyers, obtained a drug prescription from private doctors (not from SUS), and lived in the wealthiest parts of the city.[39]

Lawsuits asking for sanitation services (collective goods) relate very

---

[36] V Silva and F Terrazas, 'Claiming the Right to Health in Brazilian Courts: The Exclusion of the Already Excluded' (2011) 36/4 *Law & Social Inquiry* 825.

[37] A Chieffi and R Barata, 'Judicialização da política pública de assistência farmacêutica e equidade'('Judicialization of public health policy for distribution of medicines') (2009) 25/8 Cadernos de Saúde Pública, Rio de Janeiro 1839–49, 1841, 1842; O Ferraz, 'The right to health in the courts of Brazil: Worsening health inequities?' (2009) 11/2 Health and Human Rights 33; O Ferraz and F Vieira, 'Direito à saúde, recursos escassos e equidade: os riscos da interpretação judicial dominante' ('The right to health, scarce resources, and equity: inherent risks in the predominant legal interpretation') (2009) 52/1 Dados 223.

[38] Advocacia Geral da União/Consultoria Jurídica/Ministério da Saúde/Brasil (Federal Public Attorney's Office/Legal Department/Ministry of Health/Brazil), *Intervenção judicial na saúde pública. Panorama no âmbito da Justiça Federal e apontamentos na seara das Justiças estaduais (Court's intervention on public health. Overview from Federal Courts and notes on States Courts)* (2012) http://portalsaude.saude.gov.br/portalsaude/arquivos/Panorama.pdf, 18.

[39] See O Ferraz, 'Brazil: Health Inequalities, Rights, and Courts: the Social Impact of the "Judicialization of Health"' in A Yamin and S Gloppen (eds),

differently to the inequality issue, when compared with lawsuits requesting pharmaceuticals (private goods). Granting medicines free of charge to a single plaintiff will only benefit that patient and his or her family, unless the patient has an infectious disease. On the other hand, treating sewage, whether discharged in wealthier or poorer areas, has a positive impact on people's environmental and health rights, irrespective of whether they are wealthy or poor.

Despite this collective positive effect, some groups may benefit disproportionally depending on the scope of the lawsuit. Resolving a sanitation problem in one community and its environs will not directly impact sanitation deficiencies in a community geographically distant from the first one. If public law litigation is concerned with inequality, it is important to know what has been, and what should be, its focus.

The courts' data pertaining to environmental cases or sanitation infrastructure in public buildings did not enable an analysis of the socio-economic status of the communities that would benefit. However, in environmental cases, the plaintiffs were usually seeking a sewage treatment system, meaning that sewage collection was present. In community cases, both collection and treatment systems may be missing. Assuming that the lack of sanitation systems may be used as a proxy for the social condition of the neighborhood, and that the lack of both systems (collection and treatment of sewage) indicates a worse situation than the lack of just one (treatment), this suggests that the potential beneficiaries of community cases are in a worse situation than those in environmental cases. Therefore, it seems plausible to assume that community cases (47.40 percent of total) are dealing with poor communities, poorer than the ones that may eventually benefit from environmental cases.

The research identified that the courts considered sanitation cases in only 177 municipalities These municipalities include areas in state capitals, which do not completely lack sanitation systems.[40] Therefore, there are many people living in the 2,495 Brazilian municipalities who lack sewage systems completely, and who are not seeking their entitlements to sanitation systems from the courts. Also, public interest litigators are not seeking judicial intervention on their behalf. As shown in the results, the court cases are predominantly in the regions of Brazil that have a higher than average HDI. People without sewage systems and who are not accessing courts would appear to be among Brazil's most disenfranchised.

Litigating Health Rights. Can Courts Bring More Justice to Health? (Harvard University Press 2011), 101, 102.

[40] IBGE/Brasil, n. 6 above, 21, 35, 36, 37.

The study findings, therefore, indicate that public law litigation, although aspiring to reach the poor, has addressed so far only a very small portion of the country's sanitation needs. The reasons for this inequity in access to the judicial system require further investigation.

These results reflect the inverse equity hypothesis, which theorizes that a non-targeted initiative in public policy tends to initially increase inequality, only helping to reduce it after some time.[41] Although public law litigation cannot be described entirely as an initiative in public policy, the data supports the first part of the hypothesis: lawsuits are concentrated in the wealthier cities in each region, although it is not known whether they may be benefiting poorer areas of those municipalities. The poorest cities have not yet benefited from litigation addressing the need for sanitation. The second part of the phenomenon is yet to be observed; the oldest lawsuits were filed in the 1990s and, up to 2013, no change in this trend was observed.

Some of the factors that may explain these results could include the following. Public law litigators (who filed most cases) probably live in the wealthier cities and are more aware of local problems. People from remote and poorer cities, after failing to resolve the problem with local authorities, might not have the resources, financial and otherwise, to further explore legal options. In such instances, the costs involved in going to another city to search for a sympathetic ear for their complaints is a hurdle that should not be underestimated. People in poorer cities are less likely to be educated and less aware of their rights than people in more developed and wealthier cities. This leads to a scenario where even in the face of an opportunity to complain, people in poorer cities are less likely to do so.

CONCLUSIONS

This chapter has found that public law litigation has promoted sanitation public policies and services in Brazil. Courts have been favorable to 76 percent of claims on public health policies, granting sanitation services. There is also evidence that court decisions can help make sanitation a political priority. However, litigation has addressed only a small part of the nation's need.

---

[41] Victora et al, 'Explaining Trends in Inequities: Evidence from Brazilian Child Health Studies' (2000) 356/9235 Lancet 1093; and M Castro, *Spatio-temporal Trends of Infant Mortality in Brazil* (2010) <www.drclas.harvard.edu/sites/default/files/Lecture9_Castro-2013-IMR.pdf>.

This research has identified the potential of courts to improve public health conditions, not only in Brazil, but in any country where the law demands that public policies are developed to address public health issues. Courts in Brazil have been willing to improve access to social and economic rights; however, the scale must be increased. Hence, it is suggested that public interest litigators frame access to public health services as a human right and focus on securing collective goods claims. Broader public health policy outreach has the potential to improve the general conditions of health, particularly for poorer people, and is likely to eventually counterbalance any apparent short-term increase in inequality.

Public law litigation should also plan targeted initiatives to reach out to the most disenfranchised, so that at least the basic dimensions of human rights are fulfilled. As found in this research, the worse-off communities remain the least represented in the court cases. Thus, it is important that public interest litigators map needs across the country, so that specific initiatives to help those communities can be planned.

Brazilian courts have not ordered specific behavior of defendants, but similar to structural reform cases, they have asked defendants to provide deadlines and plans, which the courts then monitor. The monitoring is pivotal so that a positive court order promotes change and increases access to sanitation services for poor communities. This is particularly complex when the instigators of the litigation are third parties with legal expertise (NGOs, law school clinics, and public prosecutors), not the interested community itself. In this context, there is the risk that litigators will perceive the court decision—the legal outcome—as the ultimate goal of their work, while the next step—executing the decision—receives less attention.

This research found some evidence, although it was not conclusive, that this phenomenon was happening. In some cases, plaintiffs abandoned a lawsuit after a positive decision was rendered, and sought implementation only after a court inquiry. Furthermore, initial decisions were not usually put into practice if there were appeals pending, even if the appeal did not prevent implementation. Special and extraordinary appeals before the Superior Court of Justice or the Supreme Court were filed against approximately 50 percent of court decisions that granted the plaintiffs' requests. As the appeal process is known to take many years, it resulted in execution of the judicial decree being seriously delayed in more than 50 percent of cases.

Public law litigators need to remain focused on having court decisions implemented throughout this long time frame. They also need to coordinate with other social players, for example, social movements, political parties, media, and public officials, to draw attention to sanitation needs and make sanitation a public health priority. A long process takes place after a court order is rendered. A political environment sensitive to sanitation rights will make the implementation of court orders easier.

ACKNOWLEDGMENTS

This project was conducted with the support of the Takemi Program in International Health at Harvard School of Public Health. I want to thank Professors Michael Reiss, Norman Daniels, Donald Halstead, and Marcia Castro for their useful comments on previous drafts. I want also to thank Professors from Escola de Direito – Fundacao Getulio Vargas/Rio for their comments.

PART III

Economic drivers shaping the future of water

13. Demand for infrastructure investment for water services: key features and assessment methods

Sacchidananda Mukherjee and Debashis Chakraborty

1. INTRODUCTION

With the rise in the global population, the demand for safe drinking water, the lifeline of human civilization, is simultaneously growing. Projecting future demand for water has thereby become particularly important given the growing pressure on this natural resource.[1] Amarasinghe and Smakhtin have noted that the present average per capita domestic water withdrawals have already crossed the projected 'business as usual' scenario developed for 2025, which, even discounting for a possible underestimation, is a matter of serious concern.[2]

The urgency of fulfilling this ever-expanding 'thirst' have been duly acknowledged in Target 7.C of the Millennium Development Goals (MDGs), which intends to halve, by 2015, the proportion of the population without sustainable access to safe drinking water and basic sanitation. Reaching the MDG target on time is crucial in global development cooperation architecture owing to the fact that the implementation experience will provide important lessons to the countries to achieve universal access to improved water supply and sanitation (WSS) facilities by 2030, as per Goal 6 of the Sustainable Development Goals (SDGs), which aspires to, ensure availability and sustainable management of water and sanitation for all.

---

[1] See generally J Chaisse and M Polo, 'Globalization of water privatization: Ramifications of investor-state disputes in the "blue gold" economy', (2015) 38 *Boston College International and Comparative Law Review* 1.

[2] Upali A. Amarasinghe and Vladimir Smakhtin, 'Global water demand projections: past, present and future', Research Report 156, Colombo: International Water Management Institute.

However, the chances of reaching the target by the stipulated deadline looks rather thin as the recent assessment on achievement of the MDGs reveals that in 2012, while 748 million people had to rely on unsafe drinking water sources, 173 million people were sourcing the same directly from rivers, streams or ponds. In other words, a significant proportion of the population had to depend on unprotected or poorly protected water sources for meeting their requirement. It was further noted that even the population depending on improved drinking water sources were often exposed to various types of microbiological contaminations.[3] Mukherjee et al. observed that due to rapid population growth, urbanization and industrialization, water supplies from local water resources are falling far short of the high and concentrated demands in most urban areas.[4] In all, several countries are poised to fail in meeting the drinking water target, as mandated by the concerned MDG, owing to insufficient finance.

The aforesaid statistics nevertheless in no way belittle the achievements so far in enhancing access to improved drinking water on two distinct fronts. First, the world population with access to improved drinking water source has increased from 76 to 89 per cent over 1990–2012. Among the additional 2.3 billion people who gained access to improved sources of drinking water over this period, only a fraction of them have received piped drinking water supply. The sheer number of additional households receiving a secure drinking water supply is commendable. Second, the proportion of people without access to an improved water source had been reduced by 50 per cent in 2010, much earlier than the original target year of 2015.

Regrettably, such improvement has not been evenly spread across regions, and similar disparity in access to improved WSS services is also witnessed between rural and urban centres within many countries. Achievements in economically weaker regions like Oceania, Caucasus and Central Asia, South Asia and Sub-Saharan Africa have been poorer on this front and they are unlikely to fulfil their respective MDG 7C targets by 2015. It deserves mention that the residents in all these regions depend heavily on surface water or other contaminated sources, which expose them to the risks of various ailments.

In short if the current scenario continues, the low- and middle-income

---

[3] Jennyfer Wolf and others, 'Assessing the impact of drinking water and sanitation on diarrhoeal disease in low- and middle-income settings: systematic review and meta-regression' (2014) 19 *Tropical Medicine and International Health* 928.

[4] Sacchidananda Mukherjee and others, 'Sustaining urban water supplies in India: increasing role of large reservoirs', (2010) 24 *Water Resources Management* 2035.

countries in several parts of the world may end up within an unending spiral, where the initial lack of access to safe drinking water leads to various ailments, lower productivity, lower income levels and in turn a weaker financial ability to ensure safe water supply subsequently. Moreover, the livelihood challenges associated with poverty and lack of adequate facilities may lead to practices that add to further contamination of the environment in general and water resources in particular. In 2012, around 2.5 billion people still lacked improved sanitation facilities, while around 1 billion people had to defecate openly. Ensuring improved sanitation facilities is therefore an important step towards securing safe drinking water.

The evident truth that MDG targets will not be reached by 2015 underlines the fact that much greater efforts and investment flows need to be urgently channelized in this front.[5] The present analysis, which attempts to estimate the demand for investment in water services infrastructure to achieve the universal access to improved WSS by 2019, is arranged in the following manner. The second section briefly reviews the global/regional literature on methodologies to estimate the demand for investment in water services infrastructure. Section 3 presents the basic data on access to WSS across regions, to understand the regional dimension of the availability challenges. The present stock and projected demand of investment in water supply and sanitation infrastructure, based on a suitable methodology, is empirically estimated and discussed in Section 4. It also provides an analysis on possible sources for financing the infrastructure augmentation requirements. Finally, based on the findings, the policy conclusions are drawn in Section 5.

2. LITERATURE REVIEW

2.1 Estimation of Demand for Investment in Water Services Infrastructure

The mainstream literature on estimation of investment demand for infrastructure often uses parametric estimation and projections based on regression models, as developed by Fay and Yepes.[6] For projecting the

---

[5] See generally Chaisse and Polo, n. 1 above.

[6] Marianne Fay and Tito Yepes, 'Investing in infrastructure: what is needed from 2000 to 2010?' *Policy Research Working Paper 3102* (Washington DC: World Bank 2003) 3.

demand for investment in water services infrastructure, World Health Organization (WHO) however follows a different methodology. The present section first reviews the studies based on the methodology developed by Fay and Yepes, before focusing on the studies in line with WHO methodology.[7]

The analysis of Fay and Yepes established econometric relationship between access to WSS (present stock of investment in water services infrastructure) with per capita income, structural composition of the economy (share of agriculture and manufacturing in GDP), population density and level of urbanization.[8] Based on the estimated relationship, the study projected the demand for access to WSS on the projected values of the independent variables for 2010. While Fay and Yepes estimated annual investment requirements during 2005–10 across regions (Table 13.1 below), their underlying methodology did not take the achievement target for universal access to water and sanitation into consideration.[9] On the contrary, the methodology implicitly assumed that the status quo in access to water and sanitation would continue with future demand for investments come from set of independent variables, given the stock of present investment in water services infrastructure. Since the methodology does not set achievement targets for access to WSS, the projected figures of the same is likely to cross the maximum achievable target of 100 per cent. The problem cannot be solved even through application of sophisticated econometric estimation techniques, e.g., censored panel regression model. Therefore, there is sufficient reason to believe that the estimates based on this methodology is prone to over-estimate the demand for investment. However, the methodology is useful to estimate demand for investment in other infrastructure sectors (e.g., road transport) and also for water services other than for domestic purposes.

The empirical estimates by Fay and Yepes presented in Table 13.1 reveals that an annual investment of US$ 22.112 million and US$ 31.148 million is required in water supply and sanitation respectively during 2005–10.[10] In other words, in percentage terms, 0.13 per cent of world GDP (0.04 per cent in new and 0.09 per cent in maintenance) is required annually for investment in these two areas. It deserves mention that this

---

[7] Ibid; Guy Hutton and Jamie Bartram, 'Regional and Global Costs of Attaining the Water Supply and Sanitation Target (Target 10) of the Millennium Development Goals' *WHO/HSE/AMR/08/01* (Geneva: World Health Organization 2008) 2.
[8] Fay and Yepes, ibid., 5.
[9] Ibid., 8.
[10] Ibid., 19.

Table 13.1 Expected annual investment needs (2005–10) on water and sanitation, $ million and percentage of GDP (all prices in constant 1995 US$)

Countries	Water		Sanitation	
	New	Maintenance	New	Maintenance
East Asia & Pacific	1,799 (0.07)	3,602 (0.13)	2,608 (0.1)	4,202 (0.15)
South Asia	1,912 (0.21)	3,286 (0.36)	1,707 (0.19)	2,417 (0.26)
Europe & Central Asia	235 (0.02)	1,436 (0.1)	750 (0.05)	2,616 (0.18)
Middle East and North Africa	399 (0.06)	629 (0.1)	691 (0.11)	1,030 (0.16)
Sub-Saharan Africa	689 (0.15)	949 (0.2)	1,256 (0.27)	1,619 (0.35)
Latin America & Caribbean	645 (0.03)	1,245 (0.05)	1,147 (0.05)	1,989 (0.08)
High Income	565 (0)	4,719 (0.01)	982 (0)	8,133 (0.02)
Low Income	2,974 (0.19)	5,036 (0.32)	3,706 (0.24)	5,462 (0.35)
Middle Income	2,707 (0.04)	6,111 (0.09)	4,454 (0.06)	8,410 (0.12)
Developing Regions	5,681 (0.07)	11,147 (0.13)	8,160 (0.1)	13,872 (0.16)
WORLD	6,246 (0.02)	15,866 (0.04)	9,143 (0.02)	22,005 (0.05)

Note: Figures in the parenthesis show the percentage of GDP.

Source: Fay and Yepes (2003: 11).

analysis considers only 3 per cent of the replacement cost of the capital stock as maintenance requirement for water services, which is actually the minimum annual average expenditure, below which the network's functionality will be threatened.[11] It is observed that this replacement cost figure is quite low in comparison with the corresponding number considered in other similar studies.[12]

A series of studies have attempted to estimate the demand for investment in infrastructure services in various regions of the world in line with the

---

[11] Ibid., 10.
[12] World Health Organization, 'Global costs and benefits of drinking-water supply and sanitation interventions to reach the MDG target and universal coverage' *WHO/HSE/WSH/12.01* (Geneva: WHO 2012) 54; Hutton and Bartram, n. 7 above, 14.

Table 13.2 *Sub-region wise investment needs in water services infrastructure in Asia, 2010–20 (2008 US$ billions)*

Region	Water	Sanitation	Water and Sanitation	
			2008 US$ Billion	% of GDP
East and Southeast Asia	58.37	112.88	171.25	0.17
South Asia	46.12	38.97	85.09	0.39
Central Asia	8.6	14.8	23.4	0.42
The Pacific	0.14	0.36	0.51	0.30
Total	113.22	167.02	280.24	0.22

Source: Bhattacharyay (2010: 13).

methodology developed by Fay and Yepes in subsequent period. The current analysis presents here the results of the studies specific to demand for water services infrastructure.[13] It particularly focuses on those studies, where the year of analysis goes beyond 2012, so as to view such estimates in light of the concerns raised by UN-Water and WHO (2014) study. For instance, Bhattacharyay have estimated the demand for water services infrastructure investment in Asia and the Pacific during 2010–20, which stands at US$ 280.24 billion (Table 13.2).[14] Of this estimated demand for investment, US$ 113.22 billion and US$ 167.02 billion are originating from water supply and sanitation respectively, which on average represents 0.22 per cent of GDP of the region. Bhattacharya et al. have estimated that the annual investment requirements for water supply in the developing world would range over 15–30 per cent of total annual investment required in infrastructure of US$1.8–2.3 trillion (in 2008 constant prices).[15]

While there exist a sizable literature on estimates of demand for investment in water supply infrastructure, none of them are free from criticisms. The current analysis presents some of the critical issues involved in

---

[13] Fay and Yepes, n. 6 above, 11.

[14] B N Bhattacharyay, 'Estimating demand for infrastructure in energy, transport, telecommunications, water and sanitation in Asia and the Pacific: 2010–2020' *ADBI Working Paper Series No. 248* (Tokyo: Asian Development Bank Institute 2010) 13.

[15] Bhattacharya and others, 'Infrastructure for development: meeting the challenge' *Policy Paper* (Centre for Climate Change Economics and Policy and Grantham Research Institute on Climate Change and the Environment in collaboration with Intergovernmental Group of Twenty Four 2012) 13.

estimation of global demand for investment in water services infrastructure in the following:

a) *Per Unit Cost of Water Supply and Sanitation*: There is no standard estimate on the cost of supplying safe drinking water supply and sanitation for each household, which ideally represents the actual cost of providing these services across continents and income groups. Moreover, as is evident from the published papers,[16] the available estimates do not specify the technology and make no distinction between the possible costs prevailing in rural and urban belts in their analysis. Often the information on access to WSS is provided in terms of percentage of population, whereas in reality, the cost estimates are generated for per household. It is widely known that the household size varies not only across countries but also across location of residence (rural and urban) within countries. Therefore, any assumption on household size may not always reflect actual reality and the estimated demand for investment could be either under-estimated or over-estimated, depending on the assumption on average household size. Finally, reliable postulations on per unit cost of WSS services play a crucial role in estimation of the present stock of investment.

b) *Supply versus Demand for Water*: Ironically, most of the estimates on demand for investment in water services do not consider availability of water in their models. This separation of demand and supply forces in the analysis could pose a serious limitation to the cost estimates, as the incremental costs of water supply (both resource and environmental costs) are not taken into account in such models.[17] For example, improvement in access to modified sanitation services is not only dependent on investment in sanitation infrastructure but also on access to safe and sufficient water supplies. Therefore, access to safe sanitation cannot logically be de-linked from access to water supplies. Hence, the demand estimates for investment in sanitation services should take into account not only the cost of augmenting water supply but also cost of water and sewage treatment as well (including sewerage disposal).

c) *Projection of Variables*: There are several development-related factors, on which the demand for investment in water services in a country is heavily dependent (e.g., per capita income, structural composition

---

[16] Fay and Yepes, n. 6 above, 10.
[17] Dinesh M Kumar, 'Growing environmental costs of urban water supplies: Overcoming the challenges in implementing integrated urban water management', in Mukherjee and Chakraborty (eds), *Environmental Scenario in India: Successes and Predicaments* (Abingdon: Routledge 2012) 151.

of the economy, state of technology adoption, population density, urbanization, level of human development). However, obtaining a realistic long time-series projection of all these variables, so as to correctly estimate the demand for a specific future date, is an area of major concern. While short- and medium-term projections are easier to arrive at, adequate care needs to be taken for generating long-term projections.

d) *Operation and Maintenance Costs*: In addition to the capital investment requirement in new projects, the costs of maintaining existing projects (e.g., replacement and repairing expenses), while running expenditure for the new and old projects is equally important. In addition, the cost of maintenance depends on several factors, namely – vintages of stock of investment/projects, state of technologies, source of water supply, regulatory standard for drinking water, etc. Depending on the relative importance of these factors, cost of operation and maintenance of WSS projects varies widely across countries. It deserves mention that unlike sanitation services, where most of the operation and maintenance (O&M) cost is borne by the beneficiaries, responsibility of operating and maintaining water supply projects lies with the community and / or water supply authorities.

Offering an alternative methodology, WHO (2012: 7) estimates the total financial capital cost to expand coverage to achieve universal access of improved drinking-water sources and sanitation during 2010–15 in constant 2010 prices of US$ billion (Table 13.3 below). The estimate reveals that to achieve universal access in water supply across all the regions, an investment of US$ 203.3 billion is required during 2010–15. To achieve universal access in sanitation services, an investment of US$ 332.4 billion is required over the same period. In contrast to the Fay and Yepes methodology, the WHO study has been based on population projection of 2015 and unit cost of providing access to safe drinking water and sanitation.[18] One distinct edge of this analysis over the Fay and Yepes study is that the region-wise investment requirement has been presented here separately for rural and urban areas.[19] However, this methodology does not consider other demand for water services (other than domestic demand).

Following the WHO methodology of unit cost, Hutton and Bartram estimates sub-region wise cost of providing water and sanitation coverage

---

[18] Fay and Yepes, n. 6 above, 17; WHO 2012, n. 12 above, 21.
[19] Fay and Yepes, ibid., 19.

Table 13.3 Total financial capital costs to expand coverage to achieve universal access of improved drinking-water sources and sanitation, from 2010–15 (2010 US$ billions)

Region	Water supply			Sanitation		
	Urban	Rural	Total	Urban	Rural	Total
Caucasus and Central Asia (CCA)	2.0	1.8	3.8	2.7	0.8	3.6
North Africa	8.8	3.1	11.9	5.0	1.3	6.4
Sub-Saharan Africa (SSA)	13.6	16.0	29.6	47.0	48.2	95.2
Latin America and the Caribbean (LAC)	24.7	4.4	29.1	29.1	10.2	39.3
East Asia	48.9	21.3	70.2	50.8	16.6	67.4
South Asia	4.2	3.6	7.8	43.7	45.5	89.2
Southeast Asia	22.8	6.7	29.5	8.3	7.6	15.9
West Asia	15.7	4.6	20.4	11.0	3.8	14.8
Oceania	0.2	0.7	0.9	0.2	0.5	0.7
All	141.0	62.3	203.3	197.9	134.5	332.4

Source: WHO (2012: 7).

to meet MDG target (Table 13.4 below).[20] It is observed that an estimated amount of US$ 363.5 billion is required to achieve the water supply coverage goal of MDGs, while an investment of US$ 357.5 billion is required for fulfilling the access to sanitation related objective. For providing water supply services, 16 per cent of the total expenses would be in terms of capital cost while the remaining expenses are in terms of recurrent cost. The corresponding figures for sanitation services are 43 per cent and 57 per cent respectively. The investment estimates strongly underline the higher capital cost requirement for expanding the coverage of sanitation services, vis-à-vis the water services. The study further noted that a major proportion of the estimated investment is required on provision of urban water supply (68 per cent) and urban sanitation facilities (59 per cent). The majority of these investments will be spent on maintaining the existing coverage, with the corresponding figures for water supply and sanitation projected as 88 per cent and 60 per cent respectively.

---

[20] WHO 2012, n. 12 above, 41; Hutton and Bartram, n. 7 above, 9.

Table 13.4 Total spending on new and existing water and sanitation coverage to meet MDG target 10,[a] excluding programme costs (US$ millions)

WHO Sub-Region	Water Coverage	Cost item (%)		Sanitation Coverage	Cost item (%)	
		Capital	Recurrent		Capital	Recurrent
AFR-D	17,296	37	63	27,272	58	42
AFR-E	19,852	32	68	29,700	63	37
AMR-B	64,042	6	94	44,303	19	81
AMR-D	10,034	11	89	7,575	32	68
EMR-B	10,960	6	94	3,300	10	90
EMR-D	28,087	13	87	24,124	34	66
EUR-B	13,640	9	91	11,242	18	82
EUR-C	18,685	6	94	15,622	19	81
SEAR-B	8,397	32	68	16,550	31	69
SEAR-D	37,013	14	86	76,141	56	44
WPR-B	135,440	18	82	101,656	44	56
All	363,447	16	84	357,485	43	57

Notes:
[a] Total spending includes operation and maintenance of existing supply; periodic replacement of existing infrastructure; and the costs of increasing coverage to existing and increased future populations so as to meet the MDG target. Totals may not add up as a result of rounding.
WHO Sub-Regions: AFR: WHO African Region; AMR: WHO Region of the Americas; EMR: WHO Eastern Mediterranean Region; EUR: WHO European Region; SEAR: WHO South-East Asia Region; WPR: WHO Western Pacific Region.
B = low adult, low child mortality; C = high adult, low child mortality; D = high adult, high child mortality; E = very high adult, high child mortality.

Source: Hutton and Bartram (2008: 12).

Several other studies in the existing literature have attempted to estimate demand for investment in water services infrastructure. Lloyd-Owen estimates region-wise cost of capital and operating costs water and wastewater for 67 countries (Table 13.5).[21] As per the estimates, the capital spending requirement (medium scenario) would reach US$ 2.9 trillion, while the operating cost would be US$ 6.8 trillion during the period 2010–29 (OECD, 2011: 43).[22]

---

[21] D Lloyd-Owen, *Tapping Liquidity: Financing Water and Wastewater to 2029, a Report for PFI Market Intelligence* (London: Thomson Reuters 2009).
[22] OECD, 'Meeting the Challenge of Financing Water and Sanitation: Tools and Approaches' *OECD Studies on Water* (Paris: OECD 2011).

Table 13.5 Forecast operating and capital spending in countries covered, 2010–29 (US$ billions)

Region	Operating costs	Capital spending (capex)			% capex by region
		Low	Medium	High	
North America	1,821	525	630	940	23%
Europe	2,133	612	838	991	28%
Developed Asia	1,018	461	550	640	19%
Latin America	796	119	164	194	5%
Rest of World	992	472	713	1,027	24%
Overall	6,760	2,213	2,880	3,792	100%

Source: Lloyd-Owen, D. (2009) as cited in OECD (2011: 44).

According to the GLAAS report, the global cost estimates for meeting the drinking water and sanitation-related MDG targets range from US$ 6.7 billion to US$ 75 billion per year, i.e., an expenditure of US$ 33.5 billion to US$ 375 billion needs to be borne by 2015.[23]

As evident from the discussions so far, due to the challenges associated with availability of statistical data and projections on key variables at regional and country levels and access to such information, variations in estimates on investment demand tend to vary widely. Given the constraints in estimating global demand for investment in water services infrastructure, there are two alternative methods – bottom-up approach and top-down approach. It deserves mention that the purpose of global approach based studies (top-down approach) is different from those based on bottom-up approach. On one hand, the main objective of global approach is to highlight the projected demand for investment at a macro level to initiate dialogue for mobilizing the necessary finances for investment. Conversely, the importance of project level estimation (bottom-up approach) of demand for investment is concerned more with facilitating financial planning for execution of investment in efficient manner.

In addition to the global/regional studies, the literature includes several country/province level estimation of demand for investment in water services infrastructure.[24] However, it is observed that compiling all

---

[23] UN-Water and WHO, *Global Annual Assessment of Sanitation and Drinking Water (GLAAS)* (Geneva: United Nations and World Health Organization 2010) 20.

[24] Isabel Chatterton and Olga Susana Puerto, 'Estimation of infrastructure investment needs in South Asia region' *Working Paper No. 62608* (World Bank, Washington DC 2011) 4; Tito Yepes, *Expenditure on Infrastructure in East Asia Region, 2006–2010*

such studies to get the global estimate is cumbersome for two reasons. First, arriving at a policy conclusion from them is difficult, as the time period of these studies as well as the respective development scenario vary widely. Second, all such studies are scattered and their accessibility is often limited. For example, even multilateral development banks (MDB) like the World Bank and Asian Development Bank (ADB) do not provide detailed information for WSS projects in the public domain. Provision of access to basic information (e.g., unit costs by technology, locations), at least at the regional level, would facilitate future research on this field.

2.2 Benefits and Co-benefits of Improved Access to Water and Sanitation

Given the importance of ensuring safe drinking water and sanitation services in augmenting human health and well-being, a branch of the literature focuses on the potential benefits of such policies. WHO-UNICEF notes that, 'much of the health benefit of water supply and sanitation comes from the changes in hygiene they promote'.[25] WHO estimates the costs and benefits of ensuring universal access to WSS for meeting the relevant MDG target by 2015, both at regional and country levels.[26] The estimated Benefit-Cost ratio is expected to serve two purposes. First, the improved provision of WSS services lowers the risk of exposure to various diseases and hence the risk of premature mortality. This in turn curbs losses in productive time, the estimation of which would motivate governments, particularly in the low- and middle-income countries to invest more vigorously on this front. Second, the estimated potential benefits, if sufficiently lucrative, might contribute significantly in engaging external donors in the process, which can crucially supplement the efforts of the domestic governments.[27] Cronin et al. have noted that access to safe drinking water and sanitation are basic building blocks to improve health and nutrition status of people.[28]

(2004) <http://citeseerx.ist.psu.edu/viewdoc/download?doi=10 .1.1.452.9623&rep=rep1&type=pdf> accessed 18 March 2015, 10.

[25] WHO-UNICEF 'Global Water Supply and Sanitation Assessment 2000 Report' *WHO and UNICEF Joint Monitoring Programme for Water Supply and Sanitation* (United States of America: WHO-UNICEF 2000) 34.

[26] WHO 2012, n. 12 above, 57.

[27] Ibid., 3.

[28] Cronin and others, 'Water, sanitation and hygiene: moving the policy agenda forward in the post-2015 Asia' (2015) 2 *Asia & the Pacific Policy Studies* 229.

Demand for infrastructure investment for water services 269

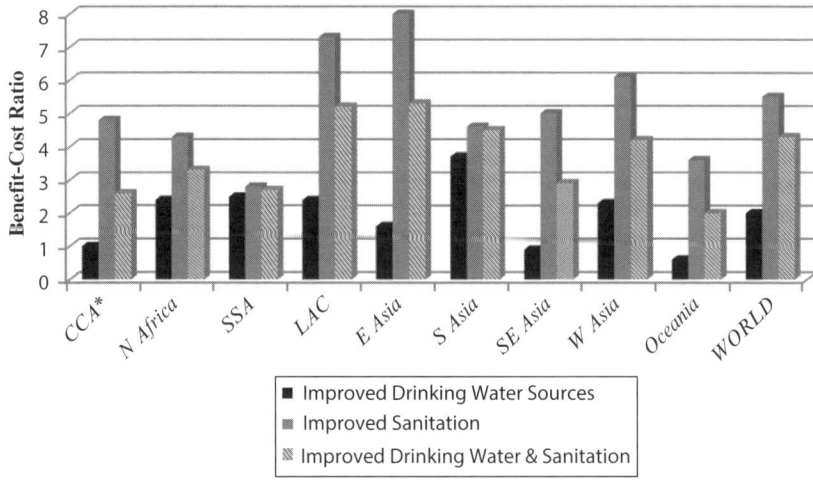

Source: Compiled from WHO (2012).

Figure 13.1 Region-wise benefit-cost ratios of interventions to attain universal access

WHO estimates region-wise benefits-cost ratios of universal access to WSS interventions, and a compilation of those estimates is presented in Figure 13.1.[29] The ratios are presented for water and sanitation services separately as well as for the collective WSS services, given the close association between the two. It is evident from the Figure that the maximum value of the ratio could be achieved for South Asia for water supply and Latin America and East Asia for sanitation services. The results are not surprising in light of the region-level accessibility challenges expressed by UN-Water and WHO, as noted earlier. In case of the WSS services, for the world on average, the expected benefits will be four times higher than the associated costs.[30]

---

[29] WHO 2012, n. 12 above, 57.
[30] UN-Water and WHO, *UN-Water Global Analysis and Assessment of Sanitation and Drinking-Water (GLAAS) 2014 Report: Investing in Water and Sanitation: Increasing Access, Reducing Inequalities* (Geneva: United Nations and World Health Organization 2014) 66.

3 ESTIMATION OF DEMAND FOR INVESTMENT IN WATER SERVICES INFRASTRUCTURE

3.1 Present State of Water and Sanitation Services

Before analysing the investment requirement for WSS services, the evolving pattern of access to improved water source and sanitation facilities over the last two decades is discussed in this sub-section with the help of World Development Indicators data. Table 13.6 demonstrates that access to improved water source is still lacking in Sub-Saharan Africa (SSA) and South Asia (SA). People's access to improved water sources has increased by more than 4 per cent in every five years in SSA region since 1990 and for South Asia the corresponding figure was more than 5 per cent. As compared to the scenario on water supply, access to improved sanitation facilities is still lacking substantially for SSA, SA, East Asia and Pacific (EAP) and Latin America and Caribbean (LAC). In case of access to sanitation services, substantial improvement for the SA region is noticed (i.e., more than 5 per cent in every five years since 2001). On the other hand, the improvement in access to sanitation for SSA has been considerably slower, particularly in the post-2005 period. In comparison with these regions, the performance of countries located in Europe and Central Asia (ECA), Middle East and North Africa (MENA) and North America (NA) have been far superior.

For analysing the scenario at a more disaggregated level, disparity in access to improved water sources between rural and urban areas over the last two decades across the regions has been compared in Table 13.7. Though the disparity was very high for EAP, LAC and SA regions during 1990s, the same has gone down substantially over the years. However for the SSA region, despite improvements in the rural belt, a similar decline in rural and urban disparity in access to improved water sources has not been observed yet.

The cross-region scenario for rural-urban disparity in improved sanitation services has been presented in Table 13.8. Unlike access to water supply, the disparity in access to improved sanitation is not declining for EAP and SSA regions. On the other hand, the pace of the corresponding decline in other low- and middle-income countries located in LAC, MENA and SA regions is found to be quite slow.

From the analysis so far, it is evident that majority of future investment for improving access to water supply is required in SSA, SA and EAP regions. Furthermore, to weed out the rural-urban disparity in access to improved water sources, a larger proportion of the new investment would be required in rural areas. On the other hand, a major proportion

Table 13.6 Region-wise average access to improved water source and sanitation (rural and urban combined)

Regions	Improved water source (% of population with access)					Improved sanitation facilities (% of population with access)				
	1990–95	1996–2000	2001–05	2006–10	2011–12	1990–95	1996–2000	2001–05	2006–10	2011–12
East Asia & Pacific (EAP)	80.8	82.0	85.0	87.9	89.5	61.7	63.1	66.4	70.3	72.4
Europe & Central Asia (ECA)	95.5	95.5	96.1	96.6	97.0	95.4	94.4	94.9	95.9	96.9
Latin America & Caribbean (LAC)	85.8	88.4	90.3	92.0	93.0	70.7	75.1	77.9	79.6	80.9
Middle East & North Africa (MENA)	88.4	89.1	91.0	91.8	92.4	83.7	86.3	88.5	90.2	91.1
North America (NA)	99.2	99.3	99.4	99.4	99.5	99.7	99.8	99.3	99.9	99.9
South Asia (SA)	68.2	73.2	78.9	84.6	88.4	36.7	41.2	46.3	51.8	55.2
Sub-Saharan Africa (SSA)	56.4	60.7	64.7	68.9	71.7	26.9	29.2	31.2	32.6	34.1

Source: Computed by authors based on World Bank (undated a).

Table 13.7 Region-wise average access to improved water source in rural and urban areas

Regions	Improved water source, rural (% of rural population with access)					Improved water source, urban (% of urban population with access)				
	1990–95	1996–2000	2001–05	2006–10	2011–12	1990–95	1996–2000	2001–05	2006–10	2011–12
EAP	75.0	76.7	80.3	83.8	85.7	91.5	91.7	93.4	95.0	96.1
ECA	92.2	92.3	93.5	94.5	95.3	98.7	98.9	99.0	99.0	99.0
LAC	73.9	78.1	81.1	84.4	86.8	94.1	94.8	95.4	95.9	96.2
MENA	81.2	82.4	84.8	86.3	87.2	94.0	93.9	95.2	95.3	95.5
NA	96.7	97.2	97.7	98.1	98.5	99.9	99.8	99.8	99.7	99.7
SA	63.1	69.9	75.9	82.1	86.4	85.1	85.7	88.9	92.0	94.2
SSA	47.3	50.7	54.5	58.7	62.1	80.6	81.7	84.0	86.6	88.4

Source: Computed by authors based on World Bank (undated a).

Table 13.8 Region-wise average access to improved sanitation in rural and urban areas

Regions	Improved sanitation facilities, rural (% of rural population with access)						Improved sanitation facilities, urban (% of urban population with access)					
	1990–95	1996–2000	2001–05	2006–10	2011–12		1990–95	1996–2000	2001–05	2006–10	2011–12
EAP	56.0	56.8	60.2	64.5	66.5		73.4	75.0	78.4	81.5	83.3
ECA	92.8	91.5	92.3	93.8	95.2		97.4	96.7	96.9	97.4	97.9
LAC	57.8	64.0	67.7	70.0	71.6		79.3	81.4	83.1	84.3	85.3
MENA	73.8	77.2	80.9	84.0	85.4		92.2	93.1	94.2	95.0	95.3
NA	98.9	99.1	99.2	99.4	99.5		99.9	99.9	100.0	100.0	100.0
SA	29.5	34.4	39.8	45.7	49.4		59.2	60.9	63.1	65.6	67.3
SSA	19.9	22.3	23.8	24.3	25.2		41.6	42.6	44.0	44.7	45.2

Source: Computed by authors based on World Bank (undated a).

Table 13.9 Region-wise population without access to improved source of water and sanitation (2012)

Regions	Average Access to Improved Water Source (% of Population)	Average Access to Improved Sanitation (% of Population)	Total Population (Billion)	Population Without Access to Improved Water Source (Billion Population)	Population Without Access to Improved Sanitation (Billion Population)
EAP	89.62	72.64	2.20	0.23	0.60
ECA	97.08	96.94	0.75	0.02	0.02
LAC	93.11	81.19	0.60	0.04	0.11
MENA	92.43	91.19	0.40	0.03	0.03
NA	99.50	99.90	0.35	0.00	0.00
SA	88.95	55.53	1.65	0.18	0.73
SSA	72.51	34.41	0.90	0.25	0.59
Total			6.85	0.75	2.10

Source: Computed by authors based on World Bank (undated a).

of the investment in sanitation services is required in SSA, SA and EAP regions. A similar presence of rural-urban disparity underlines the higher investment requirement in rural areas on this front as well. Though better equipped compared to SSA, SA and EAP regions, investment is also required in countries located in MENA and LAC, particularly in the rural areas, for augmenting the existing level of access in WSS services.

It is observed from Table 13.9 that as of 2012, there are around 750 million populations from countries included in the present study who lack access to improved sources of water, while 2.10 billion people do not have access to improved sanitation facilities. The majority of these people live in SSA, EAP and SA regions. In SA alone 730 million people do not have access to improved sanitation facilities. To compound the problem, 73 per cent of the population who do not have access to improved water source live in rural areas, while the corresponding figure for improved sanitation facilities is around 67 per cent. Therefore providing access to improved WSS services to such a sizable chunk of population living in lower- and middle-income countries would be a major challenge.

3.2 Current Stock of Investment in Water Services Infrastructure

The estimation of present stock of investment to provide improved WSS across regions is important for this analysis primarily for two reasons. First, the costs of operation and maintenance of water supply infrastructure (inclusive of both water supply and sanitation services) not only depend on future investment but also on present stock of investment. Understandably, the future investment stream on maintenance is a function of the present stock of investment. Second, the estimation of stock of investment not only depends on the percentage of population having access to WSS, but also on several ground-level base factors, e.g., size of population of the regions, average family size, and choice of technology. Therefore, the estimated stock of investment across regions will enable the present study to conduct an informed analysis on the future scenario.

The estimation of present stock of investment is contingent upon two key factors. First, the unit cost of providing access to improved WSS generally varies across countries, depending on their regulatory standard for supply water quality, source(s) adjusted quality of raw water, level of treatment requirements (and choice of technology thereof), distribution system, and population density. Apart from inter-country variations in per unit costs, intra-country (e.g., rural – urban) variations could also influence the estimation process. In most of the cases the unit cost figures are presented in per household level, whereas the data on access to WSS is available as percentage of population. Variations across regions in estimates of per unit cost prevailing in their territories are presented in Table 13.10.[31] Second, there exist wide variations in average family size across regions and countries and even across locations (i.e., rural – urban) within a country, which also influence the estimation of demand for investment in WSS.

While the unit costs estimates provided by Toubkiss provide an important perspective in understanding the modified WSS operations in various segments for Asia, Africa and LAC, the analysis suffers from a major limitation as far as coverage is concerned.[32] The main problem against using the unit costs estimated by Toubkiss is that it does not provide estimates for all the sub-regions that are in dire need for investment in this sphere

---

[31] J Toubkiss, *Costing MDG Target 10 on Water Supply and Sanitation: Comparative Analysis, Obstacles and Recommendations* (2006) 29 <www.worldwatercouncil.org/fileadmin/world_water_council/documents_old/Library/Publications_and_reports/FullTextCover_MDG.pdf> accessed 4 March 2015.
[32] Ibid.

Table 13.10 Unit costs of water and sanitation improvements, excluding programme costs

Improvement type	Per capita (US$ year 2005*)					
	Initial investment costs			Annual recurrent cost		
	Africa	Asia	LAC	Africa	Asia	LAC
Water improvement						
Household connection (treated)	164	148	232	13.4	9.6	14.6
Standpost	50	103	66	0.5	1.0	0.7
Borehole	37	27	89	0.2	0.2	0.6
Dug well	34	35	77	0.2	0.2	0.5
Rainwater	79	55	58	0.5	0.4	0.4
Average of non-household connection options	50	55	72	0.4	0.5	0.5
Sanitation improvement						
Household connection (partial treatment)	193	248	258	8.2	9.1	11.0
Septic tank	185	167	258	6.2	6.1	6.8
Pour-flush	147	81	97	6.1	5.5	5.7
VIP	92	81	84	3.8	3.8	3.8
Simple pit latrine	63	42	97	3.6	3.5	3.9
Average of non-household connection options	122	93	134	4.9	4.7	5.0

Note: *Data from 2000 adjusted to 2005 prices using an average annual gross domestic product (GDP) deflator of 10 per cent.

Source: Adjusted by authors on estimation by Toubkiss (2006) (also available in Hutton and Bartram 2008: 6).

separately (e.g., South Asia and Sub-Saharan Africa).[33] In addition, even for the regions presented in the table, per unit estimates vary significantly. Therefore, to avoid the risk of under-valuation, the present analysis has relied on per unit cost estimates as provided by Fay and Yepes, where we have estimated data on sub-regions in 2005 prices by using sub-region wise GDP deflator.[34]

---

[33] Ibid.
[34] Fay and Yepes, n. 6 above, 10.

Table 13.11   Household scenario in water supply and sanitation – 2012

Regions	Population with access (in billion)		Average Household Size[a]	Households with access (in billion)	
	Improved Water Source	Improved Sanitation		Improved Water Source	Improved Sanitation
EAP	1.97	1.60	4.3	0.45	0.37
ECA	0.73	0.73	2.9	0.25	0.25
LAC	0.56	0.49	5.0	0.11	0.10
MENA	0.37	0.36	4.3	0.08	0.08
NA	0.35	0.35	2.9	0.12	0.12
SA	1.47	0.92	5.6	0.26	0.16
SSA	0.65	0.31	4.9	0.13	0.06
Total	6.10	4.75		1.42	1.15

Note: [a] obtained from Worldmapper (undated).

Source: Computed by the authors based on World Bank (undated a).

The estimation process of the current analysis is first explained with the help of Table 13.11. The data on region-wise distribution of population with access to improved WSS services has been estimated on the basis of figures obtained from UN-Water and WHO (2014). From Worldmapper (undated), the figures on average household sizes across regions are taken. By dividing the former series by the latter, the present analysis arrives at the number of average households (in billions) per region with access to improved water and sanitation facilities separately.

The estimation results of the current analysis are presented in Table 13.12. If unit cost of providing water services are considered in line with Fay and Yepes,[35] (those are US$ 400 per household for water supply and US$ 700 per household for sanitation at 2000 prices) and taken at 2005 prices by using region-wise GDP deflator, the estimated stock of investment in water services for 1.42 billion households having access to improved water source comes out at US$ 680 billion (at 2005 prices). On the other hand, the estimated stock of investment for 1.15 billion households having access to improved sanitation would reach US$ 969 billion (at 2005 prices). It is observed from Table 13.12 that the majority of these investments need to be incurred in EAP, ECA and SA.

---

[35]   Ibid.

Table 13.12 Region-wise stock of investment in water services infrastructure – 2012 (at 2005 prices, US$ billions)

Regions	Water Supply		Sanitation	
EAP	188	(27.7)	267	(27.6)
ECA	147	(21.6)	257	(26.5)
LAC	51	(7.5)	78	(8.1)
MENA	43	(6.4)	75	(7.7)
NA	54	(7.9)	95	(9.8)
SA	127	(18.7)	139	(14.3)
SSA	69	(10.2)	58	(5.9)
Total	680	(100)	969	(100)

Note: Figure in the parenthesis show the percentage share in total investment.

Source: Computed by the authors.

3.3 Future Stock of Investment in Water Services Infrastructure

Before going into an estimation of future demand for investment in water services infrastructure, it will not be irrelevant here to look into the major challenges involved in the process, as noted in the following. To begin with, the estimation exercise must take into account the evolving nature of the population, which is growing across regions at a widely differing pace. In other words, while estimating the investment demand for a future date (e.g., in 2019) one needs to look at the cost of not only providing universal access of WSS services to the existing population, but also for securing the same to the future population.

In addition to the need for making the aforesaid population growth related adjustments, several concern areas originating from supply side as well as demand side considerations also require attention. First, while focusing on the need for securing safe water for domestic purposes, the inter-sectoral dynamics, e.g., growing competition across sectors (e.g., agriculture, industry, ecology) may often be overlooked. The effort to secure universal access may possibly increase the opportunity cost of making available additional water for domestic uses. Second, the investment is not a one-off exercise and the costs associated with protection of sources of drinking water and treatment of domestic wastewater and sewerage may rise. Third, given the recent change in focus, the costs associated with additional investment in water services infrastructure is required to cope with climate change related concerns. However, incorporation of such considerations in the equation will add further to the costs. Fourth, with improvement in human development (including per capita income as well as literacy) and greater awareness for personal/ public health and hygiene, demand for safe water is likely to grow up in coming days. The higher demand from individual households, coupled with the likely population growth (and in number of households), therefore makes supply of additional water along with the desired quality of the same a major challenge. Fifth, in certain parts of the world, relatively smaller sizes of the family as well as the lower population density (e.g., vis-à-vis the scenario prevailing in South Asia) is more likely to raise the actual per unit cost of WSS services as compared to the estimated figure. Sixth, apart from the core water service, expanding the coverage of safe sanitation services would be equally important, which will also rise with improvement in human development level. However, delivery of better sanitation service is crucially linked to increased demand for water, unless there is conscious adoption of water-saving innovations and technologies.

Finally, factoring the sustainability perspective of providing universal access to WSS services and measuring the implicit costs would add further complexity to the exercise. First, with growing population pressure, urbanization drive and improvement in socio-economic status, additional pressure will be put on resources, leading to larger demand for water. In absence of regulation, major sustainability challenges may emerge (e.g., depletion of groundwater resources, contamination of drinking water sources).[36] Second, the larger demand for environment, ecosystem and ecological status of water bodies may lead to competition for water across competing sectors like agriculture, industry etc. and in absence of proper regulatory framework, massive environmental degradation may follow. Under those circumstances, corrective governmental interventions (e.g., inter-linking of rivers, creation of dams), keeping in mind the need for augmenting water access for citizens, may not necessarily improve the ecosystem.[37]

---

[36] Sacchidananda Mukherjee and Prakash Nelliyat, *Groundwater Pollution and Emerging Environmental Challenges of Industrial Effluent Irrigation in Mettupalayam Taluk, Tamilnadu, Comprehensive Assessment of Water Management in Agriculture* (International Water Management Institute, Colombo 2007) 8.

[37] Sacchidananda Mukherjee and Debashis Chakraborty, 'Environmental Governance Scenario in Asia: Lessons for the Future' in Mukherjee and Chakraborty (eds) *Environmental Challenges and Governance: Diverse Perspectives from Asia* (Abingdon: Routledge 2015) 253.

Table 13.13 Region-wise projection of households

Regions	Total Population: 2019	Average Household Size[a]	No. of Households: 2019
EAP	2.30	4.35	0.53
ECA	0.77	2.90	0.26
LAC	0.65	5.00	0.13
MENA	0.45	4.35	0.10
NA	0.37	2.94	0.13
SA	1.80	5.56	0.32
SSA	1.08	4.88	0.22
Total	7.41		1.70

Note: [a] Obtained from Worldmapper (undated).

Source: Computed by the authors.

Based on World Bank population projection for 2019,[38] and given the average size of the household (region-wise) available from Worldmapper (undated),[39] the present analysis assumes status quo of household size will persist and estimates the number of households for the year 2019. The estimates are summarized in Table 13.13. It is observed from the Table that the total estimated number of households comes to 1.70 billion.

The demand for investment in water service infrastructure by 2019, as estimated by the current analysis, has been presented in Table 13.14 below. It is observed that if all the households are provided with improved source of water and sanitation (i.e., universal access) by 2019, the total stock of investment in water services infrastructure would reach US $2,240 billion at 2005 prices by 2019. The findings of GLAAS 2014 show that global aspirations towards universal access to safe and affordable water and sanitation are supported by political processes in many countries. Two-thirds of the 94 countries covered in the report recognized both drinking water and sanitation as human right in their national legislation. Moreover, national policies for drinking water and sanitation are in place

---

[38] World Bank, *Health Nutrition and Population Statistics: Population estimates and projections* (undated) <http://databank.worldbank.org/data/reports.aspx?source=Health%20Nutrition%20and%20Population%20Statistics:%20Population%20estimates%20and%20projections> accessed 18 March 2015.

[39] WHO, *Global Water Supply and Sanitation Assessment: 2000 Report* (Geneva: WHO and UNICEF 2000) 12. Worldmapper, *Households* (undated) <www.worldmapper.org/posters/worldmapper_map191_ver5.pdf> accessed 8 March 2015.

Table 13.14 Projection of demand for investment in water services infrastructure by 2019 (in US$ billions, 2005 prices)

Regions	Capital Expenses				O&M				Total		
	Stock of Investment Required in 2019 (A)		Additional Investment Required by 2019 (B)		Additional Annual Investment Required during 2013–2019 (C)=(B/7)		Repair, Replacement and Operation Cost (D)		Annual Investment Required during 2013–2019 (E)=(C+D)		Total
	Water Supply	Sanitation	Water Supply	Sanitation	Water Supply	Sanitation	Water Supply	Sanitation	Water Supply	Sanitation	
EAP	219.6	384.4	31.4	117.3	4.5	16.8	28.2	40.1	32.7	56.8	89.5
ECA	154.8	271.0	7.7	13.8	1.1	2.0	22.1	38.6	23.2	40.5	63.7
LAC	59.2	103.6	7.9	25.3	1.1	3.6	7.7	11.7	8.8	15.4	24.2
MENA	53.0	92.7	9.5	17.7	1.4	2.5	6.5	11.3	7.9	13.8	21.6
NA	57.4	100.4	3.4	5.5	0.5	0.8	8.1	14.2	8.6	15.0	23.6
SA	155.7	272.4	28.7	133.8	4.1	19.1	19.0	20.8	23.1	39.9	63.1
SSA	114.7	200.7	45.3	143.1	6.5	20.4	10.4	8.6	16.9	29.1	46.0
World	814	1,425	133.9	456.5	19.1	65.2	102.1	145.3	121.2	210.5	331.7

Source: Computed by the authors.

for more than 80 per cent of countries covered in the survey.[40] Looking at the sectoral distribution, the stock of investment in infrastructure should reach US$ 814 billion and US$ 1,425 billion for water supply and sanitation respectively. Therefore, given the stock of investment in 2012, an additional (capital) investment of US$ 590 billion is required during 2013–19 in new water services infrastructure to reach the desired stock of investment in 2019. The additional expenditure of US$ 134 billion would be required in the area of water supply, while the residual US$ 456 billion needs to be spent in sanitation sector.

If the capital investment requirement is spread out equally over the entire period of planning (i.e., 2013–19), an equal annual investment of US$ 19 billion in water supply and US$ 65 billion in sanitation are required to achieve the universal access to improved WSS by 2019. However, to protect the value of present of stock of investment and to keep it functional, it requires repairs, replacement, etc.

Hutton and Bartram noted that, '[A]nnual operation and maintenance (O & M) costs 5–10% of capital cost for low-technology options, water source protection an additional 5–10% of capital cost per year, and education for sanitation interventions 5% of capital cost per year.'[41] In line with Hutton and Bartram, the present analysis assumes that the cost of repair and replacement would be 15 per cent of present value (stock) of investment during 2013–19.[42] This cost also includes repair and replacement cost of future investment which will be carried out during 2013–19. Therefore, in addition to new investment for infrastructure, additional investment of US$ 247 billion is required for repair and replacement, the region-wise distribution of which is also summarized in Table 13.14. Therefore taking both the capital investment and repair and replacement components into consideration, the total annual investment in water services infrastructure required during 2013–19 comes out as US$ 332 billion, of which the annual capital cost and annual O&M costs are of US$ 84.3 billion and US$ 247.4 billion respectively.

The desired annual investment in water services infrastructure required during 2013–19 is presented in successive year's percentage of regional GDP (at constant 2005 prices) in Table 13.15. The country-wise GDP projections (in current prices, US$) is available up to 2019 from IMF.[43] The present analysis considers the projected GDP estimates in 2005 prices

---

[40] UN-Water and WHO, GLAAS 2014, n. 30 above, 14.
[41] Hutton and Bartram, n. 7 above, 7.
[42] Ibid.
[43] International Monetary Fund, 'World Economic Outlook Database' <www.imf.org/external/pubs/ft/weo/2014/02/weodata/index.aspx> accessed 10 March 2015.

Table 13.15 Region-wise estimated annual investment in water services infrastructure as percentage of GDP (in constant 2005 prices)

Regions	2012	2013	2014	2015	2016	2017	2018	2019
EAP	0.72	0.71	0.68	0.63	0.60	0.56	0.52	0.49
ECA	0.39	0.37	0.36	0.35	0.33	0.32	0.30	0.29
LAC	0.69	0.68	0.68	0.65	0.61	0.58	0.55	0.52
MENA	1.07	1.04	0.99	0.94	0.89	0.84	0.79	0.74
NA	0.15	0.15	0.14	0.14	0.13	0.12	0.12	0.11
SA	3.75	3.67	3.72	3.38	3.10	2.84	2.61	2.38
SSA	5.05	4.82	4.54	4.24	3.97	3.70	3.47	3.26
World	0.63	0.61	0.59	0.56	0.54	0.51	0.48	0.46

Source: Computed by the authors.

by using region-wise GDP deflator for 2012. It is observed from the Table that in terms of relative importance (i.e., expressed as percentage of GDP), the desired investment will vary widely across regions and as compared to water supply, providing universal access to WSS would be quite costly for Sub-Saharan Africa and South Asia. Several economies located in both the regions are characterized by poverty and unemployment concerns, fiscal mismanagement, budget deficit and food security related challenges. Therefore, other pressing priorities and limited fiscal space may overshadow their ability and commitment to allocate 3–5 per cent of the GDP annually for augmenting WSS services, especially when recovery of water charges is poor.

In order to understand the long-term demand for investment in ensuring access to WSS, the present analysis has been extended further for two additional years, namely, 2025 and 2030. The results on additional investment requirements are presented in Table 13.16. It is observed from the Table that if the period for achieving universal access is spread over a longer period, the demand for absolute value of investment would naturally go up, though the average annual investment requirement will be lower. Taking into consideration the financial, fiscal and development related challenges being faced by the economies located in South Asia and Sub-Saharan Africa, it may be pragmatic to argue that the deadline for achieving universal access to WSS may be extended to 2025.

The comparative investment requirement figures over 2012–19, 2012–25 and 2012–30 for securing universal access to WSS services are summarized in Table 13.17. The last rows of each column show the average annual investment requirement over the respective periods. As expected, the

Table 13.16 Additional investment (capital) required to achieve universal access to water and sanitation (US$ billion, 2005 prices) (with reference to present 2012 stock of investment)

Regions	2025			2030		
	Water	Sanitation	Total	Water	Sanitation	Total
EAP	82.7	207.2	289.9	88.2	216.7	304.9
ECA	9.3	16.6	25.8	10.0	18.0	28.0
LAC	11.1	30.9	42.1	13.5	35.1	48.5
MENA	13.9	25.4	39.3	17.2	31.2	48.4
NA	5.8	9.8	15.7	7.8	13.2	21.0
SA	38.2	150.4	188.6	45.5	163.1	208.6
SSA	63.2	174.3	237.5	79.6	203.1	282.7
World	224.3	614.6	838.9	261.8	680.3	942.0

Source: Computed by the authors.

Table 13.17 Total and annual capital investment required to achieve universal access to water and sanitation (US$ billion, 2005 prices) by 2019, 2025 and 2030

Regions	2012–19	2012–25	2012–30
EAP	149	289.9	304.9
ECA	21	25.8	28.0
LAC	33	42.1	48.5
MENA	27	39.3	48.4
NA	9	15.7	21.0
SA	163	188.6	208.6
SSA	188	237.5	282.7
World	590	838.9	942.0
Equal Annual Investment	84.33	64.53	52.33

Source: Computed by the authors.

investment requirement is found to be highest for Sub-Saharan Africa, followed by South Asia and East Asia and the Pacific.

3.4 Cost of Adaptation to Climate Change

Climate change induced hydrological imbalances could affect the present as well as future investment in WSS infrastructure. For instance, to cope with the recurrent floods or droughts, the infrastructure in water services needs to be resilient enough to absorb such shocks. Improvement in water use efficiency in high water-consuming sectors (e.g., agriculture), integrated urban water management, etc. could be part of effective mitigation strategies. Climate change consequences may also affect the availability of water to a sizable chunk of population significantly. For instance, the melting of Himalayan glaciers in Nepal owing to global warming may increase water flows in the local rivers till 2030, but the same will eventually be reduced significantly by end of this century.[44] It however deserves mention that the cost of universal access to WSS services can increase further, if the cost of adaptation to climate change related challenges is also factored in calculations. The adaptation costs may include both hard options like building dams and dykes, and soft options like use of early warning systems, community preparedness programmes, watershed management, water pricing, supporting water-efficient technologies, urban and rural zoning and land use restrictions, etc.[45] Instead of computing the same, the current analysis briefly reports the available literature on this front.

Ward et al. estimated annual average adaptation cost of municipal and industrial waters supply during 2010–50 across regions, see Table 13.18.[46] Based on a socio-economic baseline and two climate change scenarios developed by the Commonwealth Scientific and Industrial Research Organisation (CSIRO), Australia and the National Center for Atmospheric Research (NCAR), USA, the study estimates annual average adaptation costs for individual sub-regions and for all the developing as well as non-developing countries as a whole. The total expenditure for all developing countries comes out to be US$ 66.53 billion (in 2005 prices), while the corresponding figure for non-developing countries stands at US$ 6.44 billion. Interestingly the adaptation cost estimate for Sub-Saharan African countries turn out to be moderate, while it is quite substantial for regions where emerging countries play a key role (e.g., India in South Asia, China in East Asia and the Pacific).

---

[44] Madhukar Upadhay and Bibek R Kandel, 'Understanding the state of environmental governance in Nepal', in Mukherjee and Chakraborty (eds), *Environmental Challenges and Governance: Diverse Perspectives from Asia* (Abingdon: Routledge 2015) 79.

[45] Ward and others, 'Costs of adaptation related to industrial and municipal water supply and riverine flood protection' *World Bank Discussion Paper No. 6* (Washington DC: World Bank 2010) 17.

[46] Ward and others, 'Costs of Adaptation Related to Industrial and Municipal Water Supply and Riverine Flood Protection' *World Bank Discussion Paper No. 6* (Washington DC: World Bank 2010) 35.

Table 13.18 Average annual adaptation costs in the industrial and municipal water supply sector (2010–50) (US$ billions)

Regions	Baseline
East Asia and Pacific	20.79
Europe and Central Asia	2.15
Latin America and Caribbean	3.04
Middle East and North Africa	6.74
South Asia	28.73
Sub-Saharan Africa	5.08
Total Developing Country	66.53
Total Non-Developing Country	6.44

Source: Ward et al. (2010).

4. SOURCES OF FINANCING INFRASTRUCTURE IN WATER SERVICES

4.1 Concern Areas in Financing Infrastructure Projects: Public Funds

Public goods nature of services and various uncertainties often makes it difficult to mobilize resources, especially to attract private investors, to finance the infrastructure projects (e.g., roads, railways, power, water and sanitation). As a result, public financing of infrastructure remains the main source of funds for developing countries like India and the LDCs.[47] Unlike other infrastructure, WSS services have a special political economic dimension and at times it is provided free of costs to the stakeholders/beneficiaries (electorates), however uneconomic the policy might actually be. The falling fiscal space of governments, rising demands for public financing from other social merit goods (e.g., health and education), often leaves inadequate funds for financing water services infrastructure. Low public investment and low mobilization of resources from users and beneficiaries often tend to make the public WSS projects financially unviable.

The revenue stream of the infrastructure projects in general and WSS projects in particular depends on multiple factors, namely, the nature and

---

[47] Sengupta and others, 'Financing for infrastructure investment in G-20 countries' *Working Paper No. 2015-144* (New Delhi: National Institute of Public Finance and Policy, 2015) 13.

stringency of water pricing, i.e., type of charges/fees (volumetric *vs.* lump sum), under-recovery of charges/fees due to political interventions/reluctance to charge water, humane nature of the service, etc. The examples of political promises, with the objective of rich electoral dividends, abound in reality. For instance, during the recent Assembly Election in Delhi, India, a political party promised to provide free access up to 20 kilolitre of water per month free of cost per household.[48]

Poor financial management and accounting practices are also another area of concern resulting in pilferage of public resources away from WSS projects. Apart from budgetary resources, concessional/special loans, grants-in-aid from MDBs/agencies like the World Bank, ADB remain a major source of financing of WSS projects for developing countries like India and LDCs. In 2012, the WSS sector received official development assistance and non-concessional loans of US$ 10.5 billion and US$ 4.2 billion respectively, which included bilateral and multilateral aids as well as supports from NGOs and private foundations.[49]

4.2 PPI Investment in Water Services Sector

Given the aforesaid concern areas in securing public investment in WSS services, exploring the private investment route is of utmost importance. The present quantum and distribution of private investment in WSS sector would enable regions to accordingly strategize in their future initiatives. The distribution of private investment across the sub-regions can be observed by accessing the statistics from the Private Participation in Infrastructure (PPI) database of the World Bank.[50]

The investment patterns across regions and various infrastructure sectors over 1991 to 2013 are summarized in Table 13.19. The Table reveals that, only 3.4 per cent of the total investment in PPI projects has been channelled towards water and sewerage projects. In value terms, as amount of US$ 75 billion has been invested on water and sewerage projects (in historical prices) over this period, which looks pretty meagre as compared to other three infrastructure sectors reported in the table. It is further observed that the majority of the private investments undertaken in water infrastructure have moved to Latin America and the Caribbean and East Asia and Pacific regions. The poorer show by South Asia and

---

[48] Aam Admi Party, 'Election manifesto' (2015) 13 <www.aamaadmiparty.org/AAP-Manifesto-2015.pdf> accessed 10 March 2015.
[49] UN-Water and WHO, (GLAAS) 2014, n. 30 above, 13.
[50] World Bank, 'Private Participation in Infrastructure database' (undated) <http://ppi.worldbank.org/> accessed 14 March 2015.

Table 13.19 Region-wise, sector-wise distribution of investment in projects (US$ billions) – 1991–2013 (in historical prices)

Sector	EAP	EUR	LAC	MENA	SA	SSA	Total Investment	
Energy (Electricity & Natural Gas)	152	127	282	25	159	22	767	(34.9)
Telecom	111	190	352	62	125	109	949	(43.2)
Transport	89	29	176	7	89	18	409	(18.6)
Water and sewerage	31	4	35	4	1	0	75	(3.4)
All	383	350	845	98	374	150	2,199	
	(17.4)	(15.9)	(38.4)	(4.5)	(17)	(6.8)	(100)	

Note: Figures in the parenthesis shows the percentage share in total investment.

Source: Constructed by authors from World Bank (undated c) data.

Sub-Saharan Africa in attracting PPI investment therefore remains a major concern.

In order to analyse the evolving nature of the PPI investment on water services infrastructure across regions during 1991–2013, the data is presented in Table 13.20. For a closer focus, the entire period under analysis is divided in five time slots. The Table reveals that Latin America and the Caribbean and East Asia and Pacific regions have cumulatively attracted 46.6 per cent and 41.5 per cent of the total PPI investment, respectively. On the other hand, both Europe and Central Asia and Middle East and North Africa regions have attracted PPI investment of 5.3 per cent each. However, the regions which urgently require larger investment on water services infrastructure, i.e., Sub-Saharan Africa and South Asia, have attracted only 0.5 per cent and 0.8 per cent of the total investment, in that order. Interestingly, as the data reveals, major investment on water and sewerage sector was undertaken during 1996–2000, i.e., before the launching of the MDGs in September 2000. Last but not the least, it is further observed from the Table that the PPI investment in Sub-Saharan Africa and South Asia has increased during 2011–13 over 2006–10 level only marginally. This is all the more worrying given the incomplete water, sanitation and hygiene service delivery scenario in several Asian countries, including South Asia.[51]

---

[51] Cronin and others, n. 28 above.

Table 13.20 PPI investment in projects by region and year of investment (US$ millions) (in historical prices)

Year of Investment	East Asia and Pacific	Europe and Central Asia	Latin America and the Caribbean	Middle East and North Africa	South Asia	Sub-Saharan Africa	Total Investment	
1991–95	4,183	–	5,964	–	–	–	10,147	(13.6)
1996–2000	13,462	1,318	12,257	–	–	133	27,172	(36.5)
2001–05	6,784	1,013	3,433	679	113	12	12,034	(16.1)
2006–10	5,202	1,451	3,322	3,093	242	121	13,429	(18.0)
2011–13	1,316	133	9,745	192	251	126	11,762	(15.8)
All	30,947	3,915	34,721	3,964	606	392	74,544	(100)
	(41.5)	(5.3)	(46.6)	(5.3)	(0.8)	(0.5)	(100.0)	

Note: Figures in the parentheses shows the percentage share in total investment.

Source: Constructed by authors from World Bank (undated c) data.

Table 13.21 *Number of projects and investment in water services projects by sub-sector (US$ million) (in historical prices)*

Sub-sector	Segment	Project Count	Total Investment	
Treatment plant	Potable water and sewerage treatment plant	13	292	(0.4)
	Potable water treatment plant	141	9,120	(12.2)
	Sewerage treatment plant	330	7,876	(10.6)
Total Treatment plant		484	17,288	(23.2)
Utility	Sewerage collection	2	174	(0.2)
	Sewerage collection and treatment	21	8,011	(10.7)
	Water utility with sewerage	264	35,807	(48.0)
	Water utility without sewerage	73	11,252	(15.1)
Total Utility		360	55,244	(73.9)
Total Water Transfer System		3	2,013	(2.7)
Grand Total	..	847	74,544	

Source: Constructed by authors from World Bank (undated c) data.

The observation underlines the need to augment PPI investment on WSS services in these two regions in no uncertain terms.

The anatomy of the PPI investment in water services infrastructure has been analysed with the help of Table 13.21, where the investments have been categorized across sub-sectors and segments. It is observed from the table that 73.9 per cent of the total investment has come to the utility services, 23.2 per cent has targeted the treatment plant, and the remaining (2.7 per cent) has targeted the water transfer system. It is evident that even within water services projects, the investment quantum is not evenly distributed across different functions.

Finally, the distribution of PPI investment across regions in WSS projects by establishment type has been analysed for understanding the scenario on newer capacity creation and the results are summarized in Table 13.22. The table reflects that 64 per cent of the total PPI investment has so far been on concessions (also known as brownfield), while 13 per cent has come towards divesture (privatized). Interestingly, PPI investment in greenfield projects has been quite low (21 per cent). However, a saving grace is that in Sub-Saharan Africa, investment in greenfield projects explain a significant proportion of the total investment

Table 13.22 Investment in projects by region and type (US$ millions) (in historical prices)

Region	Concession	Divestiture	Greenfield project	Management and lease contract	Total
EAP	23,315	1,321	6,167	143	30,945
ECA	731	435	1,544	1,205	3,915
LAC	22,982	7,924	3,812	5	34,724
MENA	192	0	3,772	0	3,964
SA	359	0	245	2	605
SSA	76	0	259	57	392
Grand Total	47,654	9,680	15,798	1,412	74,545
	64	13	21	2	100

Notes: *Concessions* – A private entity takes over the management of a state-owned enterprise for a given period during which it also assumes significant investment risk; *Divestitures* – A private entity buys an equity stake in a state-owned enterprise through an asset sale, public offering, or mass privatization program; *Greenfield Projects* – A private entity or a public-private joint venture builds and operates a new facility for the period specified in the project contract. The facility may return to the public sector at the end of the concession period; and *Management and Lease Contracts* – A private entity takes over the management of a state-owned enterprise for a fixed period while ownership and investment decisions remain with the state.

Source: Constructed by authors from World Bank (undated c) data.

(66.07 per cent). Conversely, in South Asia the corresponding figure stands only at 40.50 per cent.

4.3 Financing the Investment Gap in Water Services Infrastructure

OECD describes the current water service finance architecture in terms of cost and revenue components.[52] While the three major cost components are investment, operating and maintenance costs, revenues are earned through the 3Ts, namely, tariffs (i.e., user charges), taxes (budgetary resources) and transfers (e.g., overseas development assistance, grants-in-aid from higher level of government in case of federal countries, long-term loans from multilateral development institutions like World Bank, ADB, etc.). The public water supply agencies often issue bonds (guaranteed by state government), obtain loans and advances from the appropriate government, etc. to meet the finance gap in WSS projects. The countries

---

[52] OECD 2011, n. 22 above, 52.

in South Asia and Sub-Saharan Africa urgently need to enhance their budgetary provisions towards WSS services. However, as most of them suffer from limited fiscal space and are faced with competing development agendas, meeting such costs exclusively through public funds would not be possible.

Palaniappan et al. note that private capital plays a limited role in OECD countries.[53] Given the finance gap, there is however an urgent need to augment private participation in financing WSS projects in regions like Sub-Saharan Africa and South Asia. However, it is contingent on the existing legal framework on protection of investment to promote the private participation in infrastructure projects. WHO have underlined the need for promoting services that are, 'both socially efficient and financially sustainable'.[54] As most of the countries in dire need of finance in WSS services are either LDCs or developing countries, financial sustainability is not automatically assured. Moreover, it is unlikely that participation of the domestic private sector will be able to supplement the investment requirement gap. Encouraging participation of foreign investors, either from the developed countries or advanced developing countries, might be a prudent step in this background. Presently privatization of WSS is happening in developing countries, with entry of both domestic and foreign players in key markets, but the quantum has been inadequate. Fay et al. have identified several factors for this scenario, namely: substantial up-front capital investment requirement, benefits for which are spread over time, difficulties in recovering cost, need for fiscal adjustments down the line, etc.[55]

A series of reforms are required in several South Asian and Sub-Saharan African countries before the possibility of obtaining private investments from abroad can indeed become a reality. First, the legal framework on foreign investment protection needs to be strengthened in the recipient country. For instance, many developing countries/LDCs have bilateral investment treaties (BITs) with other countries, but the extent of actual market access can be quite restrictive owing to the inherent provisions in

---

[53] Palaniappan and others, 'Water infrastructure and water-related services: trends and challenges affecting future development', in OECD Report (2007) 2 Infrastructure to 2030: Telecom, Land Transport, Water and Electricity 278.

[54] WHO 2012, n. 12 above, 3.

[55] Marianne Fay, Michael Toman, Daniel Benitez, and Stefan Csordas, 'Infrastructure and sustainable development' in Fardoust, Yongbeom and Sepúlveda (eds) *Postcrisis Growth and Development: A Development Agenda for the G-20* (Washington DC: World Bank 2011) 330.

the agreement.⁵⁶ Second, protection of intellectual property rights through necessary procedural modifications would be crucial, if the foreign firm enters the market with certain key technology (e.g., water treatment technology). Third, often the market in developing countries and LDCs are considered risky by foreign investors, laden with political crisis, non-transparency, macroeconomic instability, expropriation risks, etc. A major reform of the policy environment is required for changing the perceived risk by investors. Fourth, the governments may subject foreign investment proposals in WSS service to case-by-case review, instead of putting a blanket ban.⁵⁷ Finally, accessing money from the international market might be problematic if the exchange rate of the recipient country is weak and on the decline. Therefore, there is a need to develop and strengthen the local capital market.⁵⁸

Given the public good nature of the infrastructure sector, where the problems of market failure and revenue mobilization are enormous, Public-Private Partnership (PPP) models provide a robust option for raising the necessary investment. Here both the public and private partners can focus on their respective specific competences. Stevens and Schieb noted that PPP projects are better suited for capital-intensive large projects (e.g., urban public transport, toll motorways), with possibility of controlling access.⁵⁹ There is need to identify and implement the recipe of successful PPP models, through introduction of 'user-pay' approach. The World Bank recounts the success of water service PPPs in Sub-Saharan and other African countries characterized by civil law, where 'legal systems that have statutes and codes regulating public service contracting'.⁶⁰ WSP notes

---

⁵⁶ See Julien Chaisse and Christian Bellak, 'Do bilateral investment treaties promote foreign direct investment? Preliminary reflections on a new methodology' (2011) 3 *Transnational Corporations Review* 7. See also Julien Chaisse and Christian Bellak, 'Navigating the Expanding Universe of Investment Treaties—Creation and Use of Critical Index,' 18 *Journal of International Economic Law* 79 (2015), for understanding the link between treaties that states entered into with the increase of the FDI.

⁵⁷ United Nations Conference of Trade and Development, *Investment Policy Framework for Sustainable Development* (Geneva: UNCTAD 2015) 124.

⁵⁸ Bhattacharya and others, 'Driving sustainable development through better infrastructure: key elements of a transformation program' *Global Economy & Development Working Paper No. 91*, (Washington DC: Global Economy and Development at Brookings 2015) 16.

⁵⁹ Barrie Stevens and Pierre-Alain Schieb, 'Infrastructure to 2030: main findings and policy recommendations', in OECD Report (2007) 2 Infrastructure to 2030: Telecom, Land Transport, Water and Electricity 31.

⁶⁰ World Bank, 'Public-Private Partnerships for Rural Water Services' *Briefing Note No. 4* (Washington DC 2012) 1.

that number of water-related PPP projects reaching award stage in India is on the rise in recent period.[61] While factors like availability of public support, integration of mechanisms to address revenue mobilization, ownership and expertise of state agencies add to their success, the projects sometimes suffer from limitations like inconsistent and inadequate local stakeholder support, limited awareness and technical capacity to undertake PPPs, etc. There is a need to identify the reasons behind both successes and failures and accordingly strategize for future WSS projects.

Last but not the least, the investment strategies in general and in WSS services in particular need to facilitate the on-going climate change mitigation policies through appropriate actions (e.g., deployment of climate-friendly technologies, prohibiting exploitation of natural resources). UNCTAD argues in favour of shifting towards 'sustainable-development-based' incentives for investments from a purely 'location-based' approach.[62]

5. CONCLUSIONS

It is evident from the analysis so far that different regions are at different levels of achievement in securing universal access to WSS services. Faced with the twin problems of growing population size and inadequate WSS services infrastructure, the importance of investment demand has become apparent in several parts of the globe, particularly in South Asia and Sub-Saharan Africa. Moreover, the water supply and sanitation services should not be treated in isolation, but through an integrated approach so as to enhance the livelihood security of the people on one hand and prevent contamination risks on the other. The present analysis is an attempt to estimate the demand for such investments.

The empirical estimates of the analysis reveals that to achieve universal access to improved WSS by 2019, the stock of investment in water services should reach US $2,240 billion at 2005 prices in 2019. At the

---

[61] Water and Sanitation Program, 'Trends in Private Sector Participation in the Indian Water Sector: A Critical Review' *Field Note* (2011) 5. *See also* Julien Chaisse et al., 'Deconstructing Service and Investment Negotiating Stance: A Case Study of India at WTO GATS and Investment Fora,' 14 *Journal of World Investment and Trade* 44 (2013). See also Julien Chaisse and Mitsuo Matsushita, 'Maintaining the WTO's Supremacy in the International Trade Order – A Proposal to Refine and Revise the Role of the Trade Policy Review Mechanism,' 9 (2013) 18 *Journal of International Economic Law*.

[62] United Nations Conference of Trade and Development, *Investment Policy Framework for Sustainable Development* (Geneva: UNCTAD 2015) 127.

given stock of investment in 2012, an additional investment of US$ 590 billion (US$ 134 billion in water supply and US$ 456 billion in sanitation) would be required in new water services infrastructure to reach the desired stock of investment by 2019. In addition to new investment for infrastructure, additional investment of US$ 247 billion is required for repair and replacement activities. Therefore total investment in water services infrastructure required during 2013–19 would turn out to be US$ 838 billion. If the investment is spread out evenly over the entire period of planning (i.e., 2013–19), an average annual investment of US$ 34 billion in water supply and US$ 86 billion in sanitation would be required to achieve the universal access to improved water source and sanitation by 2019. The cost estimates would go up further if the climate change aspects are to be integrated in the calculations.

In the backdrop of the huge investment requirements, and the urgency of interventions in low-income countries located in South Asia and Sub-Saharan Africa, it is unlikely that the government budgetary devolutions and the aids from bilateral and multilateral donors would be able to bridge the gap. Attracting private investment in this field needs to be part of the long-term solution. To facilitate the process, certain strong measures are urgently required. In particular there is need for improving the politico-legal framework for facilitating private sector investment protection as well as devising efficient mechanisms for cost recovery.

Given the poor access scenario, simultaneously with the process of arranging for the necessary investment, the countries must also embrace certain policies. First, UN-Water and WHO notes that 70 and 63 countries (out of 94 respondents) have incorporated the human right to water and sanitation respectively in their constitution.[63] There is a need for all the UN member countries, especially the countries presently with poorer access, to undertake similar step at the earliest to recognize human right to water and sanitation as fundamental rights. Second, so far only around 50 per cent of countries are implementing specific measures for reusing wastewater in their national plans, while the practice have become widespread only in around 2 per cent of countries.[64] Greater focus needs to be accorded in this area as well, as environmental degradation due to sewage discharges and wastewaters remains a major area of concern in several low- and middle-income countries.[65] Third, innovations are required to

---

[63] UN-Water and WHO, GLAAS 2014, n. 30 above, 14.
[64] Ibid., 26.
[65] Richard Ashley and Adrian Cashman, 'The impacts of change on the long-term future demand for water sector infrastructure', in OECD Report (2006) 1 Infrastructure to 2030: Telecom, Land Transport, Water and Electricity 253.

facilitate operation of desalination plants at lower cost in at least some of the countries with poor access to drinkable water but a sizable coastline. Fourth, increased focus to biotechnology research is required to facilitate cultivation of drought-resistant crops and other similar innovations, thereby balancing demand for water to some extent.[66] Similar water-saving innovations would be required in the industrial sector as well. Fifth, with the global climate change effects intensifying, untimely freak torrential rains and droughts are becoming a recurrent phenomenon in recent times. There is a need to integrate stormwater and rainwater management practices with the wider urban planning approach through the necessary legal and administrative steps.[67] Finally, disputes over water right issues are common when rivers flow across the border, as one country's attempt to enhance access within its territory may lower the same in neighbouring countries. For instance, the recent building of dams over the Brahmaputra River in China and the concerns in India over future water flows deserve mention. The emerging concern is that even if the country located upstream agrees to release annually an assured amount of water, the disputes may subsequently turn far more complex with varying water flows in the river as consequence of climate change. Hence it is extremely important to develop necessary bilateral institutional mechanisms for mitigating water allocation conflicts in mutual interest.[68]

The lack of access to WSS for a sizable population of the world is a global problem, which cannot be tackled through a simple localized solution. This perspective needs to be borne in mind by all the stakeholders, including the developed countries, whose supports through technology transfer (e.g., in the area of wastewater treatment, desalination) and pre-harvest techniques (e.g., supply of genetically modified drought-resistant seeds) can provide an effective short-term solution. To arrive at the long-term solution through arrangement of the necessary investments in WSS services, a consultative approach engaging all the stakeholders is the need of the hour.

---

[66] Ibid., 290.
[67] Annicka Cettner, Richard Ashley, Maria Viklander and Kristina Nilsson, 'Stormwater management and urban planning: Lessons from 40 years of innovation' (2013) 56 *Journal of Environmental Planning and Management* 790; Zrinab Yazdanfar and Ashok Sharma, 'Urban drainage system planning and design – challenges with climate change and urbanization: a review' (2015) 72 *Water Science & Technology* 169.
[68] Dinar and others, 'Climate change, conflict, and cooperation: global analysis of the resilience of international river treaties to increased water variability' *Policy Research Working Paper 6916* (Washington DC: World Bank 2014) 19.

14. Residential water charges in Ireland: policy objectives and funding models
Thomas McDonnell

INTRODUCTION

Clean water is an economic good. It is not costless. Water is heavy and difficult to transport, and the provision of water and wastewater services requires the construction, operation, maintenance and improvement of expensive network infrastructure.[1] Such costs must be financed through present or future taxation, through tariffs, or through a combination of both. In Ireland, the provision of water and wastewater services has mainly been funded through general taxation with an additional contribution from non-domestic rates. A government commissioned assessment of water services delivery in Ireland reviewed the strengths and weaknesses of the delivery of water services through the then 34 local authorities, and concluded that there was a lack of co-ordination, an aging and poor quality network, and problems in achieving economies of scale and delivering projects of national importance.[2] The report also argued that the best way to improve the efficiency and effectiveness of operations, to increase capital investment, and to access new finances for the water sector, was to combine Irish water and wastewater services into a single public utility.[3]

The market for water and wastewater services is a natural monopoly in the sense that a single provider of the service is the most efficient market structure. Natural monopolies are characterised by network infrastructure, large fixed costs, economies of scope and scale and low

---

[1] See generally J Chaisse and M Polo, 'Globalization of Water Privatization: Ramifications of Investor-State Disputes in the "blue gold" Economy', (2015) 38 *Boston College International and Comparative Law Review* 1.
[2] See generally PWC, *Irish Water: Phase 1 Report* PWC (2011).
[3] Ibid.

or zero marginal cost. A lack of competition in these markets will tend to drive a propensity to market failure and economic inefficiency in the absence of robust regulatory measures. On 1 January 2014, responsibility for water and wastewater services to homes and businesses was transferred from the then 34 local authorities to a new national water service authority called Irish Water. Statutory responsibility for the economic regulation of the water sector and protection of the interests of customers was given to the Commission for Energy Regulation (CER) with the mandate to ensure operational efficiency is achieved across the sector.

Beyond subsidies funded from general taxation there are potentially three main sources of revenue for a water utility. These are connection fees, recurrent fixed charges and volumetric charges based on usage. Water policy pursues multiple objectives and a wide variety of pricing structures are employed within the OECD. These objectives can be structured around four sustainability dimensions: environmental sustainability; financial sustainability; economic efficiency and political and social concerns including affordability. There are clear tensions and trade-offs between the different goals of conservation, economic efficiency and affordability. For example, full cost recovery through usage-based tariffs creates affordability and equity concerns, while the most economically efficient allocation of water may not be consistent with water saving and environmental concerns. The prior abolition of residential water charges in the 1990s was opposed by environmental groups as an abandonment of the 'user pays' principle. No economic or environmental rationale was given at the time for the abolition of charges. Funding became a combination of government subsidies and non-domestic rates mainly paid by businesses but also by schools, churches and other bodies.

Drinking water is non-substitutable and essential for life. The human right to water has been recognised in international law including human rights treaties, declarations and other standards. For example, the United Nations General Assembly has recognised the right of every human being to have access to sufficient water for personal and domestic uses (between 50 and 100 litres of water per person per day), which must be safe, acceptable and affordable (water costs should not exceed 3 per cent of household income).[4, 5, 6]

Access to clean and affordable water is a human right. Yet there is no

---

[4] United Nations, 'UN General Assembly Resolution A/RES/64/292' < www.un.org/ga/search/view_doc.asp?symbol=A/RES/64/292> accessed 15/10/14.

[5] See e.g. United Nations, 'The human right to water and sanitation' < www.un.org/waterforlifedecade/human_right_to_water.shtml> accessed 15/10/14.

[6] United Nations, 'Human Rights Council Resolution A/HRC/RES/15/9'

such thing as free water. On an annual basis, the provision of water and wastewater services in Ireland costs in excess of €1 billion although the precise cost can fluctuate quite widely from year to year given the large percentage of costs attributable to capital projects. Irish Water spent over €1.4 billion in 2014 and was expected to receive close to €1.3 billion in 2015.[7, 8, 9] The question therefore is not 'whether' we should pay for water and wastewater services but 'how' we should pay for them. We can pay for water through taxes or tariffs or both. Until 2014 water and wastewater services for domestic users were mainly paid for out of monies raised from general taxation, (mainly consumption and income taxes) which was then allocated as subsidies from central government to the various local governments.

Water tariffs were reintroduced for all domestic users in October 2014.[10] A portion of the cost of water will continue to be subsidised from central government (taxes) while the remaining cost of water will be paid for through commercial and non-domestic charges and household charges linked to water use (user tariffs). The affordability of water and the distribution of the cost of provision are central concerns for any politically sustainable water funding model or indeed any funding model consistent with social justice. The change in funding model has raised affordability and poverty concerns for low-income households. In addition, water charges are more salient than other taxes and charges and the regressive structure of water charges – indeed the very concept of charging for what is seen by many as a human right – has been the subject of strong political opposition including regular marches by tens of thousands of protesters.

The Irish government provided a Universal Free Allowance (UFA) alongside the introduction of water charges as a means of protecting households from water poverty. However, UFAs as a mechanism to reduce the average tariff level over the population are costly and reduce the scope for targeted affordability measures for low-income and other vulnerable groups. A small UFA cannot adequately address affordability concerns while a large UFA undermines other policy objectives – notably economic, financial and ecological sustainability.

<http://daccess-dds-ny.un.org/doc/UNDOC/GEN/G10/166/33/PDF/G1016633.pdf?OpenElement> accessed 15/10/14.

[7] Dáil Deb 6 November 2014, 42684W
[8] Dáil Deb 21 October 2014, 40441W
[9] Dáil Deb 4 November 2014, 41161W
[10] See generally Thomas McDonnell, 'Assessing Funding Models for Water Services Provision in Ireland' (2014) 21 *Nevin Economic Research Institute* 1.

Arguably the funding model best able to achieve the main policy goals is a water credit model with a zero free allowance and with tapered contributions to the water bill from a designated government department. A volume based pricing structure with a system of income related water credits would address the affordability issue at a much lower cost to the exchequer than a UFA, and if properly designed would ensure that a combination of water charges and low income does not become a barrier to vulnerable households accessing water and wastewater services.

POLICY OBJECTIVES

A water funding model should attempt to reconcile multiple policy objectives.[11] The fundamental objective is to provide a clean, reliable, and secure supply of water. Beyond this basic goal, policy can be structured around four main 'sustainability dimensions'.[12] These are: (1) affordable water for all households consistent with the concept of the human right to water; (2) financial sustainability for an efficiently operating water utility. In practice this means the resource inputs (i.e. labour and capital) should be fairly compensated, while at the same time ensuring the consumer is best protected in terms of the quality and cost of water and wastewater services; (3) economic efficiency within the water market such that water is allocated to the highest value uses; and (4) conservation and environmental sustainability such that polluters internalise the costs of their pollution. There are additional policy goals that should also be considered, for example, the need to reduce pressure on the public finances, the need to maximise economies of scale, and the need to minimise the cost of borrowing for investment in water infrastructure. Finally, the water-funding model must be seen as fair or it will not be politically sustainable.

Social justice concerns are central to any politically sustainable water-funding model particularly given the salience of water charges compared to other taxes and charges. At issue is whether vulnerable households can be fully protected without sacrificing the other policy goals. UFAs are an expensive, badly targeted and economically inefficient solution because much of the benefits will accrue to wealthier households. A UFA also reduces the potential for reaping economic and environmental benefits

---

[11] See generally Antonio Massarutto, 'Abstraction Charges: How Can the Theory Guide us?' (OECD Expert Meeting on Sustainable Financing for Water Services from Theory to Practice, 2007).
[12] Ibid.

from water charges and places pressure on other areas of public spending. The most economically efficient way to protect vulnerable households against hardship is to supplement the capacity to pay of low-income households through direct cash transfers or some other income supplement. However, because water is a merit good, direct cash transfers are not a complete solution; that is, we want all households to actually benefit from the use of water, not just to have a theoretical capacity to afford water.

Environmental sustainability refers to ecological preservation and the minimisation of waste. The quality of Ireland's water supply has been compromised in recent years, with boil water notices increasing in regularity, while it is estimated that almost 50 per cent of treated water is unaccounted for due to leakages in the deteriorating network infrastructure.[13] Stronger environmental standards and regulatory enforcement are keys to achieving acceptable ecological and public health standards. Both supply side and demand side elements are relevant to environmental sustainability.

Supply side solutions focus on fixing leaks and efficient distribution. Typical solutions include infrastructure improvements, technology upgrades, and improved network maintenance – all of which require funding for investment. Demand side solutions should also be emphasised. These include measures to encourage water saving and discourage wasteful use through usage-based pricing and more efficient water-using devices. Advertising campaigns have frequently been found to be effective. Volumetric pricing is used in a number of OECD countries as a means to internalise the environmental costs of water services to the user of the service. This is based on the 'polluter pays' principle. The volumetric charge paid by the user is simply the volume of water consumed multiplied by the volumetric rate per cubic metre (m^3) set by the regulator. Charging according to the quantity of water used should motivate consumers to conserve water by reducing demand or by switching to more efficient appliances. A general reduction in prices (e.g. from government subsidies) diminishes the economic and environmental signals, and lead to overconsumption.

Studies of household demand tend to find that price is an effective tool for residential water demand management.[14] A study of domestic water consumption in ten OECD countries found that residential consumption responds to volumetric pricing but that of all water-saving devices, only

---

[13] Engineers Ireland, *Delivering Ireland's Water Services for the 21st Century* (Engineers Ireland 2011) 7.
[14] See generally Soteroula Hajispyrou and others, 'Household Demand and Welfare: Implications of Water Pricing in Cyprus' (2002) 4 *Environment and Development Economics* 659.

a low volume/dual flush toilet has a statistically significant and negative effect on water consumption.[15] More generally, meta-analysis studies seem to confirm that residential consumption does respond to price changes. A 1997 study used 124 elasticity estimates from various studies to obtain a median short-run price elasticity of demand of −0.38.[16] A price elasticity of −0.38 suggests a 1 per cent increase in the volumetric price leads to a 0.38 per cent reduction in residential water demand. Similarly, a meta-study conducted in 2003 combined 296 price elasticity estimates to derive an overall mean price elasticity of −0.41.[17] High-income households appear to be less price sensitive than lower income households. Long-run price elasticity is generally found to be higher as households gradually change behaviour.

The economically efficient allocation of water is the one that results in the highest return for a given water resource. In welfare economics, the optimal path for development of the industry is where the marginal benefit of the next increment of water supplied equals the marginal cost of supplying that increment. From this perspective, the optimal price which will maximise society's welfare is equal to the marginal cost of production. Unfortunately marginal cost pricing neglects equity and affordability concerns.[18] We can assess affordability by comparing the water bill to the user's capacity to pay. The key consideration is not the average tariff over the population but the size of the actual tariff paid by low-income households. While there is no commonly agreed rule, the absolute level of affordability can be measured as a percentage of disposable income. In most OECD countries the average water and wastewater bill as a share of disposable income for the least well-off income decile (i.e. the poorest 10 per cent of households) is less than 2.5 per cent, and in some countries the bill is substantially lower as a share of income for this group.

There are other issues with a pricing model based purely on marginal cost. Marginal cost pricing will not recover full costs because water service

---

[15] See generally R Quentin Grafton and others, 'Determinants of Residential Water Consumption: Evidence and Analysis from a 10-Country Household Survey' (2011) 47 *Water Resources Research* W08537.

[16] See generally M Espey and others, 'Price Elasticity of Residential Demand for Water: A Meta-Analysis' (1997) 33 *Water Resources Research* 1369.

[17] See generally Jasper Dalhuisen and others, 'Price and Income Elasticities of Residential Water Demand: A Meta-Analysis' (2003) 79 *Land Economics* 292.

[18] See generally Ghazali Mohayidin and others, 'Review of Water Pricing Theories and Related Models' (2009) 4 *African Journal of Agricultural Research* 1536.

provision is a natural monopoly with high fixed costs and economies of scale. The water utility will make a loss because the marginal cost will always be lower than the average cost. Financial sustainability through a pure tariff model requires the price of the service to be large enough to generate revenues at least as high as the cost of providing that service. The full cost of water not only includes compensating resource inputs (i.e. operation costs, maintenance costs, capital costs and debt servicing costs – the full supply cost), but also economic and environmental externalities, as well as opportunity costs.[19] Financial sustainability requires that cost recovery, or total revenue, does not fall below the full-supply cost. Cost recovery can be attained through tariffs, through taxes, or through borrowings (future taxes and tariffs).

Full Supply Costs (FSC) = capital costs + operation/maintenance costs + debt servicing (14.1)

Full cost of water = FSC + economic and environmental externalities + administration and governance costs (14.2)

There are tensions between the core policy objectives. Multiple policy objectives create dilemmas for policymakers and trade-offs become inevitable (Table 14.1). Best practice for Objective A may not be consistent with achieving Objective B. Where there are multiple policy goals there is a need for multiple policy levers. This suggests a multi-structure funding model is required.

A pure volumetric tariff model for domestic water use is consistent with improved economic efficiency and water conservation, and is at least compatible with financial sustainability. The problem is that consumption charges are regressive, impact disproportionately on low-income households and raise significant affordability concerns. A pure fixed tariff model has little to recommend itself as a policy tool. Such charges are regressive and disproportionately impact upon vulnerable low-income households. Flat charges also generate no price signals and thus no environmental benefits. They may even be counterproductive as consumers will feel they have 'paid' for their water and thus will have no incentive to conserve water. Finally, flat charges provide no benefits in terms of the allocative efficiency of water usage. The 'polluter pays' principle suggests

---

[19] See generally Antonio Massarutto 'Water Pricing and Full Cost Recovery of Water Services: Economic Incentive or Instrument of Public Finance' (2007) 9 *Water Policy* 591.

Table 14.1   Policy trade-offs and dilemmas[20,21]

Trade-off	Policy dilemma
Ecological sustainability vs. Affordability concerns	Environmental 'user/polluter pays' volumetric pricing may not provide affordable water for those on low incomes.
Ecological sustainability vs. Financial sustainability	Higher environmental standards will increase the cost of water provision.
Ecological sustainability vs. Economic sustainability	The most efficient 'high value' allocation in terms of financial rate of return may not be consistent with water saving and preservation.
Affordability concerns vs. Economic sustainability	Should priority be given to merit uses (e.g. washing), or to high-value uses (e.g. industrial processes)?
Affordability concerns vs. Financial sustainability	Full cost recovery through tariffs may be inconsistent with affordability though universally low tariffs could lead to declining infrastructure and deteriorating services which may hurt the poor more in the long-run
Financial sustainability vs. Economic sustainability	Water pricing for economic efficiency suggests marginal cost pricing. From the utilities perspective marginal cost pricing is inconsistent with the accumulation of funds for investment. This suggests the need for average cost pricing or a recurrent fixed charge or subsidy.

water should be funded, at least in part, through water charges based on the level of water usage, and paid for by the user.

In order to best reconcile the core policy objectives, most OECD countries have used a combination of the following elements in their funding models:

A. connection charges;
B. fixed charges;
C. volumetric charges;
D. block charges; and
E. minimum charges.

---

[20] See generally Massarutto, n. 11 above.
[21] See generally Massarutto, n. 19 above.

Many OECD countries use a two-part tariff structure with both fixed and variable elements.[22] The fixed element protects the supplier from demand fluctuations, reduces financial risk, and provides a stable revenue base. The variable element encourages conservation and is based on consumption levels. The downside is that the pricing mechanism (whether based on marginal or average cost pricing) generates negative distributional consequences. To alleviate the negative distributional effect and ensure low-income households have affordable access to water, regulators can be instructed to employ tariff structures such as increasing block tariffs (i.e. the price is cheapest for the first 'block' of water used), or governments can provide income-related water credits paid for out of general taxation. Increasing block tariffs can be progressive to the extent the tariff structure provides a minimum necessary amount of water at a reduced price, which is then paid for by higher prices beyond this minimum. Even so, to offset regressive distributional outcomes it is often necessary to (A) subsidise water provision, (B) use different pricing mechanisms for different household income levels, or (C) provide income supplements for low-income households.[23, 24]

REGULATORY FRAMEWORK

The optimal distribution of goods and services (allocative efficiency) occurs where prices correspond with the firm's long-run marginal cost of supply. Market failure through allocative inefficiency is likely where there is an inadequate level of competition in a market and is commonplace in markets that are natural monopolies. Examples of natural monopolies include ports, water services and other network infrastructures. Left to its own devices the monopolist will fix the price of its good or service above the marginal cost of production in order to maximise its profits. This leads to an economically inefficient level of consumption and production and a deadweight welfare loss to society. The monopolist's super-normal profits also impose competitiveness costs on the rest of the economy.

In some cases, an analysis of the determinants of supply and demand in the market will show it is possible to rectify the market failure through the

---

[22] See generally Peter Rogers and others, 'Water is an Economic Good: How to Use Prices to Promote Equity, Efficiency, and Sustainability' (2002) 4 *Water Policy* 1.

[23] See generally Ariel Dinar and others, *Water Allocation Mechanisms: Principles and Examples* (World Bank 1997) 1.

[24] See generally Mohayidin and others, n. 18 above.

application of regulatory measures, for example, ordering and enforcing the breakup of a dominant market player. However, some markets are characterised by network infrastructure, high fixed costs, and economies of scope and scale. Economies of scale refer to the presence of efficiency gains in the production process (reductions in unit cost) as the usage of inputs increases. If there are significant economies of scale the most efficient market structure may be one containing just a single supplier. This makes competition very difficult to achieve. Indeed competition may actually be destructive in these markets. In such cases, state intervention in the market or even direct state provision of the good or service may be appropriate. The test is whether by intervening in the market the state can achieve the same allocative efficiency at a lower cost than the market – and not whether the state intervention is costly per se.[25] A recent review of the literature on privatisation concluded that the empirical evidence on the comparative performance of public and private sector enterprises fails to provide a clear-cut consensus regarding the superiority of either form of ownership once regulatory environment, size, market structure and incentive structure are all controlled for.[26]

Irish Water is a public monopoly. Public monopolies can suffer from soft budget constraints and unwillingness to minimise their cost bases. This leads to technical inefficiency in the sense that the firm is failing to maximise output for a given level of resources. In principle the inefficiencies caused by the exploitation of a monopoly position can be ameliorated or eliminated by implementing appropriate regulatory mechanisms. The economic goal of regulation is to remedy the inefficiencies of monopoly by simulating the conditions of competition. The regulator must endeavour to minimise the cost to the rest of the economy by ensuring the monopolist cannot earn more than the normal rate of profit, and by ensuring the monopolist is operating to minimise its own internal costs. Regulation itself imposes costs (for example, administration costs for the state, compliance costs for the firm, and legal costs for both) and these costs must be balanced against the gains from regulation. The regulator must also endeavour to avoid setting prices so low that it eliminates the commercial case for future investment.

The regulatory framework and its enforcement form the 'rules of the game' by which the economic actors must operate. These rules are present in some form in all markets. Along with economic efficiency the regulator may

---

[25] See generally Bernadette Andreosso and David Jacobson, *Industrial Economics and Organisation: A European Perspective* (McGraw Hill 2005) 1.

[26] See generally Donal Palcic and Eoin Reeves, *Privatisation in Ireland: Lessons from a European Economy* (Palgrave Macmillan 2011) 1.

also be asked to consider broader issues of equity, distribution, environment, security of supply and service, social cohesion, human rights, and a variety of other strategic economic development concerns. There will be tensions and trade-offs between these goals making an appropriate regulatory regime complex to design. Regulation is often ineffective as a means of exercising authority and can lead to the phenomenon of 'as bad as the law allows'. The regulated firm will be incentivised to obstruct the regulator through whatever means possible (for example legal challenges) unless the costs of doing so outweigh the benefits. There is also a need for clear distance between regulator and operator to avert the risk of regulatory failure or regulatory capture. Full transparency can reduce the risk of regulatory capture.

As the revenue framework for Irish Water, the regulatory authority proposed price/revenue capping with periodic reviews as a form of incentive regulation.[27] The RPI – X formula that was proposed is sometimes described as an efficiency glide path because it gives the utility time to reduce its operational costs to efficient levels. Stephen Littlechild (1983) argued in favour of a price or revenue cap (RPI – X) as the appropriate regulatory tool for a natural monopoly.[28] He realised that a price cap – unlike profit controls – could combine price regulation with preserving incentives for efficiency (Table 14.2 below). Price cap regulation typically specifies an average rate at which the prices that a regulated firm charges for its services must decline annually after adjusting for inflation (RPI). The specified annual rate of decline (X) should reflect the extent to which the regulated industry is capable of achieving more rapid productivity growth than other firms in the economy. In this way the price cap replicates the discipline of competitive market forces by compelling firms to realise productivity gains and pass these on to customers.[29] The price inflation of inputs relative to price inflation in the economy as a whole will also play a role in the appropriate determination of the revenue cap. However, the actual choice of where to reduce costs is left to the regulated company and this helps minimise the regulatory burden.

---

[27] Commission for Energy Regulation 'Economic Regulatory Framework for the Public Water Services Sector in Ireland Consultation Response Paper CER/14/075' <www.cer.ie/docs/000832/CER14075%20Economic%20Regulatory%20Framework%20for%20the%20Public%20Water%20Services%20Sector%20in%20Ireland%20-%20Response%20to%20Consultation%20Paper.pdf> accessed 15/7/14.
[28] See generally Stephen Littlechild, *Regulation of British Telecommunications' Profitability. Report to the Secretary of State* (Department of Industry 1983) 1.
[29] See generally Jeffrey Bernstein and David Sappington, 'How to Determine the X in RPI – X Regulation: A User's Guide' (2000) 24 *Telecommunications Policy* 63.

Table 14.2 Evaluation of schemes vis-à-vis no regulation[30]

	No explicit constraints on profits	Modified profit ceiling	Output-related profit levy	RPI – X (Local) tariff reduction scheme
Protection against monopoly	Fail	Pass	Pass	Pass
Encourage efficiency and innovation	Pass	Fail	Fail	Pass
Regulatory burden	Pass	Fail	Fail	Pass
Promotion of competition	Pass	Fail	Fail	Pass
Regulatory scheme better than none?	NA	Fail	Fail	Pass

The value chosen for X is crucial. If X is too small the regulated firm will earn excessive profits at the cost of the consumer, yet if X is too large it risks threatening the financial sustainability of the firm and its ability to attract investment at a low cost. The revenue cap (RPI – X) framework is made up of three separate building blocks that allow the regulator to estimate a level of revenue sufficient to guarantee financial sustainability for an efficient well-run utility. The three building blocks are: (1) an allowance for Operational Expenditure to fund the efficient day to day running of the utility; (2) an allowance for Capital Expenditure sufficient to fund an appropriate level of investment in the water services infrastructure; and (3) an allowance for the Regulatory Asset Base or RAB. The RAB is a measure of the net value of the utility's assets used in the provision of the regulated activities. These assets (representing previous investments) should be compensated along with Opex and Capex. The estimated RAB allows the utility to receive a fair return on previous capital investments in the water infrastructure.

RPI – X should not be used in a vacuum as it is important to continually compare performance against best practice. Alternatives to RPI – X regulation include rate-of-return regulation; yardstick regulation; banded rate of return; and profit sharing regulation. Although there are no firms in Ireland against which Irish Water's cost base can be compared there

---

[30] See generally Littlechild, n. 28 above, 1.

is a reasonably compelling case that the costs of equivalent companies in the United Kingdom can be used as a yardstick to determine reasonable allowable revenue for Irish Water. Following its interim revenue review the regulator proposed allowable revenue of €2,078 million for the period 1st October 2014–31st December 2016.[31] Operating expenditure was set at €2.5 billion over 2014–16, with an average efficiency target of 2.2 per cent for 2015 and 2016. In addition, capital expenditure totalling €1.875 billion over the three-year period was allocated within the revenue allowance. The opening RAB was set at €1.037 billion.

WATER AND WASTEWATER BILLS: INTERNATIONAL COMPARISONS

Water tariffs, or charges, are the norm in most OECD countries (Table 14.3). Ireland has been a notable exception in this regard since the mid-1990s when water charges were abolished, although non-domestic charges have remained in place continuously throughout the period. Denmark (€5.43 per m^3) had the most expensive water and wastewater services, including taxes, for domestic users of 22 OECD countries surveyed in 2008 while Mexico (€0.40 per m^3) had the cheapest.[32]

Table 14.4 shows average household bills across different water service providers in Great Britain in 2011/2012. The average bill for water and sewerage was €37 per month (€444 per year) and ranged from an average of just over €30 per month in Severn Trent (Midlands) to an average of just over €50 per month in the South West. The industry average for domestic water and wastewater in 2013/2014 was €476. The average water and wastewater bill proposed by the Irish Government for 2015 was €238 or half the amount prevailing in Great Britain.[33]

There is substantial variation across the OECD countries in terms of the financial burden placed by water and wastewater charges on the lowest

---

[31] Commission for Energy Regulation 'Water Charges Plan Consultation: Executive Summary CER/14/366' < www.cer.ie/docs/000979/CER14366%20 Water%20Charges%20Plan%20CER%20Consultation%20,%20Executive%20 Summary.pdf> accessed 31/7/14.

[32] OECD 'Pricing Water Resources and Water and Sanitation Services' (OECD 2010) < www.oecd.org/env/resources/water-therightpricecanencourageef ficiencyandinvestment.htm> accessed 25 January 2016.

[33] See generally Commission for Energy Regulation 'Water Charges Plan Consultation: Executive Summary CER/14/366' <www.cer.ie/docs/000979/ CER14366%20Water%20Charges%20Plan%20CER%20Consultation%20,%20 Executive%20Summary.pdf> accessed 31/7/14.

Table 14.3 Unit price of water supply and sanitation services to households, including taxes, in OECD countries, 2008[34]

Country	Price	Country	Price
Denmark	€5.43	Poland	€1.72
Scotland	€4.63	Hungary	€1.64
Finland	€3.57	New Zealand	€1.60
Flanders (Belgium)	€3.35	Spain	€1.56
Wallonia (Belgium)	€3.18	Japan	€1.50
England and Wales	€3.09	Canada	€1.28
France	€3.03	Italy	€1.17
Sweden	€2.91	Greece	€1.13
Switzerland	€2.54	Portugal	€1.00
Austria	€1.98	Korea	€0.62
Czech Republic	€1.97	Mexico	€0.40

Notes: Prices are Euros per cubic metre. One cubic metre is 1,000 litres. Figures are converted from US Dollars at an exchange rate of 1 US Dollar equals 0.81 Euro.

Table 14.4 Annual water and sewerage household bills in Great Britain[35]

Water and sewerage company	Average bill for 2011/12	Forecast average bill for 2013/14
South West	€608	€587
Wessex	€504	€562
Dŵr Cymru	€484	€511
Anglian	€468	€511
Southern	€467	€528
United Utilities	€442	€478
Yorkshire	€399	€433
Northumbrian (North East)	€395	€422
Scottish Water	€381	€393
Thames	€375	€416
Severn Trent	€366	€394
Industry Average	€444	€476

Note: Converted from Sterling. Figures are based on an exchange rate of 1 Euro equals 0.85 British Pound Sterling.

income decile (Table 14.5). Water and wastewater bills for the lowest income decile range from a high of 10.3 per cent of disposable income in

---

[34] Ibid.
[35] Chris Wallace, *Transforming a National Water Industry: Towards a Hydro Nation. The Scottish Experience* (IIEA Seminar 2011).

Table 14.5 *Average water and wastewater bills for the poorest 10 per cent of households in 2005 (% disposable income)*[36]

Country	%	Country	%	Country	%
Turkey	10.3	Belgium	2.4	Greece	1.4
Poland	9.0	France	2.2	Switzerland	1.4
Slovakia	5.3	USA	2.2	Canada	1.3
Hungary	4.8	UK	2.1	Norway	1.2
Czech Rep.	3.9	Australia	2.1	Korea	1.1
Germany	3.5	Spain	2.0	Italy	1.1
N. Zealand	3.3	Austria	1.7	Netherlands	1.1
Mexico	3.1	Luxembourg	1.6	Sweden	1.1
Denmark	3.0	Finland	1.6	Iceland	0.8
Portugal	2.7	Japan	1.5	Average	2.3

Turkey to just 0.8 per cent in Iceland.[37] Fifteen out of 30 OECD countries, if we include Ireland, were able to keep water and wastewater bills at or below 2 per cent of disposable income for the poorest income decile in 2005 while 11 out of 39 OECD countries were able to keep bills at or below 1.5 per cent. However, eight out of 30 OECD countries would have failed to meet the minimum standards required by the UN's 2010 resolution on the right to affordable water as the cost of water and wastewater exceeded 3 per cent of disposable income for the poorest income decile.[38]

AFFORDABILITY POLICIES

Service charges are generally regressive and care must be taken to ensure low-income households and those with particular needs are protected. Yet water is an economic good and in principle there is no compelling rationale for water to be treated differently to other essential goods such as food, shelter, clothes and energy. We can consider the problem of water affordability and equity in one of two ways:

---

[36] OECD 'Managing Water for all: An OECD Perspective on Pricing and Financing' (OECD 2009) <www.oecd.org/tad/sustainable-agriculture/44476961.pdf> accessed 25 January 2016.
[37] Ibid.
[38] United Nations, 'Human Rights Council Resolution A/HRC/RES/15/9' <http://daccess-dds-ny.un.org/doc/UNDOC/GEN/G10/166/33/PDF/G1016633.pdf?OpenElement> accessed 15/10/14.

(A) We can treat the problem as being that water is too expensive, or
(B) We can treat the problem as being that certain people do not have enough resources.[39]

There are therefore two broad strategies to resolve the affordability and equity problem. The first is to reduce the price of water – potentially all the way to zero. The second is to supplement the incomes of those on low incomes. Most affordability strategies ultimately fall into one or other of these two categories. The more economically efficient way to resolve the affordability problem is to allow the price to be set as normal, and then to supplement the capacity to pay of vulnerable households. This strategy avoids price distortions and directly addresses the distributional issue. The alternative strategy of reducing the effective price of water via a UFA, universal subsidy, or some similar mechanism will lessen the degree of water conservation as well as the allocative efficiency of water use.

UFAs contravene the 'polluter pays' principle, are poorly targeted, highly costly, and unless generous, will be ineffective as a means of averting affordability concerns for low-income households. A relatively generous UFA would not encourage conservation. Low-income and other households with special needs can be better protected, and less expensively, by supplementing their income. While an income supplement will require funding from general taxation, a UFA could require even greater funding depending on the generosity of the allowance. A UFA would have to be subsidised either through higher charges on use above the UFA or through general taxation. UFAs generally represent bad value as measures to help those on low incomes because, to provide a small subsidy to low-income households, a government would have to subsidise water for all households regardless of means and circumstances. Finally, unless the UFA is differentiated by household it will disadvantage larger households relative to single person households, and will also disadvantage those with greater water requirements.

Fortunately there are a number of superior ways to reduce the burden of water charges on low-income households. Table 14.6 outlines the most commonly used affordability measures in the OECD. Income supports were common to all 30 countries in 2005, while 14 out of the 30 forbade disconnection of the water supply.[40] Substantial subsidies as a percentage of total cost for water and sanitation supply, reduced prices for certain

---

[39] TASC 'Water Charges: An Equality-Proofed Approach' (TASC 2012) <www.environ.ie/en/Environment/Water/WaterSectorReform/Submissions/Organisations/FileDownLoad,31833,en.pdf> accessed 25 January 2016.
[40] OECD *Water: The Experience in OECD Countries* (OECD 2006) 36.

Table 14.6 Common policies to support water affordability in the OECD[41]

Policy measure	Countries
Income support for low-income households	30
No disconnection of water supply for low-income households in arrears	14
Subsidies for water supply and/or sanitation over 30% of total service cost	13
Progressive water tariff in general use	13
Social water tariff (reduced price for certain groups of users)	12
VAT on water below normal rate	11
Unmetered (cheap flat rate tariff)	11
Targeted assistance (grants or forgiveness of arrears for low-income groups)	9
No fixed fee (only proportional fee)	8
Reduced waste water tax or other water charges for low-income groups	6
Provision of first block at zero price for low-income or all households	3

Note: Out of 30 OECD countries surveyed.

groups (social tariffs), progressive water tariffs (increasing blocks), and reduced VAT on water, are all commonly used by OECD countries as measures to increase affordability. A low flat rate tariff is often identified as affordability option but offers few advantages other than simplicity; these tariffs are regressive and disproportionately impact upon vulnerable low-income households. A flat annual charge also generates no environmental benefits.

The Department for Environment, Food and Rural Affairs in the UK established an Independent Review which compared a number of different charging bases (e.g. flat rates, rateable property value, property type and occupancy) against eight defined fairness principles.[42] Only volumetric charging scored well as a charging base (Table 14.7). Volumetric charging satisfied a majority of the fairness tests – although crucially the one test it did fail was the affordability test. This suggests that in order to fully address

---

[41] Ibid.
[42] See generally Anna Walker, *The Independent Review of Charging for Household Water and Sewerage Services* (Department of Environment, Food and Rural Affairs 2009) 1.

fairness principles a volumetric tariff structure needs to be combined with income supplements for low income and other vulnerable groups.

From an efficiency perspective it is preferable to charge the full cost of water including the cost of environmental externalities. Efficient water pricing mechanisms almost invariably have negative redistribution implications.[43] Negative distributional effects can be eliminated through a system of differentiated income supplements to support low-income households and other defined groups. Examples include differentiated water contributions (water credits), direct cash transfers and social tariffs.

A. Water Credits

Water is a merit good and there is a public interest in ensuring that, for public health purposes, each household uses a certain minimum amount of water. A properly designed water credit model would ensure that those on lower incomes would not have to decrease their water consumption below

Table 14.7 *Comparative assessment of charging bases against fairness principles*[44]

Principle	Fla	Vol	Rat	Hoc	Bed	Prop
Water efficiency incentive	No	**Yes**	No	No	No	No
Cost-related	No	**It can be**	No	No	No	No
Polluter pays	No	**Yes**	No	**Partly**	No	No
Affordable	No	No	No	No	No	No
Fair to companies	**Yes**	**Yes**	**Yes**	**Yes**	**Yes**	**Yes**
Simple and transparent	**Yes**	**Yes**	No	**Yes**	No	No
Administratively feasible	**Yes**	**Yes**	**Yes**	No	**Yes**	**Yes**

Notes: Fla = Flat Rate; Vol = Volumetric; Rat = Rateable Value; Hoc = Household Occupancy; Bed = Bedrooms; Prop = Property Type.

---

[43] See generally Dinar and others, n. 23 above, 1.
[44] See generally Walker, n. 42 above, 1.

a level necessary to maintain public health.⁴⁵ A designated government department could directly contribute a defined percentage of each household's water bill.⁴⁶ The size of the contribution from the government department could be anywhere from zero to 100 per cent and would be calculated based on the characteristics of the household, for example, the household's disposable income, the number of people in the household and any special need for water (e.g. related to certain medical conditions). The contribution would be zero for a substantial portion of households.

B. Direct Cash Transfers

Direct cash transfers would have the advantage of providing the recipient household with greater choice over their household expenditure. Direct cash transfers would also be more economically efficient than water credits but crucially would not deal with public health concerns as effectively as water credits. This is because the social health benefits of water usage exceed the private benefits of water usage, and therefore the willingness to pay for water services is below the socially optimal level; in other words, there is a risk that households suffering from deprivation will be forced to divert the direct cash transfers towards other essential items, and to reduce their use of water below the optimum level for their own health and for public health generally.⁴⁷

C. Cross-subsidisation

A variation on the water credit model is the cross-subsidisation model. Water credits can be funded by government subsidies or through cross-subsidisation by higher income groups. The cross-subsidisation model is similar to the basic water credit model with the important exception that higher income households would receive a negative water contribution, i.e. an additional charge, and this additional charge would offset the positive water credit allocated to lower income households. Although cross-subsidisation is more progressive than funding from general taxation there are likely to be major administrative complexities associated with developing such a model. It is also likely to be difficult to implement politically and may lead to over-consumption by subsidised consumers and under-consumption by subsidising consumers.⁴⁸

---

[45] McDonnell, n. 10 above, 25.
[46] See generally ibid., 1.
[47] See generally ibid.
[48] See generally Ming-Feng Hung and Bzi-Tzong Chie, 'Residential Water Use: Efficiency, Affordability, and Price Elasticity' (2013) 27 *Water Resources Management* 275.

D. Social Tariffs

Social tariffs entail setting different tariff structures for different categories of people. Typical categories include the unemployed, pensioners, the disabled and low-income groups. Social tariffs are intuitively attractive. However, a recent study found that existing social tariffs in Ireland (e.g. for pensioners) were actually very poorly aligned with those households likely to experience water poverty.[49] It is useful to recall the purpose of affordability measures. The justification for an affordability measure is to ensure that:

1. The overall water charging regime does not increase poverty or push households below a certain minimum standard of living, and
2. The overall water charging regime does not reduce the progressivity of the overall system of charges, taxes and benefits.

In this context, income and net wealth (post-water charges) are the only objective criteria that should be used as the basis for affordability measures. Exceptions might be appropriate for certain groups such as individuals with increased water requirements e.g. related to a medical condition.

The different affordability models are compared in Table 14.8. UFAs

Table 14.8 Comparison of water affordability models[50,51]

Model	Admin	Sub	Type	Fund	Eco	Cost
UFA	Low	All	Fixed	Taxes	Low	Highest
Direct Cash Transfer	Low/Mid	Low income or defined group	Variable	Taxes	High*	
Water Credits	Mid	Low income (tapered)	Variable	Taxes	Mid	
Cross-Subs	High	Low income	Variable	High Income	Mid	Minimal
Social Tariffs	Low/Mid	Defined groups	Fixed	Taxes	Mid	

Notes:
Admin = Administrative complexity; Sub = Group subsidised; Type = Type of subsidy; Fund = Funding source; Eco = Ecological benefits; Cost = Exchequer cost.
*Potential public health risks.
The subsidy under a cross-subsidisation model will be negative in some cases.

---

[49] Paul Gorecki and others, *Affordability and the Provision of Water Services in Ireland: Options, Choices and Implications* (Department of the Environment, Community and Local Government 2013) 51.
[50] TASC, 'Paying for Water: Equity, Efficiency and Sustainability' (2013) 14.
[51] McDonnell, n. 10 above, 22.

and social tariffs are badly targeted and inefficient mechanisms for addressing affordability concerns. Cross-subsidisation is administratively complex and politically problematic. That leaves water credits and direct cash transfers as the most promising models for addressing water poverty and affordability. Water credits best address the merit good issue (public health) whereas cash transfers are more economically efficient and are optimal from an environmental sustainability perspective.

There is no commonly agreed rule on what constitutes water affordability, but it is generally understood to refer to the user's capacity to pay or the share of household disposable income that is required to pay the bill. Measuring water affordability is difficult and complex. According to Ofwat any measure or indicator chosen will inevitably reflect a judgement about the relationship between water and other household needs.[52] For individual customers, how difficult they find it to pay their water and sewerage bills will depend on a range of factors – including other demands on their household income.[53] Ofwat settled on an indicative threshold of above 3 per cent of household income (after housing costs) spent on water and sewerage bills. In many OECD countries the average water and wastewater bills as share of income of the lowest decile of the population is below 2 per cent.

When designing an affordable funding model the precise scale of a particular household's subsidy, water credit, or income supplement could be set to vary depending on the characteristics and circumstances of the household. Examples of relevant characteristics include household income and household size. One option would be to cap water and wastewater charges for all households at a defined percentage of the household's disposable income. A reasonable affordability goal would be to ensure the cost of water does not exceed 2 per cent or even 1.5 per cent of disposable income for the poorest income decile. This seems an achievable goal. The average water and wastewater bill as a share of disposable income of the lowest income decile was just 1.1 per cent in Sweden, the Netherlands, and Italy.[54] For households already in poverty any charge at all will simply push them deeper below the poverty line. Therefore it might be unwise to rely too much on share of disposable income as the only metric for assessing water poverty. Post-water bill disposable income may be more important as a metric. In this context the required contribution

---

[52] Ofwat, 'Affordable for all: How can we help those who Struggle to pay their Water Bills' (Ofwat 2011) 8.
[53] Ibid.
[54] OECD 2009, n. 36 above.

from government could conceivably be as high as 100 per cent for those households' already experiencing poverty.

OUTCOMES AND ALTERNATIVES

Under its original plan announced in July 2014 the Irish government directed that a UFA of 30,000 litres (30m^3) per household be provided for water and for wastewater. The value of this subsidy was €146 per household using both services. There was an additional UFA of 21m^3 for each child. There was no recurrent fixed charge and the combined service would cost domestic users €4.88 per m^3. The average bill per household was expected to be a little over €238 which over a reasonably conservative base of 1.3 million or so households would generate a yield of €310 million from domestic water charges. The regulator projected that the most common household type – a two-adult household – would pay €278 per annum.[55] €278 represents 3.2 per cent of disposable household income for the bottom (poorest) decile, 1.8 per cent for the second poorest decile and 1.3 per cent for the third poorest income decile. Ireland would therefore not meet the UN's basic affordability threshold of 3 per cent for all households. Meanwhile a two-adult household in the top (richest) income decile would on average have to spend just 0.2 per cent of their disposable household income on water services.[56] The extreme regressivity of the proposed water charges regime was clear and late 2014 was characterised by anti-government demonstrations with tens of thousands of people protesting water charges.

In response the government made a series of changes to the structure of water charges in December 2014.[57] The UFA was abolished and a cap was placed on annual charges. Principal private residences with just one adult had their annual bill capped at €160 while all other households had their bill capped at €260. In addition, a prohibition was placed on reducing the supply of water to a dwelling and a water conservation grant of €100 per year – which has nothing to do with conservation – was made available to households registering with Irish Water. The combined service for domestic users was set at €3.70 per m^3. The average bottom decile household with two or more adults and availing of the conservation grant will now have to pay 1.8 per cent of disposable household income. The new regime helped dissipate public anger over the course of 2015. However, the new water

---

[55] Commission for Energy Regulation, n. 31 above.
[56] McDonnell, n. 10 above, 24.
[57] Water Services Act 2014, s. 3.

charges regime remains regressive and clearly fail when measured against three of the four core policy objectives – environmental sustainability, financial sustainability and economic efficiency – and was an extremely inefficient way of addressing the affordability issue.

An alternative water credit based scheme has been proposed that would perform better against the core policy objectives and more cost effectively deal with affordability and regressivity concerns.[58] A viable water credit based scheme needs to be consistent with the sustainability goals (conservation, economic efficiency, financial sustainability and affordability itself – which in this context is also relevant for political sustainability purposes) as well as the general tenets of design for taxes and charges (equity, efficiency and simplicity). Water charges must be affordable, must not push households into poverty and will ideally be progressive.

The most cost-effective way to deal with the issue of water affordability is to restrict household subsidies to those vulnerable to water poverty. This could be achieved through a system of income related water credits where an assigned government agency would directly make a percentage contribution to the water bill for each household. The exact size of the government's contribution to the water bill would vary from zero to 100 per cent depending on the household's disposable income. Affordability issues related to household size or a specific medical condition may best be resolved by establishing a capped annual charge. Disposable household income is the most appropriate subsidisation criteria because water affordability is best assessed by comparing the water bill to the users' capacity-to-pay and by considering the users' disposable income post-water charges. Reasonable targets would be to ensure that the cost of water for households does not exceed an average of 2 per cent, or even 1 per cent, of disposable household income for all income groups and that water charges do not push households any deeper into poverty.

The proposed model would effectively have de-commodified water for the bottom 20 per cent of households by providing a subsidy equivalent to 100 per cent of their water bill.[59] Conservation could still be encouraged for these groups by offering additional subsidies if volumetric usage falls below specified levels. The scale of the subsidy would then taper down as we move up the income deciles with the top 40 per cent of households receiving no subsidy at all. The proposed average contribution from government subsidies was 40 per cent, sufficient to yield €300 million from domestic users. This alternative structure provides for no household

---

[58] McDonnell, n. 10 above, 25.
[59] Ibid.

allowance but does retain the allowance for children. €500 million generated from 1.3 million households equates to an average bill of €385 per household. This is a lower average than the forecast average for every one of the regions in Great Britain (Scottish Water is lowest at €393). It was also proposed that the gross annual bill per household would be capped at €450 for all households. This is to protect large households and households where medical conditions require high water usage. No household would have to spend more than 1 per cent of their disposable household income on water and wastewater services while no households would be pushed deeper into poverty. The proposed model would provide strong price signals for at least two-thirds of the population and therefore contribute to conservation and to economic efficiency.

CONCLUSIONS

The market for water services is a natural monopoly and the Irish government's decision to move to a single provider has a strong economic logic. The introduction of water meters will improve water conservation by helping to identify leaks within the system. A move in the future towards a proper user-based system would contribute to the economic efficiency of water use and the financial sustainability of the water utility. Strong and independent regulation will be required to ensure that Irish Water operates with technical efficiency and that consumers are best protected in terms of both cost and quality of service. The pricing structure and Irish Water's allowable revenue should be decided by an independent regulator.

On the other hand solutions to affordability questions are more properly decided by the legislature. The change in funding model from general taxation to user-based tariffs raises potential affordability issues for low-income households and the Irish government's new funding model is still regressive. If a system of water charges is to be retained in the long term then the current model should be abolished and replaced by a system of income related water credits. This would best protect low-income households and ensure that no household is pushed deeper into poverty by the move towards a user-based funding structure.

15. The role of multinationals in providing water services – are they more efficient?

Tihomir Ancev, Samad Azad and Francesc Hernandez-Sancho*

INTRODUCTION

The growing human population on the planet, combined with the processes of economic development and urbanisation lead to a growing need for municipal water supply and wastewater treatment. One of the UN Millennium Development Goals signifies the importance of access to safe drinking water and basic sanitation for all people.[1] Despite noted progress over the last two decades, some 750 million people in the world remain without access to an improved source of drinking water, and 2.5 billion still lack access to improved sanitation facilities.[2] At the same time, the threat of climate change points to a likely increase in water scarcity in many parts of the world due to the expected changes in hydrological patterns. IPCC estimates that between 0.4 and 3.2 billion people are going to be affected by the adverse effects of climate change on water availability, with mid-range scenario of 1–2 billion people being affected.[3]

---

* Authors listed in alphabetical order.
[1] United Nations 'The millennium development goals report 2014' (2014) New York.
[2] Ibid. See also J Chaisse and M Polo, 'Globalization of water privatization: Ramifications of investor-state disputes in the "blue gold" economy', (2015) 38 *Boston College International and Comparative Law Review* 1.
[3] Intergovernmental Panel on Climate Change (IPCC) 'Climate change 2014: Impacts, adaptation, and vulnerability: Part A: Global and sectoral aspects' (2014) contribution of working group II to the fifth assessment report of the IPPC [CB Field, VR Barros, DJ Dokken, KJ Mach, MD Mastrandrea, TE Bilir, M Chatterjee, KL Ebi, YO Estrada, RC Genova, B Girma, ES Kissel, AN Levy, S MacCracken, PR Mastrandrea, and LL White (eds)]. Cambridge University Press, Cambridge and New York 1132.

In addition to the concerns posed by a growing population and possible climate change, the capacity of many countries to keep up the rates of expansion of their water infrastructure at par with the growing demand is diminishing in the light of sluggish economic growth worldwide and austerity measures taken by many governments around the world. Whereas the water sector needs to grow and provide services to an ever-growing number of users, the current economic environment does not offer favourable conditions for such growth. In the post GFC (Global Financial Crisis) world, there are tight budgetary conditions for all key economic players: governments, businesses, and individual consumers. Economic agents at all levels are becoming very conscious about spending. The situation is also very difficult on the investment front: water industries are forced to compete for investment funds on the open capital markets, in contrast to a time gone by when governments – whether local, regional or national – were willing and able to fund water infrastructure projects. Given the fiscal position of many governments around the world, this source of investment funding is not going to be very generous in foreseeable future. As a result, the overall subdued status of the economies in the world, characterised with high unemployment, government austerity measures, and low returns on capital, makes users very sensitive to pricing of all utility services, including water.

Under these difficult circumstances, the question of how to further develop the water sector so as to provide services to a growing urban population remains unanswered. Governments around the world are faced with the dilemma of investing the limited public funds into water infrastructure, or finding ways to enlist the private sector into investment in water supply and wastewater treatment projects. Many countries around the world now have a mixture of public and private ownership in their water sectors.[4] The aim of improving economic and technical efficiency has been one of the main arguments in favour of privatization of the water sector.[5]

In the light of an increasingly globalised world, including the globalisation of capital assets and investments into such assets, a viable alternative to a domestic private sector investment in water projects is to allow multinational water companies to enter the water sector in a host

---

[4] A Estache and M Rossi, 'How different is the efficiency of public and private water companies in Asia' (2002) 16 *The World Bank Economics Review* 139; M Abbott and B Cohen, 'Productivity and efficiency in the water industry' (2009) 17 *Utilities Policy* 233.

[5] S Singh and K Dickson, 'Transnational corporations and the global water industry' (2011) 53 *Thunderbird International Business Review* 601.

country.⁶ By definition, a multinational company is a business entity that operates in more than one national market.⁷ The late twentieth and early twenty-first centuries saw an advent of such multinational companies in the water sector, led by French companies CGE (Compagnie générale des eaux, currently trading as Veolia Water) and SLEE (Société Lyonnaise des eaux et de l'éclairage, currently trading as Suez Environnement), and followed by corporations from other countries (e.g. Thames Water from the UK). These corporations are seen as being able to mobilise capital on international markets and invest in, and later operate, significant water projects in variety of national markets throughout the world. As these are for-profit publically listed corporations, their objectives are to maximise shareholder return. This is not contentious, as operation of multinational corporations is widely present and accepted in many sectors of the economy.

A more pertinent question is why should any individual country allow such multinational corporations to operate in their water sectors, given the significance of the sector to the public, and the essential nature of services that the sector is providing to its users. As the water sector in an individual country is opened up to foreign investment and business operation, the questions of national security, pride and economic sovereignty inevitably arise.⁸ The relationship between multinational companies and the economies and national policies has been the subject of intense discussion and analysis – both in terms of multinational corporate strategy and the economic autonomy of nation-states.⁹

There are several reasons why individual countries might invite or allow multinational corporations to invest and operate in their water sector. One reason is that these corporations are able to bring desperately needed capital that cannot be found at home, whether from public or private sources. This capital is essential for the purposes of investing in water infrastructure. Another reason is that there is an expectation that allowing

---

⁶ See Julien Chaisse and Christian Bellak, 'Navigating the Expanding Universe of Investment Treaties—Creation and Use of Critical Index,' 18 *Journal of International Economic Law*. 79 (2015), for understanding the link between treaties that states entered into with the increase of the FDI.

⁷ B Roach, 'Corporate power in a global economy' (2007) A GDAE teaching module on social and environmental issues in economics, Global Development And Environment Institute, Tufts University.

⁸ D Hall and E Lobina, 'Profitability and the poor: Corporate strategies, innovation and sustainability' (2007) 38 *Geoforum* 772.

⁹ P Dicken, 'Global Shift' (2003) Sage Publishing, London; R Giplin, 'Global Political Economy' (2001) Princeton University Press, Princeton; JE Stiglitz, 'Globalism's Discontents' (2002) 16 The American Prospect.

multinational water companies to operate in the domestic water sector will mean that they will bring expertise and more efficient technology with them. It is further expected that the effects of this superior expertise and technology might influence the whole water sector, thus improving the efficiency of the industry as a whole and at the same time improving the quality and reliability of service.

This chapter focuses on the latter of these possible reasons of why individual countries are allowing multinationals to operate in their water sector. In particular, we will be addressing the argument that multinational water companies run more efficient operations than the indigenous companies. If this argument holds, than one should be able to observe a difference in efficiency between multinational and domestic water companies. Therefore, a key research question to be pursued is whether the multinational water companies operating in the water sector of a host nation are more technically and economically efficient than the public and/or private water companies that are operated domestically in that same sector. This question is pursued by an analysis of technical and economic efficiency using data on several countries where multinationals operate.

A secondary research question to be addressed is whether municipal water supply sectors in those countries that have a multinational presence are more efficient than sectors in those countries that do not have a multinational presence. If the argument that the multinational presence brings about an overall improvement in the water sector holds, one should be able to observe the differences in the data. We pursue this question by comparing water sectors in countries with the presence of multinationals to those in countries where there is no presence of multinational water corporations.

The overall objective of this chapter is therefore to assess the role that multinational water corporations are playing in the water sector in the host countries, with respect to the expectations of them being leaders in improving efficiency of operations. This test is an important contribution to the debate around justifying the globalisation of water services through the increased presence of multinational water corporations in various national water sectors across the globe.

The chapter is structured as follows: the subsequent section reviews the relevant literature, followed by a section that presents the conceptual framework used in the ensuing empirical analysis. This is then followed by a section on methods and data. Results from the empirical investigation are presented next, followed by discussion, policy implications and conclusions.

LITERATURE REVIEW

The role of the public sector in providing water service has been prominent throughout the world. About 95 per cent of water and sanitation services around the globe are still provided by the public sector.[10] However, since the 1990s, the argument for engagement of the private sector in provision of water and sanitation services became prominent. The key rationale for this argument was that the private enterprises would be more efficient in providing water services than the public enterprises, due to the commercial incentives.[11] Therefore, the participation of private companies in the water service sector during this period increased substantially in Asia, Africa and Latin America, but were mostly concentrated in those countries that had large-scale economies, and where the urbanisation growth rate was high.[12] In recent years the number of multinational water companies operating worldwide has remained small, and mostly originating from a handful of countries (i.e., UK, France). The largest multinational water companies are French (e.g., Suez, Veolia and SAUR) who have established their international operations in Europe and other countries. They have become the global leaders in private water supply.[13]

The entrance of multinationals into the water sector of a host nation can have a direct impact on the whole sector in terms of competiveness and efficiency. As the multinational water companies deal with local private and government-owned institutions and other interest groups, it is expected that there will be positive spillover effects on the host nation's water sector.[14] The consumers in a host country can gain improved services with a competitive price. However, whether or not these benefits will materialise depends on whether the water sector can be characterised as a competitive market with symmetric information and complete contract.[15] The existence of these benefits can be questionable if one of these attributes is absent. Although it is often expected that the entrance and expansion of multinational water companies into the international water market can generate competition between private and public water companies in the water sector, the evidence in practice is that the level of competition in the

---

[10] Hall and Lobina, n. 8 above.
[11] J Budd and G McGranahan, 'Are the debates on water privatization missing the point? Experiences from Africa, Asia and Latin America' (2003) 15 *Environment and Urbanization* 87.
[12] Ibid.
[13] Hall and Lobina, n. 8 above.
[14] Singh and Dickson, n. 5 above.
[15] Ibid.

water sector is low.[16] This is due to two main reasons – first, the number of major multinationals operating in the global water market is very small and secondly, the nature of long-term private lease and contracts often reduce market competition.[17]

Efficiency can be defined as a comparison between observed and optimal values of a production firm's output and input. Efficiency indicates a level of performance that describes a process that uses the lowest amount of production inputs to create the greatest amount of outputs. In case of the water sector, the operation of a water supply plant is considered to be economically efficient when that plant can produce and supply water at the lowest possible cost. Economies of scale and technical efficiency improvements are the mechanisms to lower production costs of water services.[18] Therefore, it is essential to improve economic efficiency of water industry to deliver high quality water services to the affordable prices.

Over the last three decades there has been a growing interest in the economic literature on productivity and efficiency in the measurement of the performance of the water industry.[19] Table 15.1 provides a list of studies that were identified for the purpose of this chapter.

Most studies that engage in measuring the performance or comparing the efficiency of water utility companies focused on two particular types of ownership structures in the water industries — public or private.[20] For instance, using the standard data envelopment analysis approach, Romano and Guerrini estimated the efficiency of Italian water utility companies to assess the impact of ownership structure on the performance of water companies.[21]

Based on 43 water utility companies it was found that the publicly owned water companies have higher efficiency scores compared to private or mixed ownership firms. In contrast, a recent study examines whether the ownership structure and the size of the corporation may affect the

---

[16] F Gonzalez-Gomez and others, 'Beyond the public-private controversy in urban water management in Spain' (2014) 31 *Utilities Policy* 1.
[17] Hall and Lobina, n. 8 above.
[18] Budd and McGranahan, n. 11 above.
[19] M Abbott and B Cohen, 'Productivity and efficiency in the water industry' (2009) 17 *Utilities Policy* 233.
[20] CL Storto, 'Are public-private partnerships a source of greater efficiency in water supply? Results of a non-parametric performance analysis relating to the Italian industry' (2013) 5 *Water* 2058.
[21] G Romano and A Guerrini, 'Measuring and comparing the efficiency of water utility companies: A data envelopment analysis approach' (2011) 19 *Utilities Policy* 202.

Table 15.1 List of studies investigating the effect of the ownership structure on the efficiency in the water sector

Authors and years of publication	Country	Results
Mann and Mikesell, 1976	USA	Public water sector more efficient than private sector
Morgan, 1977	USA	Private sector operators more efficient than public sector
Crain and Zardkoohi, 1978	USA	Private water sector more efficient than public sector
Bruggink, 1982	USA	Public water sector more efficient than private water operators
Feigenbaum and Teeples, 1983	USA	No significant different in efficiency between public and private
Byrnes et al., 1986	USA	No evidence of difference in efficiency between public and private
Teeples and Glyer, 1987	USA	No evidence of difference in efficiency between public and private
Raffiee et al., 1992	USA	Private water sector more efficient than public sector
Lambert et al., 1993	USA	Public water sector more efficient than private water operators
Shaoul, 1997	England and Wales	Improvements in productivity occurred before not after privatisation
Estache and Rossi, 2002	Asia Pacific countries	No significant difference in efficiency between private and public water sector
Garcia-Sanchez, 2006	Spain	No evidence of efficiency difference between public and private companies
Da Silva and Souza et al., 2007	Brazil	No evidence of difference in efficiency between public and private sector
Munisamy, 2009	Malaysia	No conclusive evidence of difference in performance between public and private
Romano and Guerrini, 2011	Italy	Public owned companies more efficient than mixed-owned companies
Storto, 2013	Italy	Involvement of private sector in the water service management may contribute to the improvement of water industry efficiency
Peda et al., 2013	Estonia	No significant difference in efficiency between Estonian water utilities companies with differing ownership forms

Source: Authors' literature survey; Abbott and Cohen (2009: 239).

efficiency of water companies.²² The results from the study suggest that there is no significant difference in efficiency among Estonian water utility companies along varying ownership structures (i.e., public, private or mixed). However, it was found that the efficiency of water utilities increases with the size of the corporation – large water utilities outperform small utilities.

Another recent study evaluated the performance of water companies in terms of operational efficiency in water service provision, with particular focus on the influence of ownership pattern on the overall efficiency level.²³ The findings of the study suggest that the involvement of private water companies in managing infrastructure assets and operating and delivering water services, with or without partnership with the public water sector, may contribute to increased overall efficiency of the water service industry in Italy.

In determining the relationship between ownership structure and performance of the water utility sector, a number of studies have compared the efficiency of publicly versus privately owned water utility companies. However, this literature seems to reach conflicting conclusions about the effect of ownership structure on the efficiency. Some studies find that privately owned water utilities are more efficient, while other studies find the opposite: publically owned water utilities are more efficient.²⁴ In order to examine whether the choice of models might be the source of this conflicting results, Estache and Rossi applied two different econometric approaches to justify the models and robustness of the results.²⁵ Using a sample of 50 water companies from 29 countries in the Asia-Pacific region, this study found that there is no significant difference between private and public companies in terms of efficiency or performance measures.

Abbott and Cohen conducted a review survey of articles that investigated the effects of the ownership structure on the efficiency of water utilities.²⁶ One indication that comes out of that work is that the ownership structure in a water sector is not as important as perhaps other factors are: e.g., economies of scale of a particular utility, or the level of competition in the industry in driving overall efficiency of the water sector.²⁷

---

[22] P Peda and others, 'Do ownership and size affect the performance of water utilities? Evidence from Estonian municipalities' (2013) 17 *Journal of Management and Governance* 237.
[23] Storto, n. 20 above.
[24] Estache and Rossi, n. 4 above.
[25] Ibid.
[26] Abbott and Cohen, n. 19 above.
[27] Ibid.

As is evident from the conducted literature review, there have been a number of studies reported in the literature that have investigated the effects of the three key categories of ownership structure in the water sector: public, private and mixed/public-private ownership. However, no study has focused on examining the efficiency of multinational companies within national water sectors in relation to domestic or indigenous water companies. Very few studies that tackle this question in contexts other than the water sector. For example, Benfratello and Sembenelli, 2006 and Harris and Robinson, 2002 and 2003 investigated the differences in productivity of manufacturing plants along the multinational versus domestic distinction.[28] To the best of our knowledge the present study is the first empirical study that investigates the efficiency of water utilities across multinational, public and private categories.

ANALYTICAL FRAMEWORK

Measuring the efficiency with which productive enterprises are able to turn inputs into outputs has long been a focus of significant economics literature.[29] Efficiency of individual enterprises within a sector of an economy, such as the water sector, can be conceptualised in a fairly straightforward way. This involves collecting data on key inputs and outputs from a number of individual enterprises that constitute the sector. Further, consider plotting these data on a scatter plot, as presented in Figure 15.1(a) below. The next step is to fit an outer envelope through these data. The fitting of such an envelope (a curve, or a frontier) can be done using two main groups of methods: (1) parametric methods, where the envelope is estimated using statistical/econometric methods, with the most widely used approach being the stochastic production frontier estimation;[30] and (2) non-parametric methods, where the envelope is estimated using methods of mathematical optimisation (maximise output for

---

[28] L Benfratello and A Sembenelli, 'Foreign ownership and productivity: Is the direction of causality so obvious?' (2006) 24 *International Journal of Industrial Organisation* 733; R Harris and C Robinson, 'The impact of foreign acquisitions on total factor productivity: plant level evidence from UK manufacturing, 1987–1992' (2002) 84 *Review of Economics and Statistics* 562; R Harris and C Robinson, 'Foreign ownership and productivity in the United Kingdom. Estimates for U.K. manufacturing using the ARD' (2003) 22 *Review of Industrial Organization* 207.

[29] Abbott and Cohen, n. 19 above.

[30] G Battese and T Coelli, 'A model for technical inefficiency effects in a stochastic frontier production function for panel data' (1995) 20 *Empirical Economics* 325.

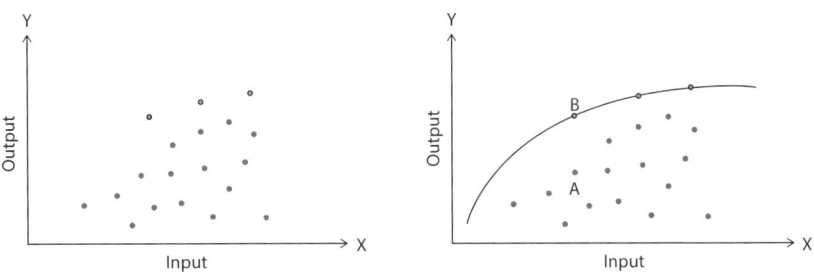

Figure 15.1(a)   Scatter plot of observations on inputs and outputs
Figure 15.1(b)   A fitted outer envelope

a given quantity of input, or minimise input for a given output), with the most widely used approach being the data envelopment analysis (DEA).[31]

Whichever method of estimation is used, those observations on the frontier are deemed to be efficient: they produce the maximum output, given the level of input they are using, or vice versa they use minimum input for the level of output that they are producing. This is depicted in Figure 15.1(b).

For example, the observation on production unit B in Figure 15.1(b) indicates that this unit is efficient, whereas the observation on unit A indicates that it is not, because it uses the same amount of input as unit B, but produces substantially less output. The inefficiency of unit A can be measured by calculating the Euclidean distance between the observation for it and the frontier. This distance can be measured by a distance function, as introduced by Farrell.[32] The distance can be measured in any direction towards the frontier. Efficiency is then expressed as a proportion of the measured distance. It is a score between zero and one, with a score of one indicating that the observed production unit operates on the frontier. The general approach described above has been extensively used in many studies and under many methodological variations.[33]

Based on the above, one can distinguish between two separate faces of

---

[31]   A Charnes and others, 'Measuring the efficiency of decision making units' (1978) 2 *European Journal of Operational Research* 429.

[32]   MJ Farrell, 'The measurement of productive efficiency' (1957) 120 *Journal of the Royal Statistical Society* Series A 253.

[33]   A Emrouznejad and others, 'Evaluation of research in efficiency and productivity: A survey and analysis of the first 30 years of scholarly literature in DEA' (2008) 42 *Socio-Economics Planning Sciences* 151.

the concept of efficiency measurement: technical efficiency and economic efficiency. Simply put, technical efficiency is defined as the ratio of the physical quantity of output(s) to the physical quantity of inputs (as discussed above), while economic efficiency is defined as the ratio of the value of output(s) to the value of inputs. It is immediately apparent that economic efficiency means that the physical quantities used in the measurement of the technical efficiency are weighted by output price(s) and by the costs of inputs. Within this regulated sector, water tariffs that utilities are able to charge to their customers could be a significant factor explaining the differences between technical and economic efficiency for each water company. In general, if technical efficiency scores are found to be lower than the economic efficiency scores, it is an indication that those utilities may be relying on higher tariffs as a mechanism to compensate for their technical inefficiency.[34] In the ensuing economic analysis, we will investigate both technical and economic efficiency of water utilities operating in several countries.

DATA AND METHODS

Detailed data on input use, output production, costs, revenues, as well as ownership and management information were needed in order to measure technical and economic efficiency of individual water utilities and test whether those water utilities owned and operated by multinational corporations attain higher efficiency scores. Databases that readily provide such information do not exist. We have therefore relied predominantly on two separate databases: IBNET and ORBIS. Data collected from these databases were supplemented with data published in other sources. The International Benchmarking Network for Water and Sanitation Utilities (IBNET) is an international database that provides a set of core cost and performance indicators of water and sanitation utilities from various countries. It is a part of the Water and Sanitation Program of the World Bank. ORBIS is a global company database, produced by Bureau van Dijk, with financial and ownership information from public and private companies throughout the world, including water utilities.

As IBNET provides data by country, an additional difficulty was to determine the ownership structure of individual water utilities in specific countries. In particular, there are relatively few countries for which data

---

[34] F Hernandez-Sancho and others, 'Tariffs and efficient performance by water suppliers: An empirical approach' (2012) 14 *Water Policy* 854.

exist in IBNET apart from OECD countries and non-OECD European countries, where there is multinational presence in the water sector. We had to use the ORBIS database to determine the ownership structure of individual water utilities. Moreover, not all countries represented in IBNET have relatively recent information on them. We aimed to use the most recent data reported by IBNET which was for the year 2008. These constraints led us to gather data from IBNET for Chile and Portugal, where there is multinational presence in the water sector. In addition, we were able to get data on the water sector for England and Wales, where several multinational companies also operate in the water sector. Data on input and output variables for the water companies operating in England and Wales were garnered from the relative efficiency assessment document 2008–09 published by the Office of Water (OFWAT, 2009).

The list of countries with multinational presence in the water sector that were analysed also included Italy. In the case of Italy, we conducted an analysis based on the results published in an article by Romano and Guerrini.[35] This paper conducted an efficiency analysis of 43 water utilities in Italy and looked at the ownership structure (public vs. private) but did not account for the multinational presence. We used the reported efficiency scores in that paper, derived the ownership structure information from the ORBIS database, and related the two to test whether utilities in Italy that are owned and operated by multinational corporations have higher efficiency scores.

We also used data on several other countries where there was no presence of water utilities owned and operated by multinational corporations. These countries were Egypt, Mexico, Philippines, Serbia and Lithuania. These countries were selected taking into account considerations for a broad geographical/regional representation, as well as based on the availability of recent data for those countries. The purpose of looking at the data on these countries was to be able to make comparisons of the overall efficiency of the water sector (measured by Total Factor Productivity (TFP)), between countries where there was a multinational presence and countries where there was no multinational presence.[36] These countries

---

[35] G Romano and A Guerrini, 'Measuring and comparing the efficiency of water utility companies: A data envelopment analysis approach' (2011) 19 *Utilities Policy* 202.

[36] Total factor productivity index is a productivity measurement index that is used to assess the productivity of using all the inputs in a production system. It is defined as the ratio of the total output to the total inputs used in an individual production process or within a sector of the economy. It can be used to investigate cross-country differences in sectoral productivity, or in productivity growth.

were selected taking into account considerations for a broad geographical/ regional representation, as well as based on the availability of recent data for those countries.

Data on the following variables were obtained from IBNET database: average quantity of water delivered to consumers (m³/connection/month) was treated as an output for the technical efficiency model, and average revenue from water sales ($US/connection/month) was treated as an output for the economic efficiency model; average operations cost ($US/ connection/month) – representing the variable costs –, and the average value of gross fixed assets ($US/connection) – representing the capital costs –, were treated as input variables in both the technical and economic efficiency model. The use of the same type of inputs in terms of costs in both models was necessary due to data availability. However, this also allows us to put an emphasis on the role of water tariffs as a factor explaining the differences between the efficiency scores obtained under the two models. As IBNET does not directly report data in the format stated above, we had to undertake calculations to bring all data in a common format, expressed per connection per month. Summary statistics for the input and output variables for all countries under investigation are provided in Table 15.2 below.

The data garnered in this way were then used in a Data Envelopment Analysis (DEA) model to derive the efficiency score for each individual water utility in each of the countries that we investigated. The DEA model, formally introduced by Charnes et al. is a technique for measuring the relative efficiency of a productive unit.[37] This model is well established and it offers a theoretically sound framework for conducting performance analysis for a set of decision-making units (DMUs). In the current context, individual water utilities are the DMUs. If a production process involves a single input and single output, the efficiency of a DMU can be simply defined as the ratio of that output to the input. However, when there are more than one input and/or output in the production system (as is the case in the present case where we have a single output but at least two inputs), the efficiency can be defined as:

Efficiency = weighted sum of outputs/weighted sum of inputs   (15.1)

The choice of weights in the above relationship is a problem that can be solved using DEA by introducing a particular weighting system for each DMU. Given the output denoted by y, and the inputs denoted by x, the

---

[37] Charnes and others, n. 31 above.

334 Charting the water regulatory future

Table 15.2 Descriptive statistics for the input and output variables included in the efficiency model

Country	Water quantity delivered to consumers (m^3/conn/month)	Total operational cost (US$/conn/month)	Value of gross fixed assets (US$/water connection served)	Total revenue (US$/conn/month)
Chile				
Mean	44.92	19.47	44.49	35.79
Maximum	203.90	61.20	218.13	144.77
Minimum	14.80	5.50	13.07	12.28
Portugal				
Mean	28.40	46.63	159.14	53.26
Maximum	108.50	110.16	412.13	143.22
Minimum	8.90	10.36	23.13	20.03
Egypt				
Mean	56.92	7.50	10.46	8.28
Maximum	67.30	7.69	75.43	8.46
Minimum	56.40	4.08	0.34	4.71
Serbia				
Mean	30.40	17.31	52.15	20.40
Maximum	60.40	35.38	180.06	42.74
Minimum	8.90	2.31	3.97	5.52
Philippines				
Mean	36.03	7.28	6.72	14.89
Maximum	94.00	13.34	14.75	43.24
Minimum	11.30	1.92	0.97	2.03
Mexico				
Mean	36.41	13.96	48.66	25.92
Maximum	60.00	33.48	314.09	51.00
Minimum	18.90	3.69	2.10	6.94
Lithuania				
Mean	34.94	44.16	83.01	34.67
Maximum	88.40	115.60	193.43	76.54
Minimum	7.90	10.40	22.44	9.48

maximum efficiency for one of the DMU's in the sample, denoted here by k_c can be calculated as follows:[38]

---

[38] Assume that the total number of DMU's in the sample is equal to z, where

$$\max \theta = \frac{\sum_{j=1}^{n} w_j y_{jk_o}}{\sum_{i=1}^{m} v_i x_{ik_o}}$$

subject to:

$$\frac{\sum_{j=1}^{n} w_j y_{jk}}{\sum_{i=1}^{m} v_i x_{ik}} \leq 1 \quad k = k_0, \ldots, z$$

$w_j, v_i \geq 0$

where z = Number of DMUs
m = Number of inputs
n = Number of outputs
w_j = Weight given to output j
v_i = Weight given to input i

(15.2)

This model involves finding values for w and v, such that the efficiency measure of the k_0^{th} DMU is maximised, subject to the constraint that all efficiency measures must be less than or equal to one. The value of θ obtained from the above equation is the efficiency score for the k_0^{th} DMU. If the efficiency score of the k_0^{th} DMU equals one, then it implies that the maximum efficiency is achieved relative to the other DMUs. On the other hand, if the efficiency score of k_0^{th} DMU is less than one, then it indicates that this production unit is inefficient compared to the frontier DMU (the best DMUs in the sample). For each water utility (DMU) for which we collected data, the above equation was solved using the General Algebraic Modelling System (GAMS) software. This produced efficiency score for each water utility.

The derived efficiency scores are reported in the subsequent section. The efficiency scores were then related to a binary variable that took a value of zero for those utilities that were not owned and managed by a multinational corporation, and unity for those utilities that were. This was conducted under the framework of the Kruskal-Wallis test.[39]

---

each DMU is denoted by the letter k and the consecutively numbered subscripts, starting with k_0.

[39] WH Kruskal and WA Wallis, 'Use of ranks in one-criterion variance analysis' (1952) *Journal of the American Statistical Association* 47.

This non-parametric test uses comparisons between the medians of two samples to determine if those samples have come from different populations. In the current context, for each country we had a sample of water utilities that were owned/managed by a multinational corporation, and a sample of water utilities that were not. The hypothesis tested using the Kruskal-Wallis test was that there is no significant difference in efficiency scores between these two samples.

To address the postulated research question of whether the multinational presence in a water sector of a given country contributes to an overall more efficient performance of that sector, relative to the water sector in those countries that did not have multinational presence, we calculated TFP indexes of the water sector for each country under investigation.[40]

Given the data that we had collected from IBNET database, where the physical quantity of inputs were not available, we used a price/cost weighted TFP index, which was a ratio of the total revenues from water sales for the whole water sector in a particular country to the total operation costs. The calculated value of the TFP for each country was related to a binary variable that took a value of zero if there was not multinational presence in the water sector in that country and unity if there was a multinational presence. This was again done using the Kruskal-Wallis test, where countries with multinational presence were treated as one sample, and countries without multinational presence as another sample. The hypothesis tested was that there is no significant difference in the value of TFP indexes between these two samples.

RESULTS

The results from the analysis of technical and economic efficiency in those countries where there is presence of multinational corporations in the water sector are presented in Figures 15.2 and 15.3. These figures show the calculated efficiency scores of individual water utilities grouped by whether or not they are owned/managed by a multinational corporation. Under the pane for each individual country, the figures also contain the results from the conducted Kruskal-Wallis tests.

---

[40] TFP was not calculated for Italy, because we did not have input and output data for this country. We only used already published efficiency scores for Italy (Romano and Guerrini, 2011) and related them to the ownership/management information for each individual water utility.

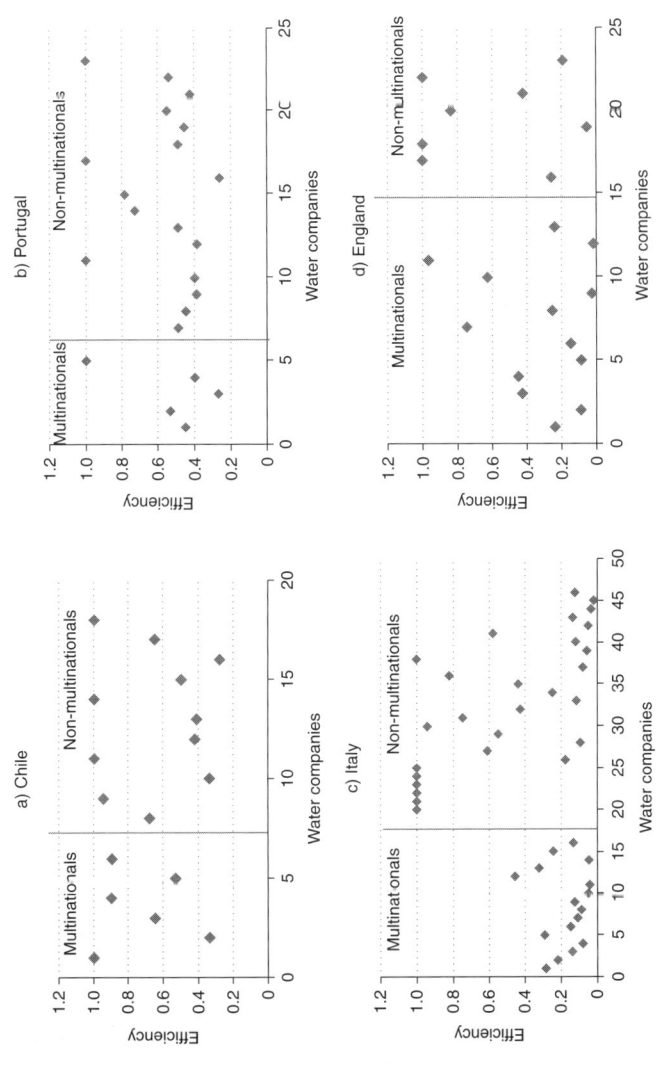

Notes:
a) Kruskal-Wallis test significance value p = 0.884 at significance level of 0.05. Fail to reject the null hypothesis.
b) Kruskal-Wallis test significance value p = 0.664 at significance level of 0.05. Fail to reject the null hypothesis.
c) Kruskal-Wallis test significance value p = 0.417 at significance level of 0.05. Fail to reject the null hypothesis.
d) Kruskal-Wallis test significance value p = 0.121 at significance level of 0.05. Fail to reject the null hypothesis.

Figure 15.2   Technical efficiency scores in countries with presence of multinationals in the water sector

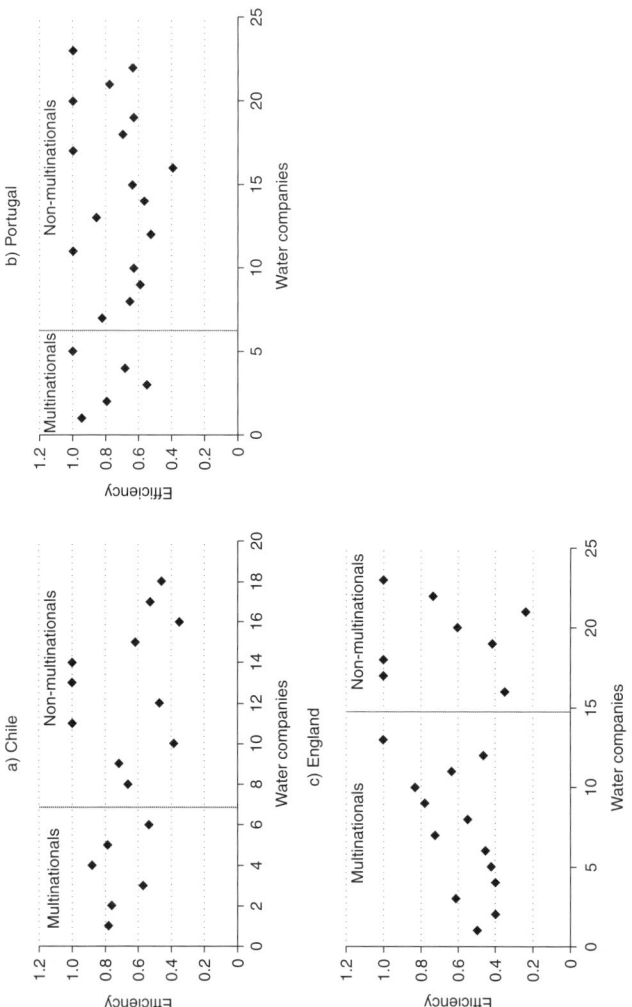

Notes:
a) Kruskal-Wallis test significance value p = 0.404 at significance level of 0.05. Fail to reject the null hypothesis.
b) Kruskal-Wallis test significance value p = 0.651 at significance level of 0.05. Fail to reject the null hypothesis.
c) Kruskal-Wallis test significance value p = 0.750 at significance level of 0.05. Fail to reject the null hypothesis.
Only technical efficiency information was available for Italy, as we did not have access to the data by individual utilities, but have instead relied on technical efficiency scores published in Romano and Guerrini (2011).

Figure 15.3   Economic efficiency scores in countries with presence of multinationals in the water sector

Only technical efficiency information was available for Italy, as we did not have access to the data by individual utilities, but have instead relied on technical efficiency scores published in Romano and Guerrini (2011).

As can be seen from the figures, there is no apparent systematic difference in efficiency scores between water companies that are owned/managed by multinational corporations and those that are not. This is confirmed with the Kruskal-Wallis test results, which indicate that the hypothesis that the efficiency scores for water companies owned/managed by a multinational are not significantly different from the efficiency scores for water companies that are not owned/managed by a multinational, cannot be rejected at the 5 per cent level of significance. These findings are consistent across the countries that were examined, as well as for both technical and economic efficiency. The implication from the findings is that there is no empirical evidence to suggest that those water companies that are owned/managed by a multinational operate more efficiently than water companies that are owned/managed domestically.

The fact that these findings are consistent across technical and economic efficiency scores is encouraging. Had we found that there was no difference between multinational owned/managed and domestically owned/managed water utilities based on technical efficiency scores, but that there was a difference based on economic efficiency scores, that would had led us to infer that there is perhaps some systemic difference in the structure of water rates between these two types of water utilities. In particular, there has been some evidence of operations with relatively low technical efficiency scores that can have high economic efficiency scores. This directly points to the possibility that the water rates these operations have been able to charge (i.e., have been approved by a regulator) are higher than they should have been. In this way, the higher water rates can cover the relative inefficiency of an operation. It is satisfying to see that such instances do not systematically appear in the data that we have analysed.

The results from the analysis of total factor productivity (TFP) indexes are provided in Table 15.3. The table contains an index calculated based on delivery of water, and an index calculated based on economic return. These indexes were calculated for the countries where there was a presence of multinational corporations in the water sector (except for Italy, for which there were no elementary data available that would allow calculating the TFP), and for countries where there was no presence of multinational corporations in the water sector.

There does not seem to be any particular pattern when comparing the TFP indexes between these two types of water sectors: with and

Table 15.3 Total Factor Productivity (TFP) for the water sector across countries

Country	Type	TFP (based on water delivered)	TFP (based on economic return)
England	Multinational presence	3.992	0.631
Chile	Multinational presence	1.341	1.075
Portugal	Multinational presence	0.138	0.259
Egypt	No presence of multinationals	3.170	0.461
Serbia	No presence of multinationals	0.438	0.294
Philippines	No presence of multinationals	2.574	1.064
Mexico	No presence of multinationals	0.581	0.414
Lithuania	No presence of multinationals	0.275	0.273

Notes:
TFP (based on water delivered) = Total quantity of water delivered to customers/total cost to the water sector.
TFP (based on economic return) = Total revenue from water sales/total cost to the water sector.

without multinational presence. This is confirmed with the results from the Kruskal-Wallis test for the TFP indexes based on water delivery, which indicated that the hypothesis that there is no significant difference between the TFP indexes for those countries where there is multinational presence in the water sector, and those countries where there is not, cannot be rejected based on a significance value of $p = 0.881$. Similar results are obtained for the TFP indexes based on economic return, where the tested hypothesis cannot be rejected based on a significance value of $p = 0.655$.

CONCLUSION

Multinational corporations have been increasing their presence in the water sector of many countries around the world over the last 20 years or so. There are several arguments that have been used to justify this increased involvement of multinationals in the international water sector. Among those arguments, the one that points to the efficiency gains from having multinational corporations establishing operations within the water sector in a host nation has been prominent.

In this chapter we conducted analyses to look for empirical evidence for this argument. The analyses were based on collating data from variety of sources. This posed a considerable challenge, as data on the key indicators of interest reported in the leading international databases are non-consistent across time or across countries. The additional difficulty was to access the information on the ownership/management of individual water utilities. All this meant that we had to restrict our attention to relatively small number of countries for which we could collate sufficient data. Ideally, we would have liked to include a much larger number of countries. Nevertheless, the findings from the empirical analysis are telling, and can be used to draw meaningful conclusions. In addition, the procedures described in this book chapter can be easily used in the future to expand the scope of the empirical analysis to other countries as more data become available.

Based on the findings from the conducted empirical analyses, one can conclude that the evidence is not supportive of the argument that multinational presence in the water sector enhances the efficiency of the individual water utilities, or of the sector overall. Calculated efficiency scores for the water utilities that are owned/managed by multinational corporations in a given country were found not to be significantly different from the efficiency scores for the water utilities that are domestically owned/managed. This finding was consistent across technical efficiency scores – which account for the efficiency in physical transformation of inputs into outputs –, and economic efficiency scores – which account for the efficiency in terms of revenues and costs. These findings suggest that multinational corporations are not running water utility operations that are more or less efficient than the rest of the water sector in individual countries.

A related argument in justifying the engagement of multinationals in the water sector of individual countries is that they might help lift the overall productivity of the water sector. We empirically tested this argument by looking at total factor productivity of the water sector in countries where multinational corporations are present in the water sector, and countries where they are not. The empirical findings indicated that there is no difference in total factor productivity of the water sectors across countries, whether there is a presence of multinational corporations or not. These findings lead to a conclusion that the multinational presence in the water sector does not necessarily lead to increased productivity of the sector.

The two key conclusions derived based on empirical evidence suggest that the argument for opening up a domestic water sector to multinational corporations based on expected productivity and efficiency gains may not be as strong as usually purported. Further empirical work should test whether other points of justification for opening the domestic water sector to multinational presence can be supported by observed data.

16. Microfinance in water and sanitation services: identifying best practices

Jonatan A. Lassa and Allen Yu-Hung Lai

INTRODUCTION

It is projected that the majority of urban population growth in the next quarter of a century will occur in developing countries.[1] Increasing numbers of people, assets, infrastructure and other types of wealth will, therefore, be located in cities. Unfortunately, these cities are likely to face additional stressors from future climate extremes. Ports may face higher exposure to extreme hazards, with one of the likely scenarios being exposure to sea level rise (SLR), but their general level of climate preparedness may be low and the degree of their vulnerability varies greatly.[2] Research communities over the last 20 years have been focusing on mega cities[3] and primary cities and the mega-risks they may be exposed to.[4]

Recently, however, the Rockefeller Foundation has initiated a new focus on secondary cities in Asia such as India, Indonesia, Thailand and Vietnam. Unfortunately, there is still little research being done on the secondary cities even though it is clear that some of them will be transformed into 'mega' cities within this century. It is suggested that

---

[1] S Hallegatte and J Corfee-Morlot, 'Understanding climate change impacts, vulnerability and adaptation at city scale: an introduction' (2011) 104(1) *Climatic Change* 1–12. DOI 10.1007/s10584-010-9981-8.

[2] S Hanson and others, 'A global ranking of port cities with high exposure to climate extremes' (2011) 104 *Climatic Change* 89, DOI 10.1007/s10584-010-9977-4. Hallegatte and Corfee-Morlot, ibid.

[3] D Parker and JK Mitchell, 'Disaster vulnerability of megacities: an expanding problem that requires rethinking and innovative responses' (1995) 37 *GeoJournal* 295; S Hochrainer and R Mechler, 'Natural disaster risk in Asian megacities: A case for risk pooling' (2011) 28 *Cities* 53.

[4] F Wenzel and others, 'Megacities – megarisks' (2007) 42 *Natural Hazards* 481.

building resilience into these current secondary cities might help reduce their vulnerability to disaster as they become the future mega cities.

In Indonesia, the two cities were selected to be part of the Asian Cities Climate Change Resilience Network (ACCCRN), namely Bandar Lampung City and Semarang City. The latter, in particular, has been increasingly affected by different sets of risks including climate risks. Some parts of the city are 'sinking' due to land subsidence coupled with an early indication of mean sea level rise.[5] Land subsidence is occurring in areas with compressible deposits and scientists suggest that the subsidence is primarily caused by ground water extraction for industrial and agricultural use.[6]

The ACCCRN project has helped the cities to conduct several necessary steps in building disaster resilient communities. Several multi-stakeholder forums known as Shared Learning Dialogues have been created. These lead to the production of vulnerability assessments (VA) and city resilience strategies have been drafted and adopted into local development planning and implementation. In Semarang, two types of VAs were conducted: the first was a science-based assessment on the technical dimensions of climate science;[7] the second was a community-based vulnerability assessment.[8]

In Semarang City, the ACCCRN Project has created a City Team that functions as a new mode of urban climate governance where decision-making processes are made collectively by the City Team through multi-stakeholder platforms made up of relevant departments, civil society organisations, local universities and representatives.[9] Figure 16.1 represents the ACCCRN processes and activities in Semarang City over the last five years. The overall framework has been named as Urban Climate Change Resilience Planning Framework (UCCRPF) which functions as

---

[5] E Chaussard, F Amelung, H Abidin and S-H Hong, 'Sinking cities in Indonesia: ALOS PALSAR detects rapid subsidence due to groundwater and gas extraction' (2013) 128 *Remote Sensing of Environment* 150.

[6] See generally J Chaisse and M Polo, 'Globalization of water privatization: Ramifications of investor-state disputes in the "blue gold" economy' (2015) 38 *Boston College International and Comparative Law Review* 1.

[7] CCROM and others, *Vulnerability and adaptation assessment to climate change in Semarang City* (2010). Unpublished Report, Mercy Corps – ACCCRN Indonesia Project.

[8] J Taylor, 'Community-Based Vulnerability Assessment: Semarang, Indonesia' in K Otto-Zimmermann (ed.) *Resilient Cities: Cities and Adaptation to Climate Change Proceedings of the Global Forum 2010* (Local Sustainability 1 2010) 329. DOI 10.1007/978-94-007-0785-6_34.

[9] JA Lassa and E Nugraha, 'From shared learning to shared action in building resilience in the city of Bandar Lampung, Indonesia' (2015) 0 *Environment & Urbanization* 1. DOI: 10.1177/0956247814552233.

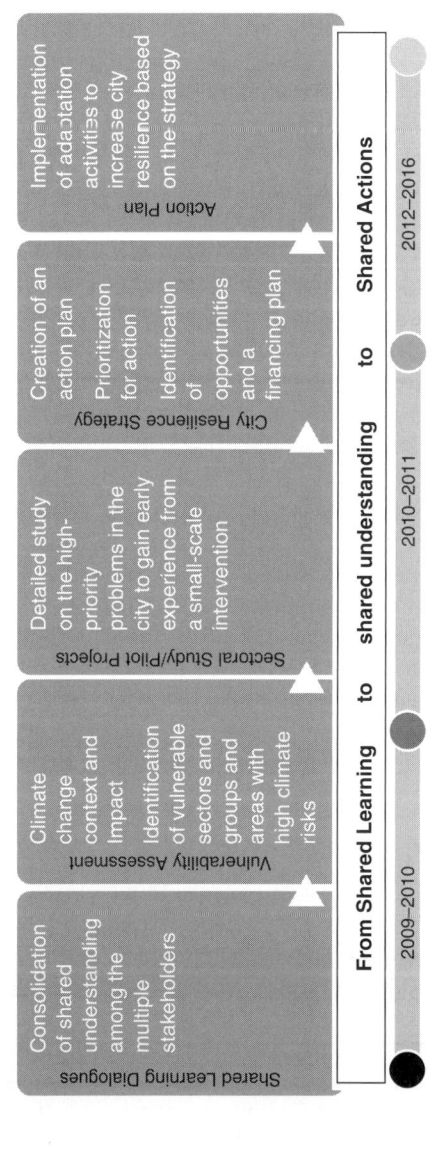

Source: Internal ACCCRN Project Report/Presentation (Lassa and Nugraha 2015).

Figure 16.1 ACCCRN typical process in Semarang City

a process-oriented framework as it 'incorporates a specific set of process considerations and supports activities that can assist urban areas in planning, capacity building, implementing and supporting the continuous process of learning that is central to the growth and maintenance of urban resilience'.[10]

In 2010, the City Team suggested that relevant agents, especially non-governmental organisations (NGOs), propose innovative ideas that could be piloted in the most vulnerable communities in the city. All the applicants should consider the VAs of Semarang City jointly produced by ISET (Institute of Social and Environmental Transition), the Center for Climate Risk and Opportunity Management in South East Asia, Pacific *Institut Pertanian Bogor*, Mercy Corps, and URDI, an Indonesian based NGO.[11]

The city team decided to approve four pilot schemes to address different hazards: landslides, flooding, coastal erosion, and storms. The VA suggested that community adaptation activities have been taking place among households in coastal areas where areas affected by flooding raised the level of their floors and fishermen had had to enter the city's informal sector, borrowing from informal lenders (loan sharks) to fulfil their daily needs. In order to protect fishponds and reduce coastal erosion, the City Team supported by a local NGO and municipal government, initiated and continued mangrove planting. The pilot projects in Semarang initially planned to be implemented from March until June 2010, but some of the projects' completion was delayed until August 2010.

This chapter investigates the sustainability of the ACCCRN intervention in a pilot project, namely community-based finance for urban sanitation. We argue that there is an existing research gaps within both climate change adaptation and risk management communities especially on the role of micro finance for urban climate adaptation and urban risk reduction.

PROJECT LOCATION

The location of the pilot project was in Kemijen Village, east of Semarang City, where local communities have been exposed to multi-risks including: a high rate of land subsidence (about -10 to -15cm annually); rising sea

---

[10] Moench and others, *Catalyzing Urban Climate Resilience. Applying Resilience Concept to Planning Practice in the ACCCRN Program 2009-2011* (ISET: Boulder Colorado 2011), 3.; S Tyler and M Moench, 'A framework for urban climate resilience' (2012) 4 *Climate and Development* 311.
[11] Delima Sari, A. 2012. Mercy Corps - ACCCRN Project Staff, personal face to face interview by Lassa on 28 June 2012.

levels; and river flooding (worsened by high sediment transport in the rivers). Based on an existing loss model (due to floods and SLR), Kemijen Village has experienced an increase in economic loss (e.g. housing, productivity, education and health) over the last five years.

Kemijen has also experienced an increase in average annual household loss (in US$ equivalent) from US$ 600 in 2007 to US$ 1,800 in 2010, most significantly in housing and productivity. Over the last three years economic losses have also been increasing.[12] Kemijen's vulnerability is also demonstrated by the fact that the village lacks services provided by the city water company; many households have to buy water from private sources which are often costly.[13]

FINANCING URBAN SANITATION ADAPTATION OPTION

The original idea of exercising a community-based revolving fund to support urban sanitation among the poor who suffer most from the urban flooding and land subsidence in Semarang City came from an NGO, Perdikan, in 2010 (the NGO no longer exists). The project took place in Kemijen Sub-District where it engaged households headed by poor females to improve their sanitation system through a revolving loan fund and to raise awareness among participants of climate issues. It looked for a model of microfinance that would improve the livelihood of poor female-headed families while simultaneously improving their sanitation conditions.

A community-based VA in Kemijen Village in October 2009[14] identified households headed by females as the group most vulnerable to climate risks. Perdikan, with the support from Mercy Corps (ACCCRN Indonesia), therefore, designed a pilot project that sought to reduce the level of vulnerability of these households in the low-lying coastal area of East Semarang City that was often affected by sea inundation and flooding. The project tried to simultaneously improve the livelihood and

---

[12] Institute for Social and Environmental Transition (ISET), Semarang's adaptation plan in responding to climate change. City Resilience Strategy of Asian Cities Climate Change Resilience Network, (ISET), 22; Bintari, 'Economic loss assessment of floods in Kemijen, Semarang City' (Kajian Kerugian Ekonomi akibat Banjir di Kelurahan Kemijen Kota Semarang) (Semarang 2007) unpublished report.
[13] Taylor, n. 8 above.
[14] Ibid.

sanitation level of female-headed families inhabiting the low-lying coastal areas by implementing a credit scheme.

The idea was to work through an existing village financing institution, the BKM Kemijen, to reach the most vulnerable group. The sanitation fund was complemented by community educational awareness. It was believed that identified vulnerable groups can be encouraged to be adaptive to climate change. In addition, the model was expected to be replicated in other areas in a similar situation as it created incentives to build the vulnerable community's resilience towards climate change through access of water and sanitation.

The pilot project was implemented during March–May 2010. Activities carried out to meet the objectives included:

- participatory planning and assessment of female-headed families;
- discussions with the Community Resilience Board (BKM Kemijen) and neighbourhood groups;
- meetings between Community Resilience Board (BKM Kemijen) and the female-headed families about the VA result and discussing the concept and impact of climate change and sanitation-saving activity implementation;
- regular meetings of female-headed families;
- serial thematic focus group discussion in ten sub-villages (neighbourhood groups);
- sanitation-saving activity (implementation of revolving fund) for 25 female-headed families;
- a Climate Change Festival at village level.

This approach was perceived by Perdikan as something new which could integrate a curative approach, organising the community and the institutionalization of ideas about climate-change adaptation at a local level.[15]

From the outset, Perdikan applied a participatory planning approach in order to improve legitimacy of the intervention. The NGO gave a presentation on the situation of women's group and the idea of sanitation credit scheme. The participants then identified potential beneficiaries of the programme. Perdikan and BKM's board members and BKM discussed and drafted a plan concerning sanitation credit model. The participants finally drafted a memorandum of understanding (MoU) between BKM and Perdikan.

---

[15] Perdikan, 'Final project of pioneering community based sanitation micro credit to anticipate climate change at urban village of Kemijen, Semarang Timur sub-district, city of Semarang' *Project Report June 2010* (2010), 5-10.

The stakeholders involved in the process included the Kemijen village institutions such as BKM (the community-based micro-finance institution), sub-village structures, the village's women's organisation (PKK) and village legislation body (LPMK). Given that the size of the fund of the pilot project was too small, the stakeholders agreed to specifically target the most vulnerable group. Agreement on the criteria of the target groups was reached on the basis that sanitation credit should be prioritised for female-headed households (FHHHs) whose houses either had poor sanitation facility or no sanitation at all. The reason why FHHHs were chosen was explained in a participatory vulnerability analysis: women:

> have been deprived of a complete education and expected to maintain household duties, women, particularly in urban poor communities, are often those least able to avoid the impacts of climate events. Women may also lack the knowledge necessary in order to prepare for such events. If, for example, they are unable to work or study, women may not have the economic means or the information necessary to cope with change.[16]

A cooperation mechanism between the community-based organisation (CBO) and Perdikan was arranged based on a MoU. The community organisations such as PKK, BKM and LPMK were committed to promoting the sanitation credit to the FHHHs in Kemijen and also to be ready to answer questions as to why only FHHHs would receive the scheme. Community consultation meetings clarified some problems of the terminology being used for the target groups. However, for the community, the term 'widow' was more acceptable without further interpretation. Also the term 'sanitation credit' was seen as misleading in the sense that communities might perceive the project as a typical government project where the aid did not need to be repaid; the BKM suggested a clearer term such as 'loan to be repaid'.

BKM and Perdikan conducted specific meetings to finalise details of the scheme of the 'sanitation loan', the *terms and conditions for loan,* and to identify the names of the FHHHs. The Sanitation micro-finance or Sanitation micro-credit was to be a joint initiative from *Badan Keswadayaan Masyarakat* (BKM), a CBO in Kemijen Village, and Perdikan Foundation. One of the conditions was that 'there should be two guarantors' (or from families). In addition, the candidate of recipient should receive a 'free arrears' status from BKM. Box 16.1 details the activities carried out by Perdikan during the course of the pilot project.

---

[16] Taylor, n. 8 above, 344.

> **BOX 16.1 SUMMARY OF ACTIVITIES BY PERDIKAN**
>
> (a) *Planning and assessment:* representatives from the Community Resilience Board (BKM Kemijen) and neighbourhood groups participated in the meeting. On the agenda was the selection of the project development of the revolving fund's mechanism and requirements. The proponent of the project also used the notion of credit to refer to the fund so that the community would understand that the funds should be repaid; hence the terminology being used for the scheme is sanitation credit.
> (b) *Sub-district awareness meeting:* this meeting was attended by 21 female-headed families, BKM Kemijen and other community groups to discuss the Semarang KM Kemijen and other community groups to discuss the Semarang used for ant impacts and measures to improve sanitation condition within their area.
> (c) *Sanitation Credit:* implementation of the revolving fund was the main activity of the pilot project. It was conducted in cooperation with BKM Kemijen. Implementation of the sanitation credit scheme was anticipated as it allowed access for the direct beneficiaries to sanitation services and improvement of their sanitation facilities.
> (d) *Regular meetings of female-headed families:* the regular meetings aimed at enhancing the direct beneficiaries' adaptive capacity towards climate change impacts and simultaneously drive cooperation between them in order to improve social control of the implementation. The meetings were also expected to enhance wider community's voluntary participation in the project.
> (e) *A Focus Group Discussion* (FGD): was conducted with the intention to disseminate information on general concepts of climate change and its subsequent impact as well as information on the on-going sanitation credit scheme. The activity was conducted in cooperation with the local neighbourhood groups with housewives and the female community as the top priority participants. Throughout the project period, FGD had been undertaken in ten neighbourhood groups with a total attendance of 371 people. It is expected that the activity would help in enhancing the community's awareness on climate change impacts and also the potential vulnerable groups.
> (f) *A Climate Change Festival:* was undertaken with the intention of raising the awareness in general of Kemijen Sub district's community on climate change impacts, especially its effects on vulnerable area such as Kemijen. A wide array of stakeholders was involved within the activity, namely: government representatives (sub district-level and sub-district level); youth groups; female groups; and children.

BKM MICRO-FINANCE: HISTORY AND POLICY APPROACH

The total population of Kemijen is about 13,500 people (or about 4,100 households). Density of population in Kemijen in 2010 was

9,599 person/km². BKM Kemijen had been considered as the most capable community-based micro-finance institution in Semarang City as it was the most sustainable of the rest of BKMs in 177 urban villages in the city.[17] It had a good reputation as it had been a self-help CBO with 12 years' experience in loan and saving services and in managing micro-finance funds from both the government and the World Bank urban project. BKM is not part of the village structure and it sees itself as an equal partner with village government of Kemijen.

By the end of 2002, BKM Kemijen's total capital was only US$ 10,000. By June 2012, the total account available for loan was about US$ 60,000. BKM Kemijen has been able to sustain itself as a microfinance institute. At June 2012 it had four permanent staff dealing with its daily operations. This was the equivalent of US$ 6,600 annual spending for the staffing. The board member (comprising sub-village representatives) meeting is the highest decision-making body at BKM Kemijen.

Key milestones of BKM Kemijen 2000–12 [BKM Kemijen 2006, 2009, 2010, 2011, 2012]:

- Established on 2 Feb 2000 facilitated by Urban Poverty Eradication Project (P2KP) of the World Bank. It received about US$ 28,000 in grants for the revolving fund and only community self-help groups (KSMs) were eligible for access to the loan. At the start, BKM Kemijen provided support to 30 KSMs (later rising to 50 KSMs). However, by 2002, there was only left US$ 10,000 for revolving funds. The gap was considered a non-performing loan (NPL); one reason why there was such a large NPL was due to the fact that the BKM policy during 2000–02 was to only allocate funds for group loan (namely community self-help groups or KSM); a policy based on the requirements of the project.
- In 2002, only nine KSMs were being sustained; BKM board members evaluated the NPL as they tried to rescue the BKM from the same sort of collapse as other BKMs in Semarang. Following a decision by board members it was decided that BKM Kemijen could diversify its options including a decision to give credit not only to groups but also to personal requests. The decision was not fully approved by the project rules because 'all the support was meant

---

[17] Delima Sari, A. 2012. Mercy Corps - ACCCRN Project Staff, personal face to face interview by Lassa on 28 June 2012.

to be "stimulant fund"'. However, it was considered to be a necessary step to sustain the BKM: 'A few members started to pay their instalments but some others did not.'[18]
- In 2003, BKM decided to start a new exercise, namely daily loans for poor neighbourhoods. Anyone who met the criteria could borrow money for one to three months. The flat rate of this 'daily loan' was 8 per cent, the reason for this percentage being both social and economic. Social because it had a social goal giving financial access to the poor people in need and economic because BKM must sustain itself by calculating the risk and benefit based on past experience. In general, existing micro-finance and micro-credit institutions often offer a monthly rate above 15 per cent.
- In 2008, BKM Kemijen received additional US$ 6,000 grants from PNPM-Kota (National Project on Urban Community Empowerment) funded by the World Bank.
- In 2010, BKM Kemijen started the 'sanitation fund' initiative supported by Perdikan-ACCCRN with additional funds of only US$ 4,000. The loan rate is 1.5 per cent, where 5 per cent will be returned to the debtor after they finish the scheme.
- As at June 2012, BKM Kemijen had US$ 60,000–65,000 in their account. Their total assets (including office building) equals US$ 45,000.

The Perdikan-ACCCRN pilot project, the Sanitation Credit, is one of the outputs of the shared decision-making process made by City Team in Semarang supported by Asian Cities ACCCRN.

The successful outcome of the idea was not guaranteed due to a very real risk as seen by the local communities. Some admitted that, 'If there are many widows here in Kemijen, it means that we have to pay back the loan so others can also get access to the credit.' When the pilot project ended, 26 FHHHs had received sanitation loan. In April 2010, 15 FHH received the support, in May, five FHH received support, and in June, six FHH received support. 22 of the FHHHs used the loan to renovate their toilets and bathroom (MCK); while the other five FHHs used the loan to install clean water systems from the city water company. Ten out of 15 FHH who received their loan in April 2010 had made their first instalment payment by the end of June 2010. The other five FHHs decided to use their loan for installing water pipes from the PDAM of Semarang city. However, the normal procedure for water installation often takes

---

[18] Sarwono and others, *Focused group discussion on Sanitation Credit* (BKM Village, 28 June 2012). [FGD by Lassa]

months. Perdikan and BKM therefore facilitated two meetings with the PDAM East Semarang, following which the water installation process was successfully made because PDAM, a member of City Team, facilitated the process. The payment for the interest rate was adjusted with the timeline of the water installation. Since the instalment payment is on a 20-month basis, BKM and Perdikan committed to working beyond the formal project timeframe.

PERDIKAN'S PERCEPTION ON CONSTRAINTS IS BASED ON THEIR PARTICIPATORY EVALUATION PROCESS

It was noted that the expansion of sanitation credit was necessary as BKM needed to move from FHHHs in order to reach other vulnerable groups in the village. Replication in other places is possible but lessons need to be made explicit as Kemijen had a different social context. Other vulnerable areas can be considered to access similar initiatives despite the limited funds available for this pilot project.

- A lack of a formal mechanism to support the present climate resilience building through sanitation credit. For a long-term view, the present investment only solves some limited pressing (immediate) problems.[19] However, given the extreme rate of land subsidence in the village, present investment is likely to be submerged in the next ten years;
- the participation from BKM and local communities allowed the local community to increase their control over the project, as a long-term view, the present investment only solves some limited pressing (immediate) problems and BKM will be watching to see if there are non-performing loans (Collective statement from FGDs among FHHHs).

CHALLENGES AND REFLECTION

Data suggests that BKM Kemijen have been able to sustain its function to serve the most vulnerable groups in need. Table 16.1 below suggests

---

[19] See J Chaisse and C Bellak, 'Navigating the Expanding Universe of Investment Treaties—Creation and Use of Critical Index' 18 *Journal of International Economic Law*. 79 (2015), for understanding the link between treaties that states entered into with the increase of the FDI.

Table 16.1 BKM Kemijen Performance 2006–11

Targeted beneficiaries	2006	2007	2008	2009	2010	2011
KSM (group)	2	4	11	17	14	13
KSM (personal)	26	19	120	89	67	61
Micro loan (personal)	46	56	55	90	77	67
Daily loan (personal)	224	185	205	219	182	149
Total reached person	296	260	380	398	326	277

Source: Authors – based on BKM Kemijen 2006, 2009–12.

that there is an increase in targeting group loan over the last six years since 2006. It has been able to increase its reach to poor households in terms of micro-loan. Our interviews suggest that most of the micro-loans during the last ten years have been associated with household adaptation to sanitation and some primary public health concerns such as water access and health concern.

It is clear that community-based micro-finance can be used to address climate and disaster vulnerabilities in the city as exemplified over the last six years by Kemijen. The total allocated funds of US$ 4,000 had been used to support 26 FHHHs by June 2010. In June 2012, some of the returned funds had been used to give support to another four FHHHs. However, the idea of up-scaling and extending the 'sanitation credit' have been extremely limited due to the lack of funds created by city government and other parties. Therefore, success has been largely isolated in the form of pilot project.

It was also found that the degree of vulnerability among the FHHHs also varied. The most vulnerable FHHHs turned out to be the candidates of NPL. Therefore, the most vulnerable of the vulnerable group need to be supported by a different mechanism. BKM accepted the offer from the NGO by setting a low interest rate (at 1.5 per cent) for the FHHHs – this was both an institutional exercise and to increase coverage of their services. However, it has been admitted that the interest rate for 'sanitation credit' was too low and BKM therefore saw it as unsustainable. Thus, they had limited incentives to promote the scheme after the intervention finished.[20] NGOs' credit schemes such as Perdikan-ACCCRN were made based on a lack of empirical exercises. BKM's board perceived sustainability differently as they believe that sustainability can be achieved if right from the

---

[20] Sarwono and others, *Focused group discussion on Sanitation Credit* (BKM Village, 28 June 2012). [FGD by Lassa]

outset, existing scheme is being used. Therefore, scaling-up and extending the support can be a possibility as long as there is a clear metric calculation on the long-term benefit for BKM. BKM maintained that 10 per cent of their annual benefit is often allocated for social activities.

The present 'sanitation credit' in Kemijen Village has proved useful in addressing some immediate problems. However, given the high rate of land subsidence, as witnessed directly by local communities, all the investment (from BKM and also physical infrastructure project from the city government) is likely to eventually be submerged, possibly in the next decade. Therefore, it is necessary for the city government to plan for the longer term.

Bibliography

JOURNAL ARTICLES

Abbott, Malcolm and Bruce Cohen, 'Productivity and efficiency in the water industry' (2009) 17 *Utilities Policy* 233.

Agrawal Anju et al, 'Water Pollution with Special Reference to Pesticide Contamination in India' (2010) 2 *Journal of Water Resource and Protection* 432.

Aldaya, Maite M and others, 'Strategic importance of green water in international crop trade' (2010) 69 *Ecological Economics* 887.

Allan, John Anthony, 'Virtual water – the water, food, and trade nexus. Useful concept or misleading metaphor?' (2003) 28 *Water International* 107.

Allan, John Anthony, 'Virtual water: A strategic resource – global solutions to regional deficits' (1998) 36 *Ground Water* 545.

Aly, K Abu-Akeel, 'Definition of trade in services under the GATS: Legal implications' (1999–2000) 32 *George Washington Journal of International Law and Economics* 189.

Armstrong, Scott C, 'Water is for fighting: transnational legal disputes in the Mekong River Basin' (2015) 17(1) *Vermont Journal of Environmental Law* 1.

Bartels, Lorand, 'The relationship between the WTO Agreement on Agriculture and the SCM Agreement: An analysis of hierarchy rules in the WTO legal system' (2016) 50 *Journal of World Trade* 7.

Bartholomeusz, Lance, 'The *amicus curiae* before international courts and tribunals', (2005) 5 *Non-State Actors and International Law* 209.

Bates, Rebecca, 'The trade in water services: How does GATS apply to the water and sanitation service sector' (2009) 31 *Sydney Law Review* 121.

Battese, George and Coelli, Timothy, 'A model for technical inefficiency effects in a stochastic frontier production function for panel data' (1995) 20 *Empirical Economics* 325.

Benfratello, Luigi and Alessandro Sembenelli, 'Foreign ownership and productivity: Is the direction of causality so obvious?' (2006) 24 *International Journal of Industrial Organisation* 733.

Benoliel, Daniel and Bruno Salama, 'Towards an intellectual property bargaining theory: The post-WTO era' (2010) 32(1) *University of Pennsylvania Journal of International Law* 265.

Benvenisti, Eyal, 'Collective action in the utilization of shared freshwater: The challenges of international water resources law' (1995) 90(3) *American Journal of International Law* 384.

Bernstein, Jeffrey and David Sappington, 'How to determine the X in RPI – X regulation: AA user's guide' (2000) 24 *Telecommunications Policy* 63.

Biro, Andrew, 'Water wars by other means: Virtual water and global economic restructuring' (2012) 12 *Global Environmental Politics* 86.

Bluemel, Erik B, 'The implications of formulating a human right to water' (2004) 31 *Ecology Law Quarterly* 957.

Bond, Patrick and Jackie Dugard, 'The case of Johannesburg Water: What really happened at the prepaid parish pump' (2008) 12 *Law Democracy & Development* 1.

Bouguignon, 'Francois, Inequality and globalization: How the rich get richer as the poor catch up' (2016) 95(1) *Foreign Affairs*11.

Bower, Herman, 'Integrated water management: Emerging issues and challenges' (2000) 45 *Agricultural Water Management* 217.

Bowmer, Kathleen H, 'Ecosystem effects from nutrient and pesticide pollutants: Catchment care as a solution' (2013) 2 *Resources* 439.

Bray, Heather L, 'ICSID and the right to water: an ingredient in the stone soup' (2013) 29(2) *ICSID Review* 474.

Brichieri-Colombi, Stephen, 'Hydrocentricity: A limited approach to achieving food and water security' (2004) 29 *Water International* 318.

Briscoe, John, 'Water security in a changing world' (2015) 144 *Dædalus – Journal of the American Academy of Arts & Sciences* 27.

Budds, Jessica and Gordon McGranahan, 'Are the debates on water privatisation missing the point? Experience from Africa, Asia and Latin America' (2003) 7 *Environment & Urbanisation* 11.

Bulto Takele Soboka, 'The emergence of a human right to water: Invention or discovery?' (2011) 12(2) *Melbourne Journal of International Law* 290.

Burke-White, William, and von Staden, Andreas, 'Investment protection in extraordinary times: The interpretation and application of non-precluded measures provisions in bilateral investment treaties' (2008) 48 *Virginia Journal of International Law* 308.

Cadeau, E and F Duhautoy, 'Le droit à l'eau, soluble dans le droit international de l'investissement?' (2013) 216 *Droit de l'environnement* 338

Carias, C and others, 'Medicamentos de dispensação excepcional: histórico e gastos do Ministério da Saúde do Brasil (Exceptional circumstance drug dispensing: history and expenditures of the Brazilian Ministry of Health)' (2011) 45 *Revista de Saúde Pública* 2.

Cettner, Annicka, Ashley, Richard, Viklander, Maria and Nilsson, Kristina, 'Stormwater management and urban planning: Lessons from 40 years of innovation' (2013) 56 *Journal of Environmental Planning and Management* 790.

Chaisse, Julien, 'Assessing the relevance of multilateral trade law to sovereign investments: Sovereign wealth Funds as "investors" under the General Agreement on Trade in Services' (2015) 3(2) *International Review of Law* 30.

Chaisse, Julien, 'Exploring the confines of international investment and domestic health protections—is a general exceptions clause a forced perspective?' (2013) 39 *American Journal of Law & Medicine* 332 (2013).

Chaisse, Julien, 'Assessing the exposure of Asian states to investment claims' (2013) 6 *Contemporary Asia Arbitration Journal* 187.

Chaisse, Julien and Christian Bellak, 'Navigating the expanding universe of investment treaties—creation and use of critical index' (2015) 18 *Journal of International Economic Law* 79.

Chaisse, Julien and Christian Bellak, 'Do Bilateral Investment Treaties promote foreign direct investment? Preliminary reflections on a new methodology' (2011) 3 *Transnational Corporations Review* 7.

Chaisse, Julien and Marine Polo, 'Globalization of water privatization: Ramifications of investor-state disputes in the "blue gold" economy', (2015) 38 *Boston College International and Comparative Law Review* 1.

Chaisse, Julien and Mitsuo Matsushita, 'Maintaining the WTO's Supremacy in the international trade order– a proposal to refine and revise the role of the trade policy review mechanism' (2013) 18 *Journal of International Economic Law* 9.

Chaisse, Julien and Puneeth Nagaraj, 'Changing lanes: Intellectual property rights, trade and investment' (2014) 37 *Hastings International & Comparative Law Review* 251.

Chaisse, Julien, Debashis Chakraborty, and Jaydeep Mukherjee, Deconstructing service and investment negotiating stance' (2013) 14 *Journal of World Investment & Trade* 44.

Chapagain, Ashok K and others, 'The water footprint of cotton consumption: An assessment of the impact of worldwide consumption of cotton products on the water resources in the cotton producing countries' (2005) 60 *Ecological Economics* 186.

Chakraborty, Debashis, Julien Chaisse, and Jaydeep Mukherjee, 'Deconstructing services and investment negotiations – A case study of India at WTO GATS and investment fora' (2013) 14(1) *Journal of World Investment and Trade* 44.

Charnes, A and others, 'Measuring the efficiency of decision making units' (1978) 2 *European Journal of Operational Research* 429.

Chaussard, E, Amelung, F, Abidin, H, and Hong, S-H, 'Sinking cities in Indonesia: ALOS PALSAR detects rapid subsidence due to groundwater and gas extraction' (2013) 128 *Remote Sensing of Environment* 150.

Chayes, Abram, 'The role of the judge in public law litigation' (1975/1976) 89 *Harvard Law Review* 1281.

Chen, Huiping, 'The human right to water and foreign investment: Friends or foes?' (2015) 40 *Water International* 297.

Chieffi, A and R Barata, 'Judicialização da política pública de assistência farmacêutica e equidade' ('Judicialization of public health policy for distribution of medicines') (2009) 25(8) *Cadernos de Saúde Pública, Rio de Janeiro* 1839.

Condon, Bradley J, 'Treaty structure and public interest regulation in international economic law' (2014) 333 17 *Journal of International and Economic Law* 351.

Crampes Claude and Estache Antonio Note, 'Regulating water concession: Lessons from the Buenos Aires Concession' (1996) *Public Policy for the Private Sector*, note No. 91, 2.

Craig, Robin Kundis and Anna M Roberts, 'When will governments regulate nonpoint source pollution: A comparative perspective' (2015) 42 *Boston College Environmental Affairs Law Review* 1.

Cronin, Aidan A and others, 'Water, sanitation and hygiene: Moving the policy agenda forward in the post-2015 Asia' (2015) 2 *Asia & the Pacific Policy Studies* 229.

Cummings, Scott L and Deborah L Rhode, 'Public interest litigation: Insights from theory and practice' (2008) 36(4) *Fordham Urban Law Journal* 603.

Curry, Elliot, 'Water scarcity and the recognition of the human right to safe freshwater' (2010) 9(1) *Northwestern University Journal of International Human Rights* 103.

Daher, Bassel T and Rabi H Mohtar, 'Water–energy–food (WEF) Nexus Tool 2.0: guiding integrative resource planning and decision-making' (2015) 40 *Water International* 748.

Dalhuisen, Jasper and others, 'Price and income elasticities of residential water demand: A meta-analysis' (2003) 79 *Land Economics* 292.

Delimatsis, Panagiotis, 'Due process and "good" regulation embedded in the GATS – disciplining regulatory behaviour in services through Article VI of the GATS' (2006) 10 *Journal of International Economic Law* 13.

Diamond, Stephen, 'Water ethics and commodification of freshwater resources' (2008) 6(1) *Santa Clara Journal of International Law* 15.

Dimitrov, Radoslav S, 'Water, conflict, and security: A conceptual minefield' (2002) 15 *Society & Natural Resources* 677.

Dolzer, Rudolf, 'The impact of international investment treaties on domestic administrative law' (2004) 37 *New York University Journal of International Law and Politics* 953.

Duarte, Rosa and others, 'Understanding agricultural virtual water flows in the world from an economic perspective: A long term study' (2016) 61 *Ecological Indicators* 980.

Echaide, J, 'Sobre el derecho humano al agua y la fragmentación del derecho internacional: El régimen internacional de protección de inversiones vis-a-vis las obligaciones erga omnes en materia de derechos humanos', (2014) VIII *Revista Electrónica del Instituto de Investigaciones* 140.

Elhance, Arun P, 'Hydropolitics: Grounds for despair, reasons for hope' (2000) 5 *International Negotiation* 202.

A Emrouznejad, A and others, 'Evaluation of research in efficiency and productivity: A survey and analysis of the first 30 years of scholarly literature in DEA' (2008) 42 *Socio-Economics Planning Sciences* 151.

Espey, M and others, 'Price elasticity of residential demand for water: A meta-analysis' (1997) 33 *Water Resources Research* 1369.

Estache, A and M Rossi, 'How different is the efficiency of public and private water companies in Asia' (2002) 16 *The World Bank Economics Review* 139.

Falkenmark, Malin, 'The greatest water problem: The inability to link environmental security, water security and food security' (2001) 17 *International Journal of Water Resources Development* 539.

Farrell, MJ, 'The measurement of productive efficiency' (1957) 120 *Journal of the Royal Statistical Society Series* A 253.

Farrugia, Bree, 'The human right to water: defences to investment treaty violations', (2015) 31 *Arbitration International* 261.

O Ferraz, O, 'The right to health in the courts of Brazil: Worsening health inequities?' (2009) 11(2) *Health and Human Rights* 33.

Ferraz, O and F Vieira, 'Direito à saúde, recursos escassos e equidade: os riscos da interpretação judicial dominante' ('The right to health, scarce resources, and equity: inherent risks in the predominant legal interpretation') (2009) 52(1) *Dados* 223.

Fishelson, G, 'The allocation and marginal value product of water in Israeli agriculture' (1994) 58 *Studies in Environmental Science* 427.

Fiss, Owen, 'The social and political foundations of adjudication' (1982) 6 *Law and Human Behavior* 121.

Fiss, Owen, 'The Supreme Court 1978 Term. Foreword: The forms of Justice' (1979) 93 *Harvard Law Review* 18.

Fitzmaurice, Malgosia, 'Symposium: Environmental protection and human rights in the new millennium: Perspectives, challenges, and opportunities' (2007) *18 Fordham Environmental Law Review* 537.

Fortier, LY and SL Drymer, 'Indirect expropriation in the law of international investment: I know it when I see it, or *caveat investor*' (2004) 19 *ICSID Review* 306.

Franck, S, 'The legitimacy crisis in investment treaty arbitration: Privatizing public international law through inconsistent decisions', (2005) 73 *Fordham Law Review* 1521.

Garcia, Frank, 'The global market and human rights: Trading away the human rights principle' (1999) 25 *Brooklyn Journal of International Law* 51.

García-Bolívar, Omar E, 'G3 Agreement: A comparison of its investment chapter with the emerging international law of foreign investment' (2004) 10 *Law and Business Review of the Americas* 779.

Gerlak, Andrea K and others, 'Hydrosolidarity and international water governance' (2009) 14 *International Negotiation* 311.

Ghorbi, Darian, 'There's something in the Water: The inadequacy of international anti-dumping laws as applied to the Fukushima Daiichi radioactive water discharge' (2012) 27 *American University International Law Review* 473.

Gilles, ME, 'In defense of making government pay: the deterrent effect of constitutional tort remedies' (2000/2001) 35 *Georgia Law Review* 845.

Gleditsch, Nils Petter, 'Armed conflict and the environment: A critique of the literature' (1998) 35 *Journal of Peace Research* 381.

Goll, Markus A, 'Desalination in Texas: Struggling to cope' (2015) 45(1) *Texas Environmental Law Journal* 51.

Gonzalez-Gomez, F and others, 'Beyond the public-private controversy in urban water management in Spain' (2014) 31 *Utilities Policy* 1.

Grafton, R Quentin and others, 'Determinants of residential water consumption: Evidence and analysis from a ten-country household survey' (2011) 47 *Water Resources Research*, W08537.

Greco, R, 'The impact of the human right to water on investment disputes' (2015) 98 *Rivista di Diritto Internazionale* 444.

Habermas, Jurgen, 'The European nation-state and the pressures of globalisation' (1999) 235 *New Left Review* 46.

Hajispyrou, Soteroula and others, 'Household demand and welfare: Implications of water pricing in Cyprus' (2002) 4 *Environment and Development Economics* 659.

Hall, Noah D, 'Protecting freshwater resources in the era of global water markets: Lessons learned from bottled water' (2009) 13(1) *University of Denver Water Law Review* 1.

Hall, Noah D, Bret B Stuntz, Robert H Abrams, 'Climate change and freshwater resources' (2008) 22(3) *Natural Resources & Environment* 30.

Hall, D and E Lobina, 'Profitability and the poor: Corporate strategies, innovation and sustainability' (2007) 38 *Geoforum* 772.

Hallegatte, S and J Corfee-Morlot, 'Understanding climate change impacts, vulnerability and adaptation at city scale: An introduction' (2011) 104(1) *Climatic Change* 1.

Hamner, Jesse H, 'Patterns in international water resource treaties: The transboundary freshwater dispute database' (1997) 9 *Colorado Journal of International Environmental Law and Policy, Yearbook* 157.

Han, Mengyao, and others, 'Virtual water accounting for a building construction engineering project with nine sub-projects: a case in E-town, Beijing' (2016) 112 *Journal of Cleaner Production* 4691.

Hansen, David R, et al. 'Solving the orphan works problem for the United States' (2013) 37(1) *Columbia Journal of Law & the Arts* 1.

Hanson, S and others, 'A global ranking of port cities with high exposure to climate extremes' (2011) 104 *Climatic Change* 89.

Hardberger, Amy, 'Whose job is it anyway? Governmental obligations created by the human right to water' (2006) 41 *Texas International Law Journal* 533.

Hardberger, Amy, 'Life, liberty and the pursuit of water: Evaluating water as human right and the duties and obligations it creates' (2005) 4 *Northwestern University's Journal of International Human Rights* 331.

Harris, R and C Robinson, 'Foreign ownership and productivity in the United Kingdom. Estimates for U.K. manufacturing using the ARD' (2003) 22 *Review of Industrial Organization* 207

Harris, R and C Robinson, 'The impact of foreign acquisitions on total factor productivity: plant level evidence from UK manufacturing, 1987–1992' (2002) 84 *Review of Economics and Statistics* 562

Hawkins, Joanne, 'Fracking: Minding the gap' (2015) 17(1) *Environmental Law Review* 8.

Heller, L, 'Relação entre saúde e saneamento na perspectiva do desenvolvimento' ('Relationship between health and environmental sanitation in view of the development') (1998) 3 *Ciência e Saúde Coletiva* 73.

Hernandez-Sancho, F and others, 'Tariffs and efficient performance by water suppliers: An empirical approach' (2012) 14 *Water Policy* 854.

Hochrainer, S and R Mechler, 'Natural disaster risk in Asian megacities: A case for risk pooling' (2011) 28 *Cities* 53.

Hoekstra, Arjen Y, 'The global dimension of water governance: Why the river basin approach is no longer sufficient and why cooperative action at global level is needed' (2011) 3 *Water* 21.

Howse, Rob, Joanna Langille and Katie Sykes, 'Sealing the deal: The WTO's Appellate Body Report in EC – Seal Products' 18(2) *Insights* (2014).

Huang, Lee-Yee, 'Not just another drop in the human rights bucket: The legal significance of a codified human right to water' (2008) 20 *Florida Journal of International Law* 353.

Hung, Ming-Feng and Bzi-Tzong Chie, 'Residential water use: Efficiency, affordability, and price elasticity' (2013) 27 *Water Resources Management* 275.

Jacob, M and SW Schill, 'Going soft: Toward a new age of soft law in international investment law?' (2014) 8 *World Arbitration and Mediation Review* 1.

Jacobson, P. and S. Soliman, 'Litigation as public health policy. Theory or reality' (2002) 30 *Journal of Law, Medicine and Ethics* 224.

Juillard, Patrick, 'L'évolution des sources du droit des investissements' (1994) 250 *Collected Courses of The Hague Academy of International Law* 208.

Kieffer, T and C Brölmann, 'Beyond state sovereignty: The human right to water' 2005 5 *Non-State Actors and International Law* 184.

Kingsbury, B and S Schill, 'Investor-state arbitration as governance: Fair and equitable treatment, proportionality and the emerging global administrative law', (2009) 146 *New York University Public Law and Legal Theory Working Papers* 1, <http://lsr.nellco.org/nyu_plltwp/146>.

Klein, A, 'Judging as nudging: New governance approaches for the enforcement of constitutional social and economic rights' (2007/2008) 39 *Columbia Human Rights Law Review* 351.

Krugman, Paul, 'A model of technology transfer, and the world distribution of income' (1979) 87 *Journal of Political Economy* 253.

Kruskal, WH and WA Wallis, 'Use of ranks in one-criterion variance analysis' (1952) *Journal of the American Statistical Association* 47.

LaBorde, Lilian Del Castillo, 'Legal regime of the Rio de la Plata' (1996) 36(2) *Natural Resources Journal* 251.

Landau, David, 'The reality of social rights enforcement' (2011) 53 *Harvard International Law Journal* < http://ssrn.com/abstract> accessed 12/02/2016.

Lassa, JA and E Nugraha, 'From shared learning to shared action in building resilience in the city of Bandar Lampung, Indonesia' (2015) 0 *Environment & Urbanization* 1.

Leb, Christina, and Mara Tignino, 'Freshwater and international law' (2011) 41(4–5) *Environmental Policy and Law* 218.

Leroux, Eric, 'Eleven years of GATS case law: What have we learned?' (2007) 10 *Journal of International Economic Law* 749.

Lévesque Céline, 'Abaclat and Others v Argentine Republic: The definition of investment' (2012) 27 *ICSID Review* 247.

Levinson, DJ, 'Making government pay: markets, politics, and the allocation of constitutional costs' (2000) 67 *University of Chicago Law Review* 345.

Luan, Xinjie, and Julien Chaisse, 'The WTO seals products dispute – traditional hunting, public morals and technical barriers to trade' (2011) 22(1) *Colorado Journal of International Environmental Law and Policy* 79.

Magsig, Bjørn-Oliver, 'Overcoming state-centrism in international water law: "Regional common concern" as the normative foundation of water security' (2011) 3 *Goettingen Journal of International Law* 317.

Manciaux, Sébastien, 'La Bolivie se retire du CIRDI' (2007) 5 *Transnational Dispute Management* 1.

Manson, Laura and Tracey Epps, 'Water footprint labelling and WTO Rules' (2014) 23 *Review of European Community and International Environmental Law* 329.

Marrella, F, 'On the changing structure of international investment law: The human right to water and ICSID arbitration', (2010) 12 *International Community Law Review* 335

Massarutto, Antonio, 'Water pricing and full cost recovery of water services: Economic incentive or instrument of public finance' (2007) 9 *Water Policy* 591.

McCaffrey, Stephen C, 'Small capacity and big responsibilities: Financial and legal implications of a human right to water for developing countries' (2009) 21 *Georgetown Environmental Law Review* 679.

McDonnell, Thomas, 'Assessing funding models for water services provision in Ireland' (2014) 21 *Nevin Economic Research Institute* 1.

McIntyre, O, 'The human right to water as a creature of global administrative law' (2012) 37 *Water International* 654.

McLeod, E, et al, 'Designing marine protected area networks to address the impacts of climate change' (2009) 7 *Frontiers in Ecology and the Environment* 362.

Mechlem, Kerstin, 'Shared resources: transboundary groundwaters' (2004) 34(4–5) *Environmental Policy and Law* 162.

Mekonnen, Mesfin M, and others, 'The consumptive water footprint of electricity and heat: A global assessment' (2015) 1 *Environmental Science: Water Research & Technology* 285.

Mekonnen, Mesfin M and Arjen Y Hoekstra, 'The green, blue and grey water footprint of crops and derived crop products' (2011) 15 *Hydrology and Earth System Sciences* 157.

Mergos, George, 'Private participation in the water sector: Recent trends and issues' (2005) 9/10 *European Water* 59.

Merrett, Stephen, 'Virtual water and Occam's Razor' (2003) 28 *Water International* 103.

Meshel, T, 'Human rights in investor-state arbitration: The human right to water and beyond', (2015) 6 *Journal of International Dispute Settlement* 277

Miller, Amy K, 'Blue rush: Is an international privatization agreement a viable solution for developing countries in the face of an impending world water crisis' (2005) 16 *Indiana International and Comparative Law Review* 217.

Mohayidin, Ghazali, and others, 'Review of water pricing theories and related models' (2009) 4 *African Journal of Agricultural Research* 1536.

Moloo, R, and J Jacinto, 'Environmental and health regulation: Assessing liability under investment treaties' (2011) 29 *Berkeley Journal of International Law* 33

Mukherjee, Sacchidananda, and others, 'Sustaining urban water supplies in India: Increasing role of large reservoirs' (2010) 24 *Water Resources Management* 2035.

Murthy, Sharmila L, 'The human right(s) to water and sanitation: History, meaning, and controversy over-privatization' (2013) 31 *Berkeley Journal of International Law* 89.

Neff, E, 'From equal protection to the right to health: Social and economic rights, public law litigation, and how an old framework informs a new generation of advocacy' (2010) 43 *Columbia Journal of Law and Social Problems* 151.

Nickson, Andrew, and Claudia Vargas, 'The limitations of water regulation: The failure of the Cochabamba Concession in Bolivia' (2002) 21 *Bulletin of Latin American Research* 128.

Nkonya, Leticia K, 'Socioeconomic rights: Empowerment for global justice: Realizing the human right to water in Tanzania' (2010) 17 *Human Rights Brief* 25.

Parker, D and JK Mitchell, 'Disaster vulnerability of megacities: an expanding problem that requires rethinking and innovative responses' (1995) 37 *GeoJournal* 295

Peart, Raewyn, 'Innovative approaches to water resource management: A comparison of the New Zealand and South African approaches' (2001) 5 *New Zealand Journal of Environmental Law* 127.

Peda, P, and others, 'Do ownership and size affect the performance of water utilities? Evidence from Estonian municipalities' (2013) 17 *Journal of Management and Governance* 237.

Post, VEA, et al, 'Offshore fresh groundwater reserves as a global phenomenon', (2013) 504 *Nature* 71.

Potestà, M, 'Legitimate expectations in investment treaty law: Understanding the roots and the limits of a controversial concept', (2013) 28 *ICSID Review* 88.

Rees, Judith A, 'Regulation and private participation in the water and sanitation sector' (1998) 22 *Natural Resource Forum* 95.

Rogers, Peter, and others, 'Water is an economic good: How to use prices to promote equity, efficiency, and sustainability' (2002) 4 *Water Policy* 1.

Romano, G, and A Guerrini, 'Measuring and comparing the efficiency of water utility companies: A data envelopment analysis approach' (2011) 19 *Utilities Policy* 202.

Sabel, C, and W Simon, 'Destabilization rights: How public law litigation succeeds' (2004) 117 *Harvard Law Review* 1015

Sadler, Barry, 'Shared resources, common future: Sustainable management of Canada-United States border waters' (1993) 33(2) *Natural Resources Journal* 375.

Salacuse, Jeswald W, 'The treatification of international investment law' (2007) 13 *Law and Business Review of The Americas* 155.

Salacuse, Jeswald W and Nicholas P Sullivan, 'Do BITs really work?: An evaluation of bilateral investment treaties and their grand bargain' (2005) 46 *Harvard International Law Journal* 67.

Salman, MA Salman, 'The human right to water and sanitation; Is the obligation deliverable' (2014) 39(7) *Water International* 969.

Salman, MA Salman and Kishor Uprety, 'Hydro-politics in South Asia: A comparative analysis of the Mahakali and the Ganges Treaties' (1999) 39 *Natural Resources Journal* 295.

Salzman, James, 'Thirst: A short history of drinking water' (2006) 31 *Duke Law School Working Paper Series* 1.

Schendel, Emily Kate, and others, 'Virtual water: A Framework for comparative regional resource assessment' (2007) 9 *Journal of Environmental Assessment Policy and Management* 341.

Schreiber, W, 'Realizing the right to water in international investment law: An interdisciplinary approach to BIT obligations', (2008) 48 *Natural Resources Journal* 473.

Schneider, Susan A, 'Predicting the future: Our food system in 2025' (2015) 11(1) *Journal of Food Law and Policy* 21.

Schwartz, GT, 'Reality in the economic analysis of tort law: Does tort law really deter?' (1994/1995) 42 *University of California Law Review* 377

Silva, V, and F Terrazas, 'Claiming the right to health in Brazilian courts: The exclusion of the already excluded' (2011) 36/4 *Law & Social Inquiry* 825.

Singh, S, and K Dickson, 'Transnational corporations and the global water industry' (2011) 53 *Thunderbird International Business Review* 601.

Sojamo, Suvi, and others, 'Virtual water hegemony: The role of agribusiness in global water governance' (2012) 37 *Water International* 169.

Stern, B, 'Civil society's voice in the settlement of international economic disputes', (2007) 22 *ICSID Review* 280.
Stiglitz, JE, 'Globalism's Discontents' (2002) 16 The American Prospect.
Storto, CL, 'Are public-private partnerships a source of greater efficiency in water supply? Results of a non-parametric performance analysis relating to the Italian industry' (2013) 5 *Water* 2058.
Subedi, Surya P, 'Hydro-Diplomacy in South Asia: The Conclusion of the Mahakali and Ganges River Treaties' (1999) 93 *American Journal of International Law* 953
Szwedo, Piotr, 'Water footprint and the law of WTO' (2013) 47 *Journal of World Trade* 1259.
Tamea, Stefania, and others, 'Drivers of the virtual water trade' (2014) 50 *Water Resources Research* 17.
Tarlock, Dan, 'Water security, fear mitigation and international water law' (2008) 31 *Hamline Law Review* 703.
Titi, Catharine, 'Investment arbitration in Latin America: The uncertain veracity of preconceived ideas' (2014) 30 *Arbitration International* 381.
Titi, Catharine, 'The arbitrator as a lawmaker: Jurisgenerative processes in investment arbitration', (2013) 14 *Journal of World Investment and Trade* 829.
Tyler, S and Moench, M, 'A framework for urban climate resilience' (2012) 4 *Climate and Development* 311.
Van Aaken, Anne, 'International investment law between commitment and flexibility: a contract theory analysis' (2009) 12 *Journal of International Economic Law* 507.
Van Harten, G, and M Loughlin, 'Investment treaty arbitration as a species of global administrative law' (2006) 17 *European Journal of International Law* 121.
Vieira, F, 'Gasto do Ministério da Saúde com medicamentos: tendência dos programas de 2002 a 2007 (Ministry of Health's spending on drugs: program trends from 2002 to 2007)' (2009) *43 Revista de Saúde Pública* 674.
Victora, CG, et al, 'Explaining trends in inequities: evidence from Brazilian child health studies' (2000) 356/9235 *Lancet* 1093.
Waelde, T, and A Kolo, 'Environmental regulation, investment protection and "regulatory taking" in international law' (2001) 50 *International and Comparative Law Quarterly* 846
Weiss, Edith Brown 'The coming water crisis: A common concern of humankind' (2012) 1 *Transnational Environmental Law* 153.
Weiss, Edith Brown and Lydia Slobodian, 'Virtual water, water scarcity and international trade law' (2014) 17 *Journal of International Economic Law* 717.

Wenzel, F, and others, 'Megacities – megarisks' (2007) 42 *Natural Hazards* 481.

'What Price for the Priceless?: Implementing the justiciability of the right to water' (2007) 120 *Harvard Law Review* 1067.

Wichelns, Dennis, 'The virtual water metaphor enhances policy discussions regarding scarce resources' (2005) 30 *Water International* 428.

Wilson, G, 'Deepwater horizon and the law of the sea: Was the cure worse than the disease?' (2014) 41 *Boston College Environmental Affairs Law Review* 63.

Wirth, DA, 'The Rio Declaration on Environment and Development: Two steps forward and one back, or vice versa' (1995) 29 *Georgia Law Review* 611.

Wolf, Jennyfer, and others, 'Assessing the impact of drinking water and sanitation on diarrhoeal disease in low- and middle-income settings: systematic review and meta-regression' (2014) 19 *Tropical Medicine and International Health* 928.

Woodhouse, Erik J, Note, 'The "Guerra del Agua" and the Cochabamba Concession: Social risk and foreign direct investment in public infrastructure' (2003) 39 *Stanford Journal of International Law* 295

Wouters, Patricia, 'The relevance and role of water law in the sustainable development of freshwater: From "hydrosovereignty" to "hydrosolidarity"' (2000) 25 *Water International* 202.

Yazdanfar, Zrinab, and Ashok Sharma, 'Urban drainage system planning and design – challenges with climate change and urbanization: A review' (2015) 72 *Water Science and Technology* 169.

Yang, Hong, and Alexander Zehnder, '"Virtual water": An unfolding concept in integrated water resources management' (2007) 43 *Water Resources Research* 1.

Ziganshina, Dinara, 'Rethinking the Concept of the Human Right to Water' (2008) 6 *Santa Clara Journal of International Law* 113.

BOOK CHAPTERS

Achtouk-Spivak, L, and A Ben Mansour, 'Reconnaissance et execution des sentences arbitrales en matière d'investissement', in C Leben (ed.) *Droit international des investissements et de l'arbitrage transnational* (Paris: Pedone 2015).

Aguirre, B. Fernando, 'Bolivia', in Jonathan C. Hamilton, Omar E. García-Bolívar and Hernando Otero (eds) *Latin American Investment Protections: Comparative Perspectives on Laws, Treaties, and Disputes for Investors, States, and Counsel*, Leiden and Boston (Martinus Nijhoff 2012).

Allan, John Anthony, 'Beyond the watershed: Avoiding the dangers of hydro-centricity and informing water policy', in Hillel Shuval and Hassan Dweik (eds) *Water Resources in the Middle East: Israel-Palestinian Water Issues – From Conflict to Cooperation* (Springer 2007).

Allan, John Anthony, 'Fortunately there are substitutes for water otherwise our hydro-political futures would be impossible' in ODA, *Priorities for Water Resources Allocation and Management* (1993).

Allan, John Anthony, 'Can improving returns to food–water in Africa meet African food needs and the needs of other consumers?' in Tony Allan and others, *Handbook of Land and Water Grabs in Africa: Foreign Direct Investment and Food and Water Security* (Routledge 2013).

Allan, John Anthony, 'The role of those who produce food and trade it in using and "trading" embedded water: What are the impacts and who benefits?' in Arjen Y Hoekstra and others (eds), 'Proceedings of the ESF Strategic Workshop on Accounting for water scarcity and pollution in the rules of international trade' *Value of Water Research Report Series No 54* (2011).

Audit, M, 'La jurisprudence arbitrale comme source du droit international des investissements', in C Leben (ed.) *Droit international des investissements et de l'arbitrage transnational* (Paris: Pedone 2015).

Barcellos, A, 'O direito a prestações de saúde: Complexidades, mínimo existencial e o valor das abordagens coletiva e abstratas' ('The right health care services: complexities, minimum standards and the value of collective and abstract approaches'), in S Guerra and L Emerique (eds) *Perspectivas Constitucionais Contemporâneas* (*Contemporary Constitutional Perspective*) (Rio de Janeiro, RJ: Lumen Iuris Book Publisher 2010).

Barry, JP, et al, 'Effects of ocean acidification on marine biodiversity and ecosystem function' in JP Gattuso and L Hansson (eds) *Ocean acidification* (Oxford University Press 2012).

Birkland, T, 'Agenda setting in public policy' in F. Fischer, G Miller, and M. Sidney (eds) *Handbook of Public Policy Analysis. Theory, Politics and Methods* (Boca Raton, London, New York: CRC Press 2007).

Brunnée, Jutta, 'Common areas, common heritage and common concern' in Daniel Bodanski and others (eds) *Oxford Handbook of International Environmental Law* (Oxford University Press 2007).

Cossy, Mireille,'Water services at the WTO' in Edith Brown Weiss and others (eds) *Fresh Water and International Law* (Oxford University Press 2005).

Fay, Marianne, Michael Toman, Daniel Benitez and Stefan Csordas,

'Infrastructure and sustainable development' in Fardoust, Yongbeom and Sepúlveda (eds) *Postcrisis Growth and Development: A Development Agenda for the G-20* (Washington DC: World Bank 2011).

Ferraz, O, 'Brazil: Health inequalities, rights, and courts: the social impact of the "judicialization of health"' in A Yamin and S Gloppen (eds) *Litigating Health Rights. Can Courts Bring More Justice to Health?* (Harvard University Press 2011).

Hamamoto, S, 'Protection of the investor's legitimate expectations: Intersection of a treaty obligation and a general principle of law', in W Shan and J Su (eds) *China and International Investment Law* (Leiden: Brill/Nijhoff 2014).

Hamamoto, S, 'Domestic review of treaty-based international investment awards: Effects of the Metalclad judgment of the British Columbia Supreme Court', in M. Kanetake and A. Nollkaemper (eds) *The Rule of Law at the National and International Levels* (Oxford: Hart Publishing 2016).

Harrison, J, 'Human rights arguments in *amicus curiae* submissions: Promoting social justice?', in PM Dupuy, F Francioni and EU Petersmann (eds) *Human Rights in International Investment Law and Arbitration*, (Oxford University Press 2009).

Hoekstra, Arjen H, 'The relation between international trade and water resources management' in Kevin P Gallagher (ed.) *Handbook on Trade and the Environment* (Edward Elgar Publishing 2008).

Hoffman, F, and F Bentes, 'Accountability for economic and social rights in Brazil' in Gauri and Brinks (eds), *Courting Social Justice: Judicial Enforcement of Social and Economic Rights in the Developing World* (Cambridge University Press 2008).

Hunt, P, and others, 'Implementation of economic, social and cultural rights', in S Sheeran and Sir N Rodley (eds) *Routledge Handbook of International Human Rights Law* (Routledge 2013).

Jackson, L and others, 'Water policy, agricultural trade and WTO rules' in P Martinez-Santos and MM Aldaya (eds) *Integrated Water Resources Management in the 21st Century: Revisiting the Paradigm* (CRC Press 2014).

Karamanian, SL, 'Human rights dimensions of investment law', in E De Wet and J Vidmar (eds) *Hierarchy in International Law: The Place of Human Rights*, (Oxford University Press, 2012).

Kumar, Dinesh M, 'Growing environmental costs of urban water supplies: Overcoming the challenges in implementing integrated urban water management', in S Mukherjee and D Chakraborty (eds) *Environmental Scenario in India: Successes and Predicaments* (Routledge 2012).

Levitt, Theodore, 'The globalisation of markets' in AM Kantrow,

Sunrise...Sunset: Challenging the Myth of Industrial Obsolescence (John Wiley and Sons, 1985).

López-Gunn, Elena, and others, 'Rethinking integrated water resources management: towards water and food security through adaptive management' in Bárbara A Willaarts and others (eds) *Water for Food Security and Well-Being in Latin America and the Caribbean: Social and Environmental Implications for a Globalized Economy* (Routledge 2014).

Markert, L, 'The Crucial Question of Future Investment Treaties: Balancing Investors' Rights and Regulatory Interests of Host States', in M Bungenberg, J Griebel and S Hindelang (eds), *EYIEL 2011, Special Issue: International Investment Law and EU Law* (Heidelberg: Springer 2011).

Mayer, P, 'Les arbitrages CIRDI en matière d'eau', in SFDI, *L'eau en droit international: Colloque d'Orléans* (Paris: Pedone 2011).

McIntyre, O, 'Emergence of the human right to water in an era of globalization and its implications for international investment law', in JF Addicott, MJH Bhuiyan and TMR Chowdhury (eds), *Globalization, international law, and human rights* (Oxford; Oxford University Press 2011)

Miles, K, 'Blue oil: Water resources, social justice and the international law on foreign investment', in S Alam, N Klein and J Overland (eds) *Globalisation and the Quest for Social and Environmental Justice* (Routledge 2012).

Mizushima, T, 'The role of the state after an award is rendered in investor-state arbitration', in S Lalani and RP Lazo (eds) *The Role of the State in Investor-State Arbitration* (Leiden: Nijhoff 2015).

Molle, François, and Jeremy Berkoff, 'Water pricing in irrigation: Mapping the debate in the light of experience', in François Molle and Jeremy Berkoff (eds), *Irrigation Water Pricing: The Gap Between Theory and Practice* (CABI 2007).

Mourra, Mary H, 'The conflicts and controversies in Latin American treaty-based disputes', in Mary H. Mourra and Thomas E. Carbonneau (eds), *Latin American Investment Treaty Arbitration – The Controversies and Conflicts* (Kluwer Law International 2008).

Mukherjee, Sacchidananda, and Debashis Chakraborty, 'Environmental governance scenario in Asia: lessons for the future' in Mukherjee and Chakraborty (eds) *Environmental Challenges and Governance: Diverse Perspectives from Asia* (Routledge 2015).

Neubert, Susanne, 'Strategic virtual water trade – a critical analysis of the debate', in Waltina Scheumann and others (eds) *Water Politics and Development Cooperation: Local Power Plays and Global Governance* (Springer 2008).

Pertile, Marco, and Paolo Turrini, 'Virtual water and international law', (forthcoming in Brill Research Perspectives – International Water Law) Ch 4.

Riedel, EH, 'The human right to water and General Comment No. 15', in EH Riedel and P Rothen (eds) *The Human Right to Water*, (Berlin: BWV, Berliner Wissenschafts-Verlag, 2006).

Rochford, Francine, 'Water sovereignty and food security', in Quentin Farmar-Bowers and others (eds), *Food Security in Australia: Challenges and Prospects for the Future* (Springer 2013).

Roth, Dik, and Jeroen Warner, 'Food security as water security: the multilevel governance of virtual water', in Otto Hospes and Irene Hadiprayitno (eds), *Governing Food Security. Law, Politics and the Right to Food* (Wageningen Academic Publishers 2010).

Rubins, Noah, The notion of investment in international investment arbitration, in N Horn and SM Kröll (eds) *Arbitrating Foreign Investment Disputes* (Kluwer 2004).

Schefer, Krista Nakavukaren, and Thomas Cottier, 'Responsibility to protect (R2P) and the emerging principle of common concern', in Peter Hilpold (ed.), *Die Schutzverantwortung (R2P): Ein Paradigmenwechsel in der Entwicklung des Internationalen Rechts?* (Martinus Nijhoff Publishers 2013).

Smets, Henri, 'Economics of water services and the right to water', in Brown-Weiss and others (eds) *Fresh Water And International Economic Law* (OUP 2005).

Staddon, Chad, and Nick James, 'Water security: A genealogy of emerging discourses', in Graciela Schneier-Madanes (ed.), *Globalized Water: A Question of Governance* (Springer 2014).

Stephens, Tim, 'International courts and sustainable development: Using old tools to shape a new discourse', in Brad Jessup and Kim Rubenstein (eds) *Environmental Discourses in Public and International Law* (Cambridge University Press 2012).

Tanzi, A, 'Public interest concerns in international investment arbitration in the water services sector: Problems and prospects for an integrated approach', in T Treves, F Seatzu and S Trevisanut (eds) *Foreign Investment, International Law and Common Concerns*, (Routledge, 2014).

Taylor, J, 'Community-based vulnerability assessment: Semarang, Indonesia', in K Otto-Zimmermann (ed.) *Resilient Cities: Cities and Adaptation to Climate Change Proceedings of the Global Forum 2010* (Local Sustainability 1 2010).

Thielbörger, P, 'The human right to water versus investor rights: Double-dilemma or pseudo-conflict?', in PM Dupuy, F Francioni and EU Petersmann (eds) *Human Rights in International Investment Law and Arbitration* (Oxford University Press 2009).

Turton, Anthony, 'Hydropolitics: The concept and its limitations', in Anthony Turton and Roland Henwood (eds) *Hydropolitics in the Developing World: A Southern African Perspective* (African Water Issues Research Unit 2002).

Truswell, JE, 'Thirst for profit: Water privatisation, investment law and a human right to water', in C Brown and K Miles (eds) *Evolution in Investment Treaty Law and Arbitration*, (Cambridge University Press 2011)

Upadhay, Madhukar, and Bibek R Kandel, 'Understanding the state of environmental governance in Nepal', in Mukherjee and Chakraborty (eds), *Environmental Challenges and Governance: Diverse Perspectives from Asia* (Routledge 2015).

BOOKS

Abukhater, Ahmed, *Water as a Catalyst for Peace: Transboundary Water Management and Conflict Resolution* (Routledge 2013).

Allan, Tony, *Virtual Water: Tackling the Threat to Our Planet's Most Precious Resource* (IB Tauris 2011).

Andreosso, Bernadette, and David Jacobson, *Industrial Economics and Organisation: A European Perspective* (McGraw Hill 2005).

Bakker, K, *Privatizing Water: Governance Failure and the World's Urban Water Crisis* (New York: Cornell University Press 2010).

Barcellos, A, *A eficácia jurídica dos princípios constitucionais. O princípio da dignidade da pessoa humana (The legal effect of the constitutional principles. The principle of human dignity)* (3rd ed, Rio de Janeiro, RJ: Renovar Book Publisher, 2011).

Barlow, Maude, and Barry Clarke, *Blue Gold: The Battle Against the Corporate Theft of the World's Water* (New Press 2002).

Boisson de Chazournes, L, *Fresh Water in International Law* (Oxford University Press, 2013).

Bulto, Takele Soboka, *The Extraterritorial Application of the Human Right to Water in Africa* (Cambridge University Press 2013).

Butler, J, *The Ethics of Health Care Rationing. Principles and Practices* (New York, NY: Cassell 1999).

Castro, Jose Esteban and Leo Heller, *Water and Sanitation Services: Public Policy and Management* (Routledge 2009).

Daniels, N, *Just Health. Meeting Health Needs Fairly* (New York, NY: Cambridge University Press 2008).

Desta, Melaku Geboye, *The Law of International Trade in Agricultural Products* (Kluwer Law International 2002).

Dicken, P, *Global Shift* (Sage Publishing 2003).
Dolzer, R and Stevens, M *Bilateral Investment Treaties* (The Hague: Martinus Nijhoff 1995).
Finger, Matthias, and Jeremy Allouche, *Water Privatisation: Trans-National Corporations and the Re-Regulation of the Water Industry* (Spon Press 2002).
Gargarella, R, P Domingo and T Roux (eds) *Courts and Social Transformation in New Democracies. An Institutional Voice for the Poor?* (Ashgate 2006).
Garner, Bryan, *Black's Law Dictionary* (10th edn, Thomson Reuters 2014).
Giplin, R, *Global Political Economy* (Princeton University Press 2001).
Hoekstra, Arjen H, and Ashok K Chapagain, *Globalization of Water: Sharing the Planet's Freshwater Resources* (Blackwell Publishing 2008).
Horowitz, D, *The Courts and Social Policy* (The Brookings Institution 1977).
Hunter, David, and others, *International Environmental Law and Policy* (4th edn, Foundation Press 2011).
Kingdon, J, *Agendas, Alternatives and Public Policies* (2nd edn, Harper Collins 2010).
Kiss, Alexandre and Dinah Shelton, *International Environmental Law*, 2nd ed (Transnational Publishers Inc, USA, 2000), 395.
Landes, W, and R Posner, *The Economic Structure of Tort Law* (Harvard University Press 1987).
M Malek, M, (ed.) *Setting Priorities in Health Care* (John Wiley & Sons 1994).
Matsushita, Mitsuo, Thomas J. Schoenbaum, Petros C. Mavroidis, and Michael Hahn, *The World Trade Organisation: Law, Practice and Policy* (2nd ed., OUP, 2006) 619.
McIntyre, Owen, *Environmental Protection of International Watercourses under International Law* (Ashgate 2007).
Moench, M, S Tyler and J Large (eds), *Catalyzing Urban Climate Resilience. Applying Resilience Concept to Planning Practice in the ACCCRN Program 2009–2011* (ISET: Boulder, CO 2011).
Monitt, Santiago, *State Liability in Investment Treaty Arbitration: Global Constitutional Law in the BIT Generation* (Hart Publishing 2009).
Morrison, Jason and Peter Gleick, *Freshwater Resources: Managing the Risks Facing the Private Sector* (Pacific Institute 2004).
Newcombe, A, and L Paradell, *Law and Practice of Investment Treaties* (Kluwer Law International 2009).
Palcic, Donal, and Eoin Reeves, *Privatisation in Ireland: Lessons from a European Economy* (Palgrave Macmillan 2011).

Pellet, A, 'Notes sur la "fragmentation" du droit international: Droit des investissements internationaux et droits de l'homme', *Unity and Diversity in International Law. Essays in Honour of Professor Pierre-Marie Dupuy* (Martinus Nijhoff Publishers 2014).

Post, A, *Foreign and Domestic Investment in Argentina: The Politics of Privatized Infrastructure* (Cambridge University Press 2014).

Redgwell, C, *Transboundary Pollution: Evolving Issues of International Law and Policy* (Edward Elgar Publishing 2015).

Rezende, S, and L Heller, *O saneamento no Brasil. Políticas e interfaces (Sanitation in Brazil. Public policies and interfaces)* (Editora UFMG 2002).

Robert-Cuendet, S. *Droits de l'investisseur étranger et protection de l'environnement* (Leiden: Nijhoff 2010).

Rosenberg, G, *The Hollow Hope: Can Courts Bring About Social Change* (2nd edn, University of Chicago Press 2008).

Sands P and Peel J, *Principles of International Environmental Law* (3rd edn, Cambridge University Press 2012).

Sen, Amartya, *Development as Freedom* (Oxford University Press 1999).

Shiva, Vandana, *Water Wars: Privatisation, Pollution and Profit* (South End Press 2002).

Sieder, R, L Schjolden and A Angell (eds) *The Judicialization of Politics in Latin America* (Palgrave Macmillan 2009).

Solanes, Miguel and Andrei Jouravkev, *Revisiting Privatization, Foreign Investment, International Arbitration, and Water* (Santiago: UN 2007).

Special Rapporteur on the Human Right to Water and Sanitation, *Realizing the Human Rights to Water and Sanitation: A Handbook, Financing* (Portugal 2014).

Titi, Catharine, *The Right to Regulate in International Investment Law* (Nomos and Hart Publishing 2014).

Tomuschat, C, *Human Rights: Between Idealism and Realism* (3rd edn, Oxford University Press 2014).

Viñuales, E, *Foreign Investment and the Environment in International Law* (Cambridge University Press 2012).

'Vital Water Graphics - An Overview of the State of the World's Fresh and Marine Waters' (2008) 2nd Edition UNEP, Nairobi, Kenya (2008), <www.unep.org/dewa/vitalwater/article43.html> accessed 20 February 2016

Williams, I, S Robinson and H. Dickinson, *Rationing in Health Care. The Theory and Practice of Priority Setting* (The Policy Press 2012).

Wilson, EO, *Half Earth: Our Planet's Fight for Life* (Liveright 2016).

Winkler, T, *The Human Right to Water: Significance, Legal Status and Implications for Water Allocation* (Hart Publishing 2012).

Yamin, A, and S Gloppen (eds), *Litigating Health Rights. Can Courts Bring More Justice to Health?* (Harvard University Press 2011).

INTERNATIONAL REPORTS

Adelphi, 'The rise of hydro-diplomacy: Strengthening foreign policy for transboundary waters' *Climate Diplomacy report* (2014).

Ashley, Richard, and Adrian Cashman, 'The Impacts of Change on the Long-term Future Demand for Water Sector Infrastructure', in OECD Report (2006) 1 Infrastructure to 2030: Telecom, Land Transport, Water and Electricity 253.

Bintari, *Economic loss assessment of floods in Kemijen, Semarang City (Kajian Kerugian Ekonomi akibat Banjir di Kelurahan Kemijen Kota Semarang)* (Semarang 2007) Unpublished report

CCROM and others, *Vulnerability and adaptation assessment to climate change in Semarang City* (2010). Unpublished Report, Mercy Corps – ACCCRN Indonesia Project.

Committee on Economic, Social and Cultural Rights, 'General Comment No. 3: The Nature of States Parties' Obligations', UN Doc E/1991/23, Annex III (1991).

de Albuquerque, C, 'Report of the independent expert on the issue of human rights obligations related to access to safe drinking water and sanitation', A/HRC/15/31 (2010).

Darwish, Adel, Analysis: Middle East Water Wars, *BBC News*, Friday 30 May 2003 at: http:/news.bbc.co.uk/ 1 /hi/world/middle east/2949768.stm

Dinar, Ariel, and others, *Water Allocation Mechanisms: Principles and Examples* (World Bank 1997) 1.

Engineers Ireland, *Delivering Ireland's Water Services for the 21st Century* (Engineers Ireland 2011).

GATT: Belgian Family Allowances (Allocations Familiales), Report of the Panel, adopted on 7 November 1952, BISD 1S/59 (1955) 412.

Hoda, Anwarul, and Ashok Gulati, 'India's Agricultural Trade Policy and Sustainable Development' http://www.ictsd.org/themes/agriculture/research/india%E2%80%99s-agricultural-trade-policy-and-sustainable-development *Issue Paper No. 49. ICTSD* (2013).

Hoekstra, Arjen Y, 'Virtual water: An introduction' in Arjen Y Hoekstra, '*Virtual water trade: Proceedings of the International Expert Meeting on Virtual Water Trade*' Value of Water Research Report Series No 12 (2003).

Littlechild, Stephen, *Regulation of British Telecommunications' Profitability. Report to the Secretary of State* (Department of Industry 1983).

Lloyd-Owen, D, 'Tapping Liquidity: Financing Water and Wastewater to 2029, a Report for PFI Market Intelligence' (London: Thomson Reuters 2009).

Massarutto, Antonio, 'Abstraction Charges: How Can the Theory Guide Us?' (OECD Expert Meeting on Sustainable Financing for Water Services from Theory to Practice, 2007).

McLoughlin, Patrick, 'Scientists Say Risk of Water Wars Rising', Planet Ark, 23 August 2004 at. http.//www.planetark.com/dailynewsstory.cfm/newsid/26728/story.htm.

Meeting Future Food Demands of Pakistan Under Scarce Water Situations (70th Annual Session Proceedings, Pakistan Engineering Congress 2007)

Ofwat, *Affordable for all: How can we help those who Struggle to pay their Water Bills* (Ofwat 2011).

Palaniappan and others, 'Water Infrastructure and Water-related Services: Trends and Challenges Affecting Future Development', in OECD Report (2007) 2 Infrastructure to 2030: Telecom, Land Transport, Water and Electricity 278.

Perdikan, 'Final project of pioneering community based sanitation micro credit to anticipate climate change at urban village of Kemijen, Semarang Timur sub-district, city of Semarang' *Project Report June 2010* (2010).

Report of the United Nations High Commissioner for Human Rights on the Scope and Content of the Relevant Human Rights Obligations related to Equitable Access to Safe Drinking Water and Sanitation under International Human Rights Instruments', A/HRC/6/3 (2007).

Report on the United Nations Water Conference, Mar del Plata, GA Res 32/158, UN GAOR, 107thPlenMtg, UN Doc E77IIA12 (1977).

Stevens, Barrie, and Pierre-Alain Schieb, 'Infrastructure to 2030: Main Findings and Policy Recommendations', in OECD Report (2007) 2 Infrastructure to 2030: Telecom, Land Transport, Water and Electricity 31.

UN Committee on Economic, Social, and Cultural Rights, The Right to the Highest Attainable Standard of Health: General Comment No. 14: art 12 of the International Covenant on Economic, Social, and Cultural Rights, § 4, UN Doc E/C12/2000/4 (2000).

UN Committee on Economic, Social, and Cultural Rights, Substantive Issues Arising in the Implementation of the International Covenant on Economic, Social and Cultural Rights: General Comment No 15: The right to water: arts. 11 and 12 of the International Covenant on Economic, Social and Cultural Rights, UN Doc E/C12/2002/11.

United Nations, Department Of Economic and Social Affairs, Population division, *World Population Prospect: The 2006 Revision, Highlights,* Working Papers ESA//P/WP 202 (2007).

UN General Assembly, 'Report of the independent expert on the issue of human rights obligations related to access to safe drinking water and sanitation, Catarina de Albuquerque' (1 July 2009) UN Doc. A/HRC/12/24.

UN Human Rights Council, 'Report of the Special Representative of the Secretary-General on the issue of human rights and transnational corporations and other business enterprises, John Ruggie' (7 April 2008) UN Doc. A/HRC/8/5.

UN Human Rights Council, 'Report of the Special Representative of the Secretary-General on the Issue of Human Rights and Transnational Corporations and Other Business Enterprises, Guiding Principles on Business and Human Rights: Implementing the United Nations "Protect, Respect and Remedy" Framework,' (21 March 2011) UN Doc. A/HRC/17/31.

UN-Water and WHO, UN-water global analysis and assessment of sanitation and drinking-water (GLAAS) 2014 report: investing in water and sanitation: increasing access, reducing inequalities (Geneva: United Nations and World Health Organization 2014).

United Nations, 'The millennium development goals report 2014' (2014) New York.

United Nations, 1977, 'Report of the United Nations Conference on Water, Mar del Plata' 70/29 E/Conf. 14.

Wallace, Chris, *Transforming a National Water Industry: Towards a Hydro Nation. The Scottish Experience* (IIEA Seminar 2011)

WHO-UNICEF, 'Global Water Supply and Sanitation Assessment 2000 Report' *WHO and UNICEF Joint Monitoring Programme for Water Supply and Sanitation* (United States of America: WHO-UNICEF 2000) 34.

World Economic Forum (2016), 'Figure 1.2: The Top Five Global Risks of Highest Concern for the Next 18 Months and 10 Years', Global Risks Report, 13 <http://reports.weforum.org/global-risks-2016/part-1-title-tba/> accessed in February 2016.

World Ocean Council, 'Conference Report' (Belfast, United Kingdom 2010).

MISCELLANEOUS

Abacha v Fawehinmi (2000) 6NWLR (Pt 660) 228.
Abdul Latif v Additional Sessions Judge, Sahiwal (2001 CLC 1139).
Abuja Declaration by African Ministerial Conference on Water (AMCOW), April 29–30, 2002.
Aam Admi Party, 'Election manifesto' (2015).

Advocacia Geral da União/Consultoria Jurídica/Ministério da Saúde/ Brasil (Federal Public Attorney's Office/Legal Department/Ministry of Health/Brazil), *Intervenção judicial na saúde pública. Panorama no âmbito da Justiça Federal e apontamentos na seara das Justiças estaduais (Court's intervention on public health. Overview from Federal Courts and notes on States Courts)* (2012).

Agence Française de Développement, 'Water Services and The Private sector in Developing Countries, Comparative Perceptions and Discussion Dynamics' (2012).

Agreement on the Encouragement and Protection of French Investments in Indonesia (1973) 985 United Nations Treaty Series 258.

Aguas Cordobesas SA, Suez, and Sociedad General de Aguas de Barcelona SA v Argentina, ICSID Case No ARB/03/18.

Aguas del Tunari SA v Bolivia, ICSID Case No ARB/02/3.

Alam, M, and Bhutta, MN, 'Availability of water in Pakistan during 21st century' *Proceedings of the International Conference on Evapotranspiration and Irrigation Scheduling* (San Antonio, Texas, USA 1996).

Anglian Water Group (AWG) v Argentina [2010] (UNCITRAL) Decision on Liability (2010).

Annotated Laws of Massachusetts.

Appellate Body Report, *United States – Measures Affecting the Cross-Boundary Supply of Gambling and Betting Services* WT/DS285/AB/R, adopted 20 April 2005; Panel Report WT/DS285/R, adopted 20 April 2005 (*US – Gambling*).

AP Pollution Control Bd II v Prof MV Nayudu (2001) 2 SCC 62

Argentina-United Kingdom Bilateral Investment Treaty (1990).

ATA Construction, Indus. and Trading Co v Jordan, ICSID Case No ARB/08/02, Award (2010).

Attakoya Thangal v Union of India (1990) 1 KLT 583.

Azurix Corp v Argentina, ICSID Case No. ARB/01/12, Award (2006).

Baluchistan Ground Water Rights Administration Ordinance, 1978.

Barlow, Maude, *The Free Trade Area of the Americas: The Threat to Social Programs, Environmental Sustainability and Social Justice* (Council of Canadians, 2001) 2.

Barlow, Maude and Clarke, Tony, 'Water Privatization' (Polaris Institute, January 2004)

Benazir Bhutto v President of Pakistan (PLD 1998 Supreme Court 388.

Bergallo, P, 'Justice and Experimentalism: The Judiciary's remedial function in public interest litigation in Argentina' (2005).

Bhattacharyay, BN, 'Estimating demand for infrastructure in energy, transport, telecommunications, water and sanitation in Asia and the Pacific: 2010–2020' *ADBI Working Paper Series No. 248* (Tokyo: Asian Development Bank Institute 2010).

Bhattacharya, BN, and others, 'Infrastructure for development: meeting the challenge' *Policy Paper* (Centre for Climate Change Economics and Policy and Grantham Research Institute on Climate Change and the Environment in collaboration with Intergovernmental Group of Twenty Four 2012).

Bhattacharya and others, 'Driving sustainable development through better infrastructure: Key elements of a transformation program' *Global Economy & Development Working Paper No. 91*, (Washington DC: Global Economy and Development at Brookings 2015).

Biwater Gauff (Tanzania) Ltd v Tanzania, ICSID Case No ARB/05/22, Award (2008).

Bolivia-Netherlands BIT (1992)

Bolivia, Ley Nº 1544 - Ley Marco de Capitalización (1994).

Brazil Constitution 1988

Brazil Law 8080/90.

Brazilian Supreme Court, Adin 2340, Adin 2077 and Adin 1842.

'Cabinet meeting approves Food Authority Act' *The Express Tribune*.

Canada — Certain Measures Affecting the Automotive Industry, WTO Doc WT/DS139/R, WT/DS142/R, Report of the Panel (2000).

Cantonment Pure Food Act 1966.

Case concerning Gabčíkovo-Nagymaros Project (Hungary/Slovakia), Judgment (1997).

Castro, M, *Spatio-temporal trends of infant mortality in Brazil* (2010).

Central Product Classification (CPC) Version 1.0, Statistical Papers Series, M, No. 77 Ver 1.0, United Nations 1998.

Chatterton, Isabel and Olga Susana Puerto, 'Estimation of infrastructure investment needs in South Asia region' *Working Paper No. 62608* (World Bank, Washington DC 2011).

Chemtura v Canada, UNCITRAL (1976), Award, 2 August 2010.

China – Measures Affecting Trading Rights and Distribution Services for Certain Publications and Audiovisual Entertainment Products – AB-2009-3 – Report of the Appellate Body (2009).

Ch, Riaz Ahmad Yazdani v The Federation of Pakistan and 8 others 1990 CLC 1406.

Clean Water Should be Recognized as a 'Human Right' *PLoS Med* (2009).

Coca-Cola Company 10-K 2010

COHRE, WaterAid, SDC and UN-HABITAT, 'Sanitation: A Human Rights Imperative' (Geneva 2008).

COHRE, AAAS, SDC and UN-Habitat, *Manual on the Right to Water and Sanitation* (2007).
Colorado Revised Statutes.
Column in the UNCTAD's Investment Dispute Settlement Navigator <http://investmentpolicyhub.unctad.org/ISDS/FilterByEconomicSector>.
Commission for Energy Regulation 'Economic Regulatory Framework for the Public Water Services Sector in Ireland' Consultation Response Paper CER/14/075
Commission for Energy Regulation 'Water Charges Plan Consultation: Executive Summary CER/14/366'
Companhia Estadual de Águas e Esgotos – CEDAE v Ministério Público Federal, Brazilian Supreme Court, RE 417408 AgR / RJ, March 20, 2012.
Compañiá de Aguas del Aconquija SA and Vivendi Universal SA v Argentina, ICSID Case No. ARB/97/3, Award (2007).
Constitution of the Federal Republic of Nigeria 1999.
Constitution of Kenya, 2010.
Constitution of the Republic of South Africa, 1996.
Contracting out water and Sanitation Services 5 <http://wedc.lboro.ac.uk/resources/books/Contracting_Out_Water_and_Sanitation_Services_-_Vol_2_-_Complete.pdf>.
Convention on the Law of the Non-navigational Uses of International Watercourses 1997
Convention on the Protection of Use of Transboundary Watercourses and International Lakes 1992.
COP21: on a oublié d'inviter l'océan, CNRS Le Journal and Libération (Online Edition, 2015).
Côté, C, 'A chilling effect? The impact of international investment agreements on national regulatory autonomy in the areas of health, safety and the environment' (PhD Thesis, The London School of Economics and Political Science (LSE) 2014).
Cottier, Thomas, 'Renewable Energy and Process and Production Methods', E15 Think Piece August.
Coulby, Hillary, 'A Guide to Multistakeholder Work: Lessons from the Water Dialogues' (May 2009).
Crema, L, 'Tracking the origins and testing the fairness of the instruments of fairness: Amici curiae in international litigation', *Jean Monnet Working Paper* (2012)
Cuq, M, *L'eau en droit international: Convergences et divergences dans les approches juridiques* (Bruxelles: Larcier, 2013).
Dakas CJ, 'Judicial Reform of the Legal Framework for Human Rights Litigation in Nigeria: Novelties and Perplexities' (Nigerian Institute of Advanced Legal Studies, 2012).

Dáil Éireann, Debate 6 November 2014, 42684W.

De Albuquerque C, 'Water and Sanitation as Human Rights' L'eau et son droit, Rapport public, Etudes et documents du Conseil d'Etat (2010) 483.

Dinar and others, 'Climate change, conflict, and cooperation: Global analysis of the resilience of international river treaties to increased water variability' *Policy Research Working Paper 6916* (Washington DC: World Bank 2014).

Dubreuil, C, *Right to water: From concept to implementation* (World Water Council 2006).

EDF v Romania, ICSID Case No ARB/05/13, Award (2009).

El Paso v Argentina, ICSID Case No. ARB/03/15, Award (2011).

Enron v Argentina, ICSID Case No ARB/01/3, Award (2007).

Ercin, AE, MM Aldaya and AY Hoekstra, 'Corporate water footprint accounting and impact assessment: The case of the water footprint of a sugar-containing carbonated beverage' (2011) Water Resource Management (2011) 25:721.

ETC Group, 'World's 10 Largest Water Companies: Who they are, what they do and how much revenue they collected' (2012).

Ethyl Corporation v Canada, UNCITRAL (1976).

European Charter on Water Resources 2001 (17 October 2001).

European Communities – Measures Prohibiting the Importation and Marketing of Seal Products – AB-2014-1–AB-2014-2 – Reports of the Appellate Body (2014).

European Commission, *Fact Sheet: Why the new EU proposal for an Investment Court System in TTIP is beneficial to both States and investors* (Brussels 2015).

European Communities — Regime for the Importation, Sale and Distribution of Bananas, (*EC–Bananas III*) WTO Doc WT/DS27/R/USA (1997) (Report of the Panel).

European Union, *Concept Paper: Investment in TTIP and beyond – the path ahead* (Brussels 2015).

Fay, Marianne, and Tito Yepes, 'Investing in Infrastructure: What is needed from 2000 to 2010?' *Policy Research Working Paper 3102* (Washington DC: World Bank 2003).

Federal Emergency Management Agency, US Fire Admin., *Water Supply Systems & Evaluation Methods* (FDMA 2008) 2.

Flint Michigan water crises. *The Guardian* February 2, 2016.

Food and Agriculture Organization, *The State of Food Insecurity in the World 2001* (Rome 2002) 10.

Food and Agricultural Organization of the UN, *New dimensions in water security: Water, society and ecosystem services in the 21st century*, Doc No AGL/MISC/25/2000 (2000).

'Gaz de schiste: risque de pollution de l'eau potable par le méthane' Science et Avenir (Online Edition 12 2011).

Gelsenwasser AG v Algeria, ICSID Case No ARB/12/32.

General Secretary, West Pakistan Salt Miners Labour Union (CBA) Khewra, Jhelum v The Director, Industries and Mineral Development, Punjab, Lahore (1994 SCMR 2061).

Geneva Convention Relative to the Treatment of Prisoners of War (1949), 6 UST 3316, 74 UNTS 135.

Geneva Convention (IV) Relative to the Protection of Civilian Persons in Time of War (1949), 75 UNTS 287, 6 UST 3516.

Geneva Conventions 1949, Additional Protocol Relating to the Protection of Victims of International Armed Conflicts (1977), 1125 UNTS 3.

Geneva Water Hub

Gleick, Peter, and others, 'The New Economy of Water: The Risks and Benefits of Globalization and the Privatization of Fresh Water' (Pacific Institute for Studies in Development, Environment and Security 2002).

Global Equity Research, 'Watching water: A guide to evaluating corporate risks in a thirsty world' (JP Morgan Securities Inc 2008)

Goldsmith, Oliver, 'The Deserted Village'.

Gorecki, Paul, and others, *Affordability and the Provision of Water Services in Ireland: Options, Choices and Implications* (Department of the Environment, Community and Local Government 2013).

Griswold v Connecticut 381 US. 479 (1965).

Gulbenkian Think Thank on Water and the Future of Humanity, *Water and the Future of Humanity: Revisiting Water Security* (Calouste Gulbenkian Foundation/Springer 2014).

Hague Convention Respecting the Laws and Customs of War on Land (1907).

Hamamoto, S, 'Requiem for indirect expropriation: On the theoretical and practical uselessness of a contested concept', *Private International Law as Global Governance (PILAGG) e-series*, IA-1, (2013), http://blogs.sciences-po.fr/pilagg/pilagg-e-series/ also at http://ssrn.com/abstract=2666836/.

Heller, L, 'Report of the Special Rapporteur on the human right to safe drinking water and sanitation', A/HRC/30/39 (2015).

Hershkoff, H, *Public interest litigation: selected issues and examples* (2005).

Hoekstra, Arjen Y, 'The relation between international trade and freshwater scarcity' *World Trade Organization Staff Working Paper ERSD-2010-05* (2010).

Hutton, Guy and Jamie Bartram, 'Regional and global costs of attaining the water supply and sanitation target (Target 10) of the Millennium Development Goals' *WHO/HSE/AMR/08/01* (Geneva: World Health Organization 2008).

Hutchinson, A, 'Is radioactive water worth worrying about?' The New Yorker (Online Edition 2015).

Impregdilo SpA v Argentina, ICSID Case No. ARB/07/17, Award (2011).

Instituto Brasileiro de Geografia e Estatística - IBGE/Brasil (Brazilian Institute of Geography and Statistics/Brazil), *Atlas do saneamento* (*Sanitation atlas*) (Rio de Janeiro, RJ: IBGE 2011).

Instituto Brasileiro de Geografia e Estatística - IBGE/Brasil (Brazilian Institute of Geography and Statistics/Brazil), *Síntese dos indicadores (Summary of social indicators)* (Rio de Janeiro, RJ: IBGE 2010).

Instituto Brasileiro de Geografia e Estatística - IBGE/Brasil (Brazilian Institute of Geography and Statistics/Brazil), *Pesquisa nacional de saneamento básico* (*National survey for basic sanitation*) (Rio de Janeiro, RJ: IBGE 2008).

Instituto de Pesquisa Econômica Aplicada – IPEA/Brasil (National Institute of Applied Economics/Brazil), *Políticas sociais: acompanhamento e análise 20* (*Welfare public policies: monitoring and analysis 20*) (Brasília: IPEA 2012) 105.

Intergovernmental Panel on Climate Change (IPCC) 'Climate change 2014: Impacts, adaptation, and vulnerability: Part A: Global and sectoral aspects' contribution of working group II to the fifth assessment report of the IPPC, CB Field, VR Barros, DJ Dokken, KJ Mach, MD Mastrandrea, TE Bilir, M Chatterjee, KL Ebi, YO Estrada, RC Genova, B Girma, ES Kissel, AN Levy, S MacCracken, PR Mastrandrea, and LL White (eds) (Cambridge University Press 2014).

International Consortium of Investigative Journalists 'Promoting privatization' (2003).

International Monetary Fund, 'World Economic Outlook Database' (2014).

International Rice Research Institute (IRRI), From Far Eastern Agriculture Web site: <www.fareasternagriculture.com/crops/agriculture/irri-introduces-water-saving-technique-in-producing-rice>.

International Seabed Authority, ISBA/18/A/11 (2012).

International Seabed Authority, ISBA Reg/16/1/12/Rev 1 (2010).

International Tribunal for the Law of the Sea Reports (ITLOS), *Responsibilities and Obligations of States with Respect to Activities in the Area, Advisory Opinion* (February 2011).

Ioane Teitiota v The Chief Executive of Ministry of Business, Innovation and Employment, CA 50/2014; [2014] NZCA (2014).

ISO, *ISO 14046-2014 Environmental Management – Water Footprint – Principles, Requirements and Guideline* (Geneva 2014).

ISET, SEMARANG's adaptation plan in responding to climate change. City Resilience Strategy of Asian Cities Climate Change Resilience Network, (Institute for Social and Environmental Transition).

IWA Specialist Group on Statistics and Economics, *International Statistics for Water Services, Information every Water Management should know about,* International Water Association (2012).

Justice (R) Fazal Karim, [2006] *Judicial Review of Public Actions,* Pakistan Law House: Karachi.

Kessides, Ioannis. N., 'Reforming Infrastructure, Privatization, Regulation and Competition' (World Bank Policy Research Report, 2004).

Kumar v Bihar, (1991) 1 SCC 598.

Kundell, Jim (ed.), *Water Profile of Nigeria*

Caflisch L, 'Le droit à l'eau – un droit de l'homme internationalement protégé?' L'eau en droit international, Colloque d'Orléans, Pedone, (Paris, 2011).

Legality of the Threat or Use of Nuclear Weapons (Advisory Opinion, ICJ Reports 1996), para 29, 242.

'Like a deer that yearns for running streams. . .' (The Bible, Psalm 42).

Lisbon Charter: Guiding the Public Policy and Regulation of Drinking Water Supply, Sanitation and Wastewater Management Services (2015)

Louisiana Revised Statutes (Annotated).

Luizaga, C, S Gotlieb, M Jorge and R Laurenti, 'Maternal deaths: reviewing the adjustment factor for official data' (2010) 19/1 Epidemiologia e serviços de saúde.

Madani, Dorsati, 'A review of the role and impact of export processing zones' (World Bank 1999)

Mader, P, 'Water paradigms: Full cost recovery versus human rights', Working Paper (2011), Human Rights to Water & Sanitation Program, Harvard Kennedy School, Cambridge, MA, USA.

Mara, D and Bos R, 'Risk Analysis and Epidemiology: the 2006 WHO Guidelines for the Safe Use of Wastewater in Agriculture', Wastewater Irrigation and Health: Assessing and Mitigating Risk in Law-Income Countries (International Water Management Institute, Earthscan, London 2010) 51.

Marrakesh Agreement Establishing the World Trade Organization (1994), 1867 UNTS 3, annex 1B, *General Agreement on Trade in Services* (1995) 1869 UNTS 183.

Marin, Philippe, *Public-Private Partnerships for Urban Water Utilities* (World Bank Publications 2009).

Marvin Feldman v Mexico ICSID Case No ARB(AF)99/1, Award (2002).

Methanex Corporation v USA, UNCITRAL (1976), Decision of the Tribunal on Petitions from Third Persons to Intervene as 'amici curiae' (2001).
Mazibuko and Others v City of Johannesburg and Others (CCT 39/09) [2009] ZACC 28; 2010 (3) BCLR 239 (CC); 2010 (4) SA 1 (CC) (2009).
Mazibuko and Others v The City of Johannesburg and Others 2008 High Court of South Africa (Witwatersrand Local Division) Case No 06/13865 3 (S Afr).
MC Mehta v Kamal Nath, (1997) 1 SCC 388.
MC Mehta v Union of India (1988) 1 SCC 471.
MC Mehta v Union of India, (2004) 3 SCR 128.
MC Mehta v State of Orissa, AIR 1992 Ori 225.
Mekonnen, MM and Hoekstra, AY, 'Four billion people facing severe water scarcity' (2016:12) Science Advances (Online edition).
Metalclad Corporation v Mexico, ICSID Case No. ARB(AF)/97/1, Award (2000).
Methanex Corporation v USA, UNCITRAL (1976), Final Award of the Tribunal on Jurisdiction and Merits (2005).
Milman, A, 'Ocean warming and acidification needs more attention, argues US' *The Guardian* (Online Edition, 2015).
Minerva Mills Ltd v Union of India AIR 1980 SC 1789.
Ministério da Saúde and Fundação Nacional de Saúde, 'Termo de Referência para elaboração de planos municipais de saneamento básico. Procedimentos relativos ao convênio de cooperação técnica e financeira da Fundação Nacional de Saúde – FUNASA/MS, Brasília' (2012).
Ministério Público do Estado de Rondonia v Companhia de Águas e Esgotos de Rondônia – CAERD, TJRO, Appeal n. 1100324-60.2008.8.22.0018 (2010).
Ministério Público do Estado de São Paulo v Município de Sorocaba, Brazilian Supreme Court, RE 254764/SP (2010).
Ministério Público Federal v Município de Barra do Sul, Cia Catarinense de Aguas e Saneamento – CASAN, IBAMA, TRF4, Appeal n. 0002755-71.2003.404.7201/SC.
Minnesota Statutes (Annotated).
Mrs Anjum Irfan v Lahore Development Authority through Director-General and Others (PLD 2002 Lahore 555).
Mukherjee, Sacchidananda, and Prakash Nelliyat, *Groundwater Pollution and Emerging Environmental Challenges of Industrial Effluent Irrigation in Mettupalayam Taluk, Tamilnadu, Comprehensive Assessment of Water Management in Agriculture* (International Water Management Institute, Colombo 2007).
Narmada Bachao Andolan v Union of India A.I.R. 2000 SC 375.

National Geographic, 'Ocean Acidification' Pristine Seas (Online Edition).
New Jersey Constitution.
New Jersey Code.
New York Constitution.
New York General Obligations Law.
Nixon, R, 'Farm Subsidies Leading to More Water Use', *The New York Times* (2013)
OECD, '"Indirect expropriation" and the "right to regulate" in international investment law', *Working Papers on International Investment* (2004).
OECD 'Managing water for all: An OECD perspective on pricing and financing' (2009).
OECD, 'Meeting the challenge of financing water and sanitation: Tools and approaches' *OECD Studies on Water* (2011).
OECD 'Pricing Water Resources and Water and Sanitation Services' (OECD 2010).
OECD Principles on Water Governance under the Water Governance Initiative (C/MIN(2015)12).
OECD *Water: The Experience in OECD Countries* (2006).
OFWAT, 'The economic regulator of water services in the England and Wales'.
Ohio River Valley Water Sanitation Comm'n, 'Strategic Plan for the Ohio River Valley Water Sanitation Commission' (2008)
Pac Rim Cayman LLC v El Salvador, ICSID Case No ARB/09/12, Award (2009.
Pakistan Constitution.
Pakistan Environmental Protection Act 1997.
Pakistan Factories Act 1934
Pakistan National Drinking Water Policy 2009.
Pakistan Penal Code (Act XLV of 1860).
Pakistan to face 25 MAF water shortage: IRSA – Pak Observer <http://pakobserver.net/201103/10/ detailnews.asp?id=80250>.
Paris Agreement, FCCC/CP/2015/L.9/Rev.1, 12 December 2015.
Parkerings-Compagniet AS v Lithuania, ICSID Case No ARB/05/8, Award (2007).
Perlman, D, 'Scientists alarmed by ocean dead-zone growth', San Francisco Gate (Online Edition 2008).
Philip Morris Asia Ltd v Australia, UNCITRAL (1976), PCA Case No. 2012–12.
Piero Foresti, Laura de Carli & Others v South Africa, ICSID Case No. ARB(AF)/07/01, Award (2010).

Programa das Nações Unidas para o Desenvolvimento – PNUD and Instituto de Pesquisa Econômica Aplicada – IPEA (United Nations Program for Development and Institute of Applied Economics), *Atlas do Desenvolvimento Humano no Brasil 2013* (*Brazilian Human Development Atlas 2013*).

Pulp Mills on the River Uruguay (Argentina v Uruguay), Judgment, ICJ Reports 2010.

Punjab Local Government Ordinance 2001.

Punjab Pure Food Ordinance 1960.

Punjab Food Safety And Standards Act 2011.

PWC, 'Irish Water: Phase 1 Report' (2011).

Ray, G, 'The Ocean is broken', *Sydney Morning Herald* (Online Edition 2013)

Raz, Joseph, 'Human rights without foundations' *Oxford Legal Studies Research Paper No 14/2007* (2007).

Razzaque, Jona, 'Access to environmental justice: Role of the Judiciary in Bangladesh' (unpublished manuscript).

Reformulated Gasoline and Brazil: Measures Affecting Imports of Retread Tyres, Report of the Appellate Body, WT/DS332/AB/R (2007).

Residents of Bon Vista Mansions v Southern Metropolitan Local Council 2002 (6) BCLR 625 (W) (S Afr).

Roach, B, 'Corporate power in a global economy' (2007) A GDAE teaching module on social and environmental issues in economics, Global Development and Environment Institute, Tufts University.

Rodrigez, DJ, and others, 'Investing in water infrastructure: capital, operation and maintenance' *Water Paper* (The World Bank 2012).

Royal Academy of Engineering, 'Global Water Security – an engineering perspective' (2010).

Rubenstein, Edwin S, 'The untapped potential of water privatization' 2000, A Hudson Institute Report For American Water Works, Inc.

Ruggie, John, 'Business and human rights: Towards operationalizing the "protect, respect and remedy" framework', Report of the UNSRSG on human rights and transnational corporations and other business enterprises, A/HRC/11/13 (2009).

Ruggie, John, 'Report of the Special Representative of the Secretary-General on the issue of human rights and transnational corporations and other business enterprises', A/HRC/14/27 (2010).

Salini Costruttori v Jordan, ICSID Case no ARB/02/13 (2006).

Salman, MA Salman and S McInerney-Lankford, *The Human Right to Water: Legal and Policy Dimensions* (World Bank 2004).

Saluka Investments BV v The Czech Republic, UNCITRAL (1976), Partial Award (2006).

'Sanitation and Water Supply: Improving Services for the Poor' (International Development Association, The World Bank).
Sarwono and others, *Focused group discussion on Sanitation Credit* (BKM Village 2012).
SAUR International SA v Argentina, ICSID Case No ARB/04/4, Décision sur la competence et sur la responsabilité (2012).
SBNP Local Government Ordinance 2001.
Scelle, G, 'Obsession du Territoire. Essai d'étude réaliste du droit international' (Symbolae, JHW Verzijl, Martinus Nijhoff, La Haye 1958).
Scheduling of Initial Commitments in Trade in Services: Explanatory Notes, MTN.GNS/W/164, 3 September 1993.
Sengupta and others, 'Financing for Infrastructure Investment in G-20 Countries' *Working Paper No. 2015-144* (New Delhi: National Institute of Public Finance and Policy, 2015).
Services Sectorial Classification List, MTN.GNS/W/120, World Trade Organisation (W/120).
SETEP construções SA v Companhia Catarinense de Águas e Saneamento – CASAN (Brazilian Superior Court of Justice), AgReg na SS 2418 (2011).
Sewage Treatment, Foundation for Water Research.
Shehla Zia and Others v WAPDA, PLD 1994 SC 693.
Shortage of water at tail-end may hit cotton crop: *SCA- Dawn News*.
Smets, H, *De l'eau potable à un prix abordable* (Paris: Editions Johanet 2007).
Social and Economic Rights Action Centre (SERAC) v Nigeria (2001) AHRLR 60 (ACHPR 2001).
Social Justice in an Open World: The Role of the United Nations.
South Africa v Grootboom, 2000 (11) BCLR 1169 (CC)(S Afr).
South Dakota Codified Laws
Stucki, Philipp, 'Water Wars or Water Peace? Rethinking the Nexus between Water Scarcity and Armed Conflict' *Programme for Strategic and International Security Studies occasional paper no 3* (2005).
Suez, Sociedad General de Aguas de Barcelona SA, and InterAguas Servicios Integrales del Agua SA v Argentina, ICSID Case No ARB/03/17, Decision on Liability (2010).
Suez, Sociedad General de Aguas de Barcelona, SA and Vivendi Universal SA v Argentine Republic, Decision on Liability ARB/03/19 (2010).
TASC, 'Water charges: An equality-proofed approach' (2012).
TASC, 'Paying for water: Equity, efficiency and sustainability' (2013).
Total S.A v Argentina, ICSID Case No ARB/04/1 (2010).
Toubkiss, J, *Costing MDG target 10 on Water Supply and Sanitation: Comparative Analysis, Obstacles and Recommendations* (2006)
Tribunal de Contas da União (Brazilian Federal Court of Accounts),

Relatório Sistêmico de Fiscalização da Saúde (Health System Report), 2014.
Tsur, Yacov, and Ariel Dinar, 'Efficiency and equity considerations in pricing and allocating irrigation water' *World Bank Policy Research Working Paper No 1460* (1995) 33.
Ulysseas v Ecuador, UNCITRAL (1976), Final Award (2012).
United Kingdom-Tanzania BIT (1994).
'UN experts voice concern over adverse impact of free trade and investment agreements on human rights' (2015).
UN Commission on Human Rights Res A/HRC/RES/15/9.
UN Convention of the Prohibition of Military or Any Other Hostile Use of Environmental Modification Techniques, May 18, 1977, 31 UST 333, TIAS No 9614.
UN Convention on the Elimination of All Forms of Discrimination against Women, GA Res 34/180 (1979) UN Doc A/34/46.
UN Convention on the Rights of the Child, GA Res 44/25 annex, at art 24(2)(c), UN GAOR, 44th Sess, Supp No 49, UN Doc A/44/49 (1989).
UN Convention on the Settlement of Investment Disputes between States and Nationals of Other States (1965) 575 UNTS 159.
UN Convention on the Law of the Sea 1982.
UNCITRAL 'Rules on Transparency in Treaty-based Investor-State Arbitration' (2014).
UNCTAD, 'Investment policy framework for sustainable development' (2015).
UNCTAD, 'The role of international investment agreements in attracting foreign direct investment to developing countries' (2009).
UNCTAD, 'Bilateral Investment Treaties 1995–2006: Trends in investment rulemaking' (2007).
UNCTAD, 'Intellectual property provisions in international investment arrangements' (2007).
UN Declaration on the Right of Development, GA Res 44/128 (1986).
UN Declaration of the United Nations Conference on the Human Environment, Stockholm (1972).
UN, Department of Economic and Social Affairs, Population division, *World Population Prospect: The 2006 Revision, Highlights,* (Working Papers ESA//P/WP 2007).
UNDP, *Human Development Report 2006 - Beyond Scarcity: Power, Poverty and the Global Water Crisis* (New York, 2006).
UN Economic Commission for Europe, 'Guidance on water and adaptation to climate change' (2009).
UNEP, 'Global Environment Outlook – 5: Environment for the future we want', UNEP, Nairobi (2012).

UNEP, 'Keeping Track of our changing environment: From Rio to Rio+20 (1992–2012)' Nairobi (2011).

UNEP, 'Sick water? The central role of wastewater management in sustainable development: A rapid response assessment' (2010).

UN 1974, 'Third United Nations Conference on the Law of the Sea: Summary records of meetings of the First Committee' A/CONF62/C1/SR10, Vol II, 10th meeting.

UN General Assembly, 'The human right to water and sanitation' A/ RES/64/292 (2010).

UN General Assembly, 'Sustainable fisheries, including through the 1995 Agreement for the Implementation of the Provisions of the United Nations Convention on the Law of the Sea of 10 December 1982 relating to the Conservation and Management of Straddling Fish Stocks and Highly Migratory Fish Stocks, and related instruments' A/ RES/64/72 (2010).

UN General Assembly, 'Letter dated 16 March from the Co-Chairpersons of the Ad Hoc Open-ended Informal Working Group to the President of the General Assembly' A/65/68, (2010).

UN General Assembly, '108th Plenary Meeting' Official Records, A/64/PV108 (2010).

UN General Assembly, 'The right to development' A/RES/54/175 (2000).

UN General Assembly, 'United Nations Millennium Declaration' A/ RES/55/2 (2000).

UN General Assembly, 'Declaration for the establishment of a new international economic order' A/RES/S-6/3201 (1974).

UN General Assembly, Human Rights Council, Report of the United Nations High Commissioner for Human Rights on the Scope and Content of the Relevant Human Rights Obligations Related to Equitable Access to Safe Drinking Water and Sanitation Under International Human Rights Instruments, P 66, UN Doc A/HRC/6/3 (2007).

UN Global Compact, 'The CEO Water Mandate White Paper, The human right to water: Emerging corporate practice and stakeholder expectations' (UN Global Compact – The Pacific Institute, November 2010).

UN High Commissioner for Human Rights, 'Human rights, trade and investment', E/CN.4/Sub.2/2003/9 (2003).

UNHR Committee, General Comment No. 6 (CCPR/C/Rev.1 1985).

UNHR Council, 'The human right to safe drinking water and sanitation', A/HRC/RES/24/18 (2013).

UNR Council 2010, 'Human rights and access to safe and drinking water and sanitation', A/HRC/15/L14.

UNHR Sub-Commission on the Protection and Promotion of Human Rights Resolution No 2001/2 and Decision No 2002/105.

UNHR Sub-Commission on the Promotion and Protection of Human Rights 2003, 'Sub-Commission Begins Consideration of Economic, Social and Cultural Rights', *Press Release*.

UNHR First Optional Protocol to the International Covenant on Civil and Political Rights, 999 UNTS 302 (1966).

UNICEF/WHO, 'Millennium Development Goal drinking water target met – Sanitation target still lagging far behind', Joint news release (2012)

UNICEF, 'Progress on Drinking Water and Sanitation' (2014).

UN Summit on the Millennium Development Goals 20-22 (2010). <www.un.org/ millenniumgoals/>

United Utilities (Tallinn) BV and Aktsiaselts Tallinna Vesi v Estonia, ICSID Case No ARB/14/24.

UN-Water and WHO, *Global Annual Assessment of sanitation and drinking water (GLAAS)* (Geneva: United Nations and World Health Organization 2010).

United Nations Water Conference: Summary and Main Documents (1978).

United Nations World Water Development Report, 'Water and Energy: Executive Summary' (2014).

United States Environmental Protection Agency, 'Drinking Water Treatment guidance' (EPA 816-F04-034 2004).

United States – Measures Affecting the Cross-Boundary Supply of Gambling and Betting Services WT/DS285/AB/R, adopted 20 April 2005; Panel Report WT/DS285/R, adopted 20 April 2005 *(US – Gambling)*.

United States Model BIT (2012).

United States Code (the 'Wire Act').

United States Code (the 'Travel Act').

United States Code (Illegal Gambling Business Act).

United States- Import Prohibition of Certain Shrimp and Shrimp Products, 38 ILM 118 (1999).

United Utilities (Tallinn) BV and Aktsiaselts Tallinna Vesi v Estonia, ICSID Case No ARB/14/24 (2015).

Urbaser SA and Consorcio de Aguas Bilbao Bizkaia, Bilbao Biskaia Ur Partzuergoa v Argentina, ICSID Case No ARB/07/26 (2010).

Uruguay Round, 'Group of negotiations on services, service sectors classification list,' MTN.GNS/W/120 (1991).

Utah Code (Annotated).

Vellore Citizens' Welfare Forum v Union of India (1996) 5 SCC 647.

Vienna Convention on the Law of Treaties (1969), 1155 UNTS 331; 8 ILM 679.
Vienna Declaration on Program of Action on Human Rights 1993.
Vis-Dunbar, Damon, and Luke Eric Peterson, *Bolivian water dispute settled, Bechtel forgoes compensation* (Investment Treaty News 2006).
Walker, Anna, 'The independent review of charging for household water and sewerage services' (Department of Environment, Food and Rural Affairs 2009).
Walsh, B, 'Ocean acidification will make climate change worse' The Time (Online Edition 2013).
Ward and others, 'Costs of adaptation related to industrial and municipal water supply and riverine flood protection' *World Bank Discussion Paper No. 6* (Washington DC: World Bank 2010).
Water and Sanitation Program, 'Trends in Private Sector Participation in the Indian Water Sector: A Critical Review' *Field Note* (2011).
Water Framework Directive 2000/60/EC of the European Parliament and of the Council of 23 October 2000 establishing a framework for Community action in the field of water policy, OJ L 327/1, 22/12/2000.
Water Industry Act, 1991 (UK)
Water Management Practices in Pakistan Issues and Options for Productivity Enhancement-Round table Discussion on Agriculture and Water in Pakistan (2011) <http://siteresources.worldbank.org/PAKISTANEXTN/Resources/ WMPracticesinPakistan.pdf> accessed on 13/05/2011.
'Water war in Bolivia' *The Economist* (2000)
Water wars: a new reality for business and governments', The Guardian (Online Edition, 2014)
Waymouth, B, 'How to Protect the Ocean From Us?' The Huffington Post (Online Edition, 2015).
Weckel, P, 'L'eau et le droit humanitaire' L'eau en droit international, Colloque d'Orléans, Pedone (Paris 2011).
Western Front, Aerial Bombardment and Related Claims Eritrea's Claims, ICGJ 356 (PCA 2005).
World Bank and World Trade Organization, I-TIP Services.
World Bank, 'Health nutrition and population statistics: Population estimates and projections' (undated a).
World Bank, 'Introduction to Wastewater Treatment processes' (undated b).
World Bank, 'Private participation in infrastructure database' (undated c).
World Bank, 'Public-private partnerships for rural water services' *Briefing Note No. 4* (2012).
World Bank, 'Wastewater Disposal and Transport Options – Sewerage' (undated).

World Economic Forum, 'Global Risk Interconnections Map', Global Risks Report (2016).
World Health Organization, 'Global water supply and sanitation assessment: 2000 Report' (Geneva: WHO and UNICEF 2000) 12. Worldmapper, *Households* (undated).
World Health Organization, 'Global costs and benefits of drinking-water supply and sanitation interventions to reach the MDG target and universal coverage' *WHO/HSE/WSH/12.01* (Geneva: WHO 2012).
World Health Organization, Guidelines for Drinking-Water Quality (4th edn. WHO 2011).
World Trade Organization, 'Communication from the European Communities and their Member States' GATS 2000: Environmental Services S/CSS/W/38 (22 December 2000).
World Trade Organization, Disputes by Agreement (GATS).
World Trade Organization, *EU and its Member States – Certain Measures Relating to the Energy Sector*, DS476 (2014).
World Trade Organization, GATS – Fact and Fiction
World Trade Organization, 'The General Agreement on Trade in Services (GATS): objectives, coverage and disciplines'.
Yepes, Tito, 'Expenditure on infrastructure in East Asia Region, 2006–2010' (2004).
Youkhana, Eva, and Wolfram Laube, 'Cultural, Socio-Economic and Political Constraints for Virtual Water Trade: Perspectives from the Volta Basin, West Africa' *Zentrum für Entwicklungsforschung Working Paper No 13* (2006) 11.

Index

2030 Water Resources Group 233

Abacha v. Fawehinmi case 206
Abbott, M. 328
Abukhater, Ahmed 72
access to water
  affordability, and 154–5
    approaches 312–18
    challenges 311–14
    cross-subsidisation 315
    definition 317
    direct cash transfers 315
    identifying and measuring 317
    income support 312–13
    social tariffs 316–18
    subsidies 312–13
    water credits 314–15
  climate change adaptation costs, and 284–6
  daily minimum needs 298
  definition 203
  full cost recovery paradigm 218
  human rights paradigm 218
  improvement benefits 268–9
  investment treaties impact on 145
  minimum standards 200
  per unit costs 263, 275–9
  policy development 172–3
  population growth, and 275–80, 322
  provision inequality 200–201, 215, 258–9
  right to water, and 298
    affordability, and 154–5
    Dublin Principles 217–18
    investment treaties impact on 145
    policy development 172–3
  rural-urban areas disparity 270–74
  supply and demand trends 200–201, 215, 257–8, 270–74
  sustainable access trends 123, 279–80

  targets 257–8, 266
  total cost estimates 264–8, 275–84
  trends 270–74, 321
  UN initiatives 219–20
  water-related disease, and 123, 199
Agenda 21 110, 134
Agreement on Agriculture 24–7
  purpose 25
  subsidies, classification 25–6
agricultural subsidies
  actionable subsidies 28
  Agreement on Agriculture 24–7
  classification 25–6
  de minimis limits 26
  domestic subsidies 25–6
  effectiveness, improvement methods 28–30
  energy-related subsidies 27
  export subsidies 25
  GATT 25
  generally 24
  Global Subsidies Initiative 26–7
  irrigation subsidies 26–9
  prohibited subsidies 28
  Subsidies and Counterveiling Measures Agreement 24, 27–8
  Subsidies Code 1979 25
  trade distorting effects 26–8
  Trade Facilitation Agreement 29–30
  virtual water, and 66–7
agriculture, generally
  drought-resistant plants 296
  water use 221
Aguas Argentinas / Suez v. Argentina case 151, 156–7, 160, 163, 230
Aguas de Santa Fe / Suez v. Argentina case 157, 160, 163
Aguas del Tunari dispute 92, 101–2, 168, 218, 229
algal blooms 122–3
Allan, Anthony 58–60

395

Allouche, Mathias 37
alternate wetting and drying irrigation 33
Amarasinghe, Upali 257
amicus curiae briefs 155–8, 229–30
anaerobic digestion 88
animal welfare
   public moral restrictions, and 44
arbitration
   awards, enforceability 149–50
   chilling effect 150
   investment treaty disputes
      amicus curiae briefs 156–8, 229–30
      expropriation 158–9, 164
      legitimate expectations of investor 159–61
      right to water, and 145–8, 153–4, 156–66
   investor-state dispute settlement 79, 82, 89, 92
   necessity defence 97
   US-Gambling case 47–52, 54
   water services disputes
      right to regulate 99–103, 160–61, 163
      trends 79, 82, 89
Argentina
   water services, right to regulate disputes 102–3, 151, 156–7, 160–61, 163
Articles of State Responsibility for Internationally Wrongful Acts 96–7
Asian Cities Climate Change Network (ACCCRN) 344–6
Aswan Dam 2
AWG case 102–3
Azurix v. Argentina case 157, 229–30

Barry, J.P. 137
Bartram, Jamie 264–5, 282
Benazir Bhutto v. Pakistan case 186
Benfratello, L. 329
Bhattacharyay, B.N. 262
bilateral investment treaties *see* investment treaties
Biwater v. Tanzania case 156–7, 229
Bolivia
   water wars 92, 101–2, 168, 218, 229

Bonn Freshwater Conference 2001 232–3
bottled water 24
Brazil
   health services
      constitutional right to health 238
      expenditure 236–7
      litigation 236–8, 249
   legal system overview 239
   public law litigation
      costs 249
      inequality 249–51
      right to sanitation services 240–53
      trends 236–8
   sanitation services
      expenditure 237–8, 240
      human rights litigation, and 240–53
      improvements, court role in 244–7, 252–3
      scope 237
      state obligations 237–40
Bulto, T.S. 112–13

CEO Water Mandate 232–3
Chile
   multinational companies and water service efficiency 334, 340
China
   Three Gorges Dam 2
clean water
   definition 203
   distribution systems 86–7
climate change 106, 296
   marine environment, and 137–40, 343
   resilience initiatives 343–6
   sea level rises 343–4
   water resources, and 106, 284–5, 321
   water services adaptation costs, and 284–6
Coca-Cola 221–2, 224
Cochabamba case 92, 101–2, 168, 218, 229
Cohen, B. 328
common heritage of mankind 130–32
competition
   economies of scale 306
   lack of, implications 298, 305
   market failure, and 305–6

monopolies, and 297–8, 305–6
multinational companies, and 325–6
regulation, role of 305–6
water services, for 279
water services investment, and 322
composting 88
concession agreements 83–6, 226
conventions and international agreements
see also investment treaties
African Charter on Human and Peoples Rights 1987 205–6, 209–10, 216–17
African Charter on the Rights and Welfare of the Child 1990 216–17
Agreement Relating to Part XI of the 1982 Law of the Sea Convention 1994 132–3
Biological Diversity 1992 119, 133–4
Elimination of All Forms of Discrimination Against Women 1979 109, 171, 216
European Charter on Water Resources 2001 217
Fishing and Conservation of the Living Resources of the High Seas 1958 127–8
Framework Convention on Climate Change 1992 (UNFCCC) 118–19
Geneva Conventions 1948 116–17, 129, 173
Geneva Conventions 1949 173
High Seas 1958 127
Law of the Sea 1958 126–7
Law of the Sea 1982 125–7, 132
Non-Navigational Uses of International Watercourses 1997 174
Paris Agreement 2015 140, 142
Prohibition of Military or Any Other Hostile Use of Environmental Modification Techniques 1977 174
Protection of Use of Transboundary Watercourses and International Lakes 1992 109
Rights of Persons with Disabilities 2007 216
Rights of the Child 1992 109, 171
Territorial Sea and the Contiguous Zone 1958 127
corporate social responsibility 213, 231–2
Cossy, Mirelle 45, 51–2
Cronin, A.A. 268
cross-subsidisation model 315
customary international law 202
investment disputes 97
investment treaties, and 78–80
necessity defence 97
precautionary principle 117, 121, 133–4
right to water 152–3, 170, 173

de jure praedae 126
desalination policies 295–6
disease, water-related
trends 123, 199
wastewater management impacts 123–4
dispute resolution
investment treaty disputes
amicus curiae briefs 156–8, 229–30
expropriation 158–9, 164
legitimate expectations of investor 159–61
right to water, and 145–8, 153–4, 156–66
investor-state dispute settlement 79, 82, 89, 92
necessity defence 97
US-Gambling case 47–52, 54
water services disputes
right to regulate 99–103, 160–61, 163
trends 79, 82, 89
Dublin Principles on Water and Sustainable Development 1992 217–18

efficiency
definition 326, 331
efficiency glide path 307–9
measurement
approaches 329–31
challenges 331–2, 341

private sector role in water services 324–9
  multinational company studies 332–42
  virtual water policies 62–3
Egypt
  Aswan Dam 2
  multinational companies and water service efficiency 334, 340
El Salvador
  water services, right to regulate disputes 103
energy resources
  water resources, and 4, 107
environmental disasters 136, 343–4
environmental protection
  Framework Convention on Climate Change 1992 (UNFCCC) 118–19
  international law, applicable sources 69–70, 116–22
  Law of the Sea, and 128–9
  no-harm principle 116–20
  precautionary principle 121, 133–4
  right to development, and 118
  significant adverse impact 120
  significant transboundary harm 119
  Stockholm Declaration 1972 117–18
  water protection challenges 169
  water services
    impacts of 295–6
    improvement challenges for 279
environmental services
  trade liberalisation 54
environmental sustainability
  definition 301
Estache, A. 328
EU-Canada CETA 99
European Union
  GSP Plus Scheme 33–4
  state right to regulate 93
  water services classification, WTO communication regarding 53
eutrophication 122–3
expropriation 90, 158–9, 164
  broad interpretation 151
  human rights protections, and 150–51
  indirect expropriation 98–9, 159

public interest, and 158–9
right to regulate, and 98–9

fair and equitable treatment 90, 149, 158–64
Falkenmark, Malin 73
Fay, Marianne 259–62, 264, 276
Feely, Richard 137
Ferromanganese Crusts Recommendations 135
Finger, Jermey 37
fisheries
  subsidies 29–30
  sustainability 137–8
food and beverage industry
  multi-stakeholder water initiatives 232–3
  water scarcity, and 223–4
  water use trends 221–5
food production
  virtual water, and 66–8
  wastewater management impacts 124
  water security, and 68–9, 192
food security
  challenges 193
  food, definition 192
  irrigation, and 192–3
  Pakistan, legislative protection 191–3
  right to water, and 190–91
  water security, and 4, 58, 68–9, 74–5
foreign investment
  see also investment treaties
  challenges 291–4
  definition 80–82, 89–90
  incentives 78–9
  Public-Private Partnership (PPP) schemes 291–4
  risks 293–4
  water and sanitation services
    concession agreements 83–6, 226
    as form of 80–82
    purposes 81–2
    trends 77
Framework Convention on Climate Change 1992 (UNFCCC) 118–19, 139
full protection and security provisions (FPS) 90, 101

General Agreement on Tariffs and
 Trade (GATT)
 Agreement on Technical Barriers to
  Trade, and 32
 agricultural subsidies 25
 differences from GATS 41–2
 general exceptions 34
 product definition criteria 32
 services classification 45, 47–52
General Agreement on Trade in
 Services (GATS)
 background 36, 39–40
 definitions 39–40
 differences from GATT 41–2
 dispute resolution 47–52, 54
 human, animal or plant life and
  health 44
 key provisions 40–41
 limitations, exceptions and
  exclusions 40, 43–4
 market access 41–2
 most favoured nation treatment
  41–2
 national treatment obligations 41
 product classification 36–7
 public morals and public order
  44
 trade liberalisation 47–52, 54
 transparency 41
Germany
 Vattenfall dispute 92
Global Compact 232–3
Global Risks Report 2016 23
Global Subsidies Initiative 26–7
globalisation
 definition 37–8
 rate 38
 water utility companies 226–8,
  322–3
ground water
 industrial contamination 222–3
Guerrini, A. 326–7, 332, 339
Guyana
 water privatization 218

Habermas, Jurgen 38
harm
 no harm principle 116–20
 polluter pays principle 301, 303–4,
  312

 significant adverse impact 120
 significant transboundary harm 119
Harris, R. 329
health services
 priority setting challenges 247
Hoekstra, Arjen Y. 71
host state rights and obligations *see*
 state rights and responsibilities
household size
 access to water, and 275–9
 trends 280
Hulot, Nicholas 139
human rights
 see also access to water; right to
  water
 definition 201–2
 investment treaties, and
  compensation obligations, and
   148
  conflicts between 145–52
  right to water, and 145–8, 153–4,
   156–66
 litigation role in promoting 236–7
 right to health 238
 right to life 44, 98–9, 217
 right to sanitation 238
 rights of the Child 109, 171
Hutton, Guy 264–5, 282
hydro-centricity 57
hydro-diplomacy 56
hydro-governmentality 56
hydro-hegemony 56, 72
hydro-politics 56
hydro-solidarity 56

Impregilo v. Argentina case 160–63
incineration 88
India
 food and beverage industry 222–4
 ground water contamination 222–3
 right to water
  constitutional right, as 217
  justiciablity 175–8
Indonesia
 environmental resilience 343–4
 initiatives, generally 344–6
 vulnerability assessments 344
 Semarang City sanitation
  microfinance project
   background 347–50

challenges 353–5
environmental risks 346–7
limitations 353
policy development 350–53
principal activities 350
progress 351–3
purpose 347–8
stakeholders 349
industry
see also water utility companies
food and beverage industry 221–5
ground water contamination by 222–3
water footprint 221–2
water mining 222–3
water-saving innovations 296
water use trends 3, 221–5
infrastructure investment
challenges 278–84
climate change adaptation costs, and 284–6
competition, and 278, 322
critical issues 262–4
developmental assistance 287
expenditure trends 77–8
GDP, compared with 282–3
household size, and 275–80
human development improvements, and 279
influences on 275
integrated approach, need for 294
investment gap 291–5
needs 78, 259–68
operations and maintenance costs 264
per unit costs 263, 275–9
population growth, and 275–80
private participation in infrastructure (PPI) schemes 287–91
privatization 83
public funding 286–7
Public-Private Partnership (PPP) role 293–4
reforms needed 292–3
sources 286–91
total cost estimates 264–8, 275–84, 294–5
trends 275–8

variables projection 263–4
water availability, and 263
integrated water resources management 73–4
intellectual property rights 293
International Court of Justice
Advisory Opinion on Nuclear Weapons 1996 119–20
International Covenant on Civil and Political Rights 1966 109
International Covenant on Economic, Social and Cultural Rights 1966 109, 112–13, 152, 171–2, 174–5, 191, 205–6, 216
international humanitarian law
water, rights and protection under 173–4
international investment agreements
see investment treaties
international investment law, generally
purpose 78–9
sources 79–80
international law, generally
duty of preservation 117–18
duty of prevention 121, 133–4
environmental objective challenges 72–3
environmental protection role 69–70, 116–22
necessity defence 97
no-harm principle 116–20
precautionary principle 117, 121, 133–4
right to regulate 96–8
right to water 152–3
enforceability 172
historical development 170–74
progressive *vs.* immediate right 174–5
scope 174
sources 5–6
water quality protections 116–17
International Rice Research Institute 33
International Seabed Authority 132–6
international trade
see also World Trade Organisation
agricultural subsidies 24–30
generally 23

most favoured-nation treatment 33–4, 41–2
national treatment obligations 34, 41
process and production methods classification 32–4
virtual water, interaction with 64
water footprint, and 31–2
water, trade aspects 24
international treaties, generally
see also investment treaties
interpretation 112
internationally wrongful acts 96–7
investment, generally
see also foreign investment; infrastructure investment
definition 80–82, 89–90
developmental assistance 287
funding gap 291–5
private participation in infrastructure (PPI) schemes 287–91
public funding 286–7
tribunals role 89–90
investment treaties
applicability 90
arbitration
amicus curiae briefs 156–8, 229–30
coercive force 149–50
foreign investors, availability for 82
investment protection, and 82
investor-state dispute settlement 79, 82, 89, 92, 99–103
legitimate expectations of investor 159–61
necessity defence 97
public policy objectives 97
Vattenfall dispute 92
water policy, influences on 149–52
water utility companies 228–31
customary international law, and 78–80
development trends 78
enterprise focus 81
expropriation 90, 98–9, 158–9, 164
flexibility in 91
human rights, impact on 145
investment, definition 80–82

purpose 79–80, 89–90
right to regulate 104
compensation obligations, and 94, 101, 148
conflicts between 91–3, 148–52
exceptions 94–5
flexibility 91
indirect right 98–9
principle 93–4
public interest protection 92–3
restrictions 91–2
water services disputes 99–103, 160–61, 163
sources of law, as 79–80
state rights and obligations 148
compensation 94, 101, 148
human rights protection 145–52
right to regulate, and 91–103, 101, 148–52
US-Canada BIT 91
Ireland
domestic water charges
affordability, and 318–20
approaches 299–300, 312, 318–19
cross-subsidisation 315
demand management, and 301–2
direct cash transfers 315
funding models 300–301, 319–20
government review of 297–8
income-related credits 299–300, 312, 318–20
income support 312–13
introduction 299–300
policy aims 300–305, 320
RPI-X formula 307–9
social tariffs 316–18
subsidies 312–13
Universal Free Allowances 299–300, 312, 318–19
water credits 314–15
water services
annual costs 299
investment trends 299
monopoly role in 306–7, 320
policy conflicts 298
regulatory framework 307–9
responsibility for 298
revenue sources 298
social justice, and 300–301
water funding model 300–301

Irfan v. Lahore Development Authority case 186–7
irrigation
  regulatory controls, in Pakistan 192–3
  subsidies 26–9
  techniques 33
  water use trends 191
ISO 14046 (Water Footprint) 31
Italy
  water company efficiency 326–7

Law of the Sea
  Advisory Opinion on Deep-Sea Mining 121
  common heritage of mankind 130–32
  development 125–6
  ecological policy developments 132–6
  freedom of the high seas 127
  marine environment
    definition 128–9
    ecological challenges 133
  state rights and responsibilities 128–33, 138–40
  theories 126
  water, definition 127–8
  water regulation, and 127–41
    common heritage of mankind, and 130–32
    development 125–9
    ecological policy developments 132–6
    political and economic influences on 129–31
Leroux, Eric 39, 50
Levitt, Theodore 37
Lithuania
  multinational companies and water service efficiency 334, 340
litigation
  human rights promotion, and
    costs 249
    criticism 249–51
    improvements, role in 244–7, 252–3
    judicial mechanisms 247–9
    trends 236–7
Littlechild, Stephen 307

Mar del Plata Water Conference 1977 4–5, 108–9, 171
mare claudum 126
mare liberum 126
marine environment
  see also Law of the Sea
  climate change, and 137–40
  definition 128–9, 134
  environmental disaster impacts 136
  importance 136–7
market access
  trade agreement provisions 41–2
market valuation of water 3
Matsushita, Mitsuo 42–3
Mazibuko v. City of Johannesburg case 208
Metalclad v. Mexico case 159
Methanex v. United States cases 159
Mexico
  multinational companies and water service efficiency 334, 340
microfinance initiatives *see* Indonesia
Millennium Development Goals 7, 110–111, 114, 123, 219–20, 257–9, 266, 321
Minerva Mills v. Union of India case 209
Mining Code 133–4
monopolies
  disadvantages 307
  water services, role in 306–7
most-favoured-nation treatment 33–4, 41–2
Mukherjee, Sacchidananda 258
multinational companies
  corporate social responsibility 231–2
  definition 323
  food and beverage industry water use 221–5
  national policies, relationships between 323–4
  right to water, and
    human rights conflicts 228–31
    multi-stakeholder initiatives 231–5
  trends 323, 325
  water services
    efficiency, and 324–9, 332–42
    investment 323–4
    provision role 215, 323–4

Narmada Bachao Andolan case 217
national treatment obligations 34, 41
necessity defence 97
neo-liberal economic theory 38
Neubert, Susanne 71
New Economic Order 129
Nigeria
  access to water
    trends 195, 199
  African Charter on Human and
    Peoples Rights 205–6, 209–10
  economic and political overview
    195–6
  population 197
  right to water 210, 214
    challenges 210
    constitutional rights 204, 206, 209
    international law, and 204–6,
      209–10
    judicial enforcement 207–11
    social justice, and 211–13
  water and sanitation services
    challenges 210
    domestic human rights protection
      204, 206, 209
    international human rights law,
      and 204–6, 209–10
  water resources
    overview 196–7
no-harm principle 116–20
nuclear weapons, ICJ Advisory
  Opinion 1996 119–20

Ocean Day 2015 139–40

Pac Rim Cayman v. El Salvador case
  103
Pakistan
  food security policy 191–3
  irrigation, regulatory controls 192–3
  right to water, domestic
    implementation 178–93
    constitutional provisions 178–9
    enforceability 183–4
    judicial developments 185–8
    legislative provisions 179–81
    national and regional policies
      181–5
    progress and limitations 188–90,
      193–4

Palaniappan, Heather 292
Perdikan *see* Semarang City *under*
  Indonesia
Philip Morris Asia v. Australia case
  150
Philippines
  multinational companies and water
    service efficiency 334, 340
Piero Foresti v. South Africa case 150
polluter pays principle 301, 303–4, 312
pollution
  atmospheric 138
  trends 116, 124
  untreated wastewater discharges
    122–4
population growth
  access to water, and 275–80, 322
Portugal
  multinational companies and water
    service efficiency 334, 340
precautionary principle 117, 121,
  133–4
preferential trade and investment
  agreements *see* investment treaties
Private Participation in Infrastructure
  (PPI) schemes 287–91
private sector
  see also multinational companies
  right to water, and 154
  social justice, role in 213
  water services
    global stakeholders 226–8
    investment trends 287–94
    role trends 92, 325
privatization
  definition 82–3
  right to water, and 154
  of water
    Bolivia water wars case 92, 101–2,
      218, 229
    violation of human rights, as
      168–9
    water and sanitation services 38–9
      benefits 83
      concession agreements 83–6, 226
      trends 83, 92
process and production methods
  goods classification by 32–4
Protocol on Water and Health 1999
  109

public health
  expenditure distribution 249
  human right to health 238
  litigation
    costs 249
    criticism 249
    human rights promotion, and 240–53
    judicial mechanisms 247–9
    trends 236–8
  right to regulate, and 98–9
  trade liberalisation restrictions 44
  water-related disease
    trends 123, 199
    wastewater management impacts 123–4
  water treatment processes 88
public interest protection
  investment treaty disputes
    expropriation 158–9, 164
    fair and equitable treatment, and 158–64
    legitimate expectations of investor 159–61
    non-water cases 158–60
    water services cases 160–61
  right to regulate, and 92–3
  right to water, and 155–65
public international law *see* World Trade Organization
  right to water, and 152–3, 172–4, 199–200
public morals
  trade liberalisation restrictions 34, 44
public policy
  defence 97
  importance 164–5
  judicial interference mechanisms 247–9
public-private partnerships (PPP) 293–4
public services
  classification 52–3
Pulp Mills on the River Uruguay (Argentina v. Uruguay) case 120

rainwater management policies 296
regulation *see* water services regulation
right to development 118

right to health 238
right to life 44, 98–9, 217
right to regulate
  conventional law, and 94–6
    exceptions 94–5
    indirect right 98–9
    legitimate regulatory interests 94, 102–3
    positive obligations 95–6
  general principles 93–4
  human rights, and 98–9
  international law, and 96–8
  investment treaties 104
    compensation obligations, and 94, 101, 148
    conflicts between 91–3, 148–52, 160–61, 163
    exceptions 94–5
    flexibility 91
    indirect right 98–9
    principle 93–4
    public interest protection 92–3
    restrictions 91–2
    water services disputes 99–103, 160–61, 163
  public health, and 98–9
  water services
    arbitration 99–103, 160–61
    EU-Canada CETA 99
    exceptions 99–100
    indirect right 98–9
right to sanitation 238
right to water 199–201, 213–14, 217
  see also access to water
  affordability, and 154–5
  amicus curiae briefs 155–8, 229–30
  challenges 115
  classification 168–70
  customary international law, and 152–3, 170, 173
  enforceability 172
  food security, and 190–91
  General Comment 15 111–13, 115, 152, 172, 174–5, 181, 200, 216
  historical development 108–15, 170–74
  historical view 198
  human right, as 199–201, 213–14, 217

India, domestic law implementation 175–7
international conventions 108–110, 108–112, 171–4
international humanitarian law 173–4
international law, generally 152–3, 172–4, 199–200
investment treaties, and
  amicus curiae briefs 155–8, 229–30
  compensation obligations, and 148
  disputes 145–8, 153–4, 153–66, 156–66
  jurisdiction and justiciability 175–8
  limitations 174
  Pakistan, domestic law implementation 178–93, 178–94
  private sector participation, and 154
  progressive vs. immediate right 174–5
  public interest protection, and 155–65
  restrictions on 114–15
  right to regulate, and 98–9, 160–61, 163
  scope 152–5
  South Africa, domestic law implementation 177–8
  state responsibilities, and 70–71, 113–14, 161–3
  activities in breach 161–3
Sustainable Development Goals, and 114–15, 123
UN Resolution 15/L.14 113
UN Resolution 64/292 13, 98, 113, 200
UN Resolution 68/157 152–3
water privatization as violation of 168–9
rights of the child 109, 171
Rio Conference on Environment and Development 1992 110, 118
Rio Declaration on Environment and Development 1992 118, 133
Robinson, C. 329
Romano. G. 326–7, 332, 339
Rossi, M. 328
Roth, Dik 65

RPI-X formula 307–9
rural areas
  access to water trends 270–74

safe drinking water, access to see access to water
sanitation services
  access trends 270–74
  Brazil, in
    expenditure 237–8, 240
    human rights litigation, and 240–53
    improvements, court role in 244–7, 252–3
    scope 237
    state obligations 237–40
  functions 87–9
  human right to 238, 240–53
  Indonesia improvement microfinance initiative 346–55
  Indonesia, microfinance initiative in
    background 347–50
    challenges 353–5
    environmental risks 346–7
    limitations 353
    policy development 350–53
    principal activities 350
    progress 351–3
    purpose 347–8
    stakeholders 349
  international regulation 78
  investment
    concession agreements 83–6, 226
    foreign 80–82
    needs 78, 259–68
    purposes 81–2
    total costs of 264–8
    trends 77
  investment cost estimates 264–8, 275–84, 294–5
  lack of 199
  per unit costs 263
  private participation in infrastructure (PPI) schemes 287–91
  private sector role 6
  sanitation, definition 203
  treated water, final destination 88
  wastewater treatment processes 88–9

water charges, international
    comparison 309–11
WTO services classification 46–7
SAUR International v. Argentina case
    99–100, 153, 157, 160, 162
Schiebm Pierre-Alain 293–4
sea level rises 343–4
Selden, J. 126
Semarang City environmental
    resilience initiative *see*
    Indonesia
Sembenelli, A. 329
Sen, Amartya 212
SERAC v. Nigeria case 209–10
Serbia
    multinational companies and water
        service efficiency 334, 340
sewage *see* sanitation services
shared resources, definition 5
Shehla Zia v. WAPDA case 185–6
Shiva, Vandana 52
significant adverse impact 120
significant transboundary harm
    concept 119
Smakhtin, Vladimir 257
Smets, Henri 170
social justice
    citizenship participation 211–12
    corporate social responsibility 213,
        231–2
    critical social engagement 211
    definition 202
    education role 212
    law, and 212–13
    private sector role 213
    water charges, and 300–301
social tariffs 316–18
soft drinks manufacture, water use in
    221–5
sole effects doctrine 158–9
South Africa
    right to water
        court rulings 208
        justiciablity 177–8
state rights and responsibilities
    duty of preservation 117–18
    duty of prevention 121, 133–4
    fair and equitable treatment 90, 149,
        158–63
    food production 66, 70–71

international liability, distinction
    from 119
internationally wrongful acts 96–7
investment treaties
    human rights protection, and
        145–52
    right to water, and 145–58
    Law of the Sea 128–33, 138–40
    necessity defence 97
    New Economic Order, and 129
    no harm principle 116–20
    precautionary principle 117, 121,
        133–4
    right to development 118
    right to regulate
        compensation obligations, and
            94, 101, 148
        conventional law 94–6, 98–9
        indirect right 98–9
        international law 96–8
        investment agreement conflicts,
            and 91–3, 160–61.148–152,
            163
        legitimate regulatory interests 94,
            102–3
        public interest protection 92–3
        water regulation, arbitration
            99–103, 160–61, 163
        water regulation, indirect right
            98–9
    right to water, and 70–71, 113–14
        activities in breach of obligations
            161–3
        amicus curiae briefs 156–8,
            229–30
        investment treaty disputes 145–8,
            153–4, 156–8
    virtual water theory, and 66, 70–71
    water quality protections 116–17
Stevens, Barrie 293–4
Stockholm Declaration 1972 117–18
stormwater management policies 296
subsidies
    see also agricultural subsidies
    water affordability, and 312–13
Subsidies and Counterveiling Measures
    Agreement 24, 27–8
Subsidies Code 1979 25
supply and demand
    critical issues 262–4

improved services, relationship between 278–9
industry, for 3
patterns and trends 1–4, 23, 76
per unit costs 263
tourism, for 3–4
variables projection 263–4
water charges, and 301–2
sustainability
environmental sustainability, definition 301
Sustainable Development Goals 23, 30, 114–15, 123, 257
sustainable development model 139

Technical Barriers to Trade Agreement 32–4
technology transfer arrangements 296
Three Gorges Dam 2
Toubkiss, J. 275–8
tourism
water supply and demand trends 3–4
trade liberalisation 38
see also General Agreement on Trade in Services (GATS)
environmental services 54
exclusions and exceptions 34, 44
role 38–9, 52–3
unintentional 47–54
US-Gambling case 47–52, 54
water services, and 38–9, 51–4
transboundary watercourses
conflicts over 296
Convention on 109
significant transboundary harm 119
transparency
trade agreement provisions 41
transport
democratization, implications of 3–4

UNESCO Water Report 2014 107
United Kingdom
domestic water charges
affordability, and 313–14
trends 309–10, 320
multinational companies and water service efficiency 334, 340

United Nations
access to water initiatives 219–20
Charter 110
Committee on Economic, Social and Cultural Rights 111–12
Conference on Environment and Development 1992 (Rio de Janeiro) 110, 118
Framework Convention on Climate Change 1992 (UNFCCC) 118–19, 139
General Comment 15 111–13, 115, 152, 172, 174–5, 181, 200, 216
Global Compact 232–3
Guiding Principles for Business and Human Rights 220
Millennium Development Goals 110–111, 114, 123, 219–20, 257–9, 266, 321
Resolution 15/L.14 113
Resolution 64/292 13, 98, 113, 200
Resolution 68/157 152–3
Sustainable Development Goals 114–15, 123, 257
Universal Declaration of Human Rights 170
Universal Free Allowances 299–300, 312
Urban Climate Change Resilience Planning Framework 344–5
urbanization 3, 343
access to water services trends 270–74
climate change impacts 343
planning policies, need for 296
trends 343–4
US-Canada BIT 91
US-Gambling case 47–52, 54

Vattenfall dispute 92
Vienna Convention on the Law of Treaties 1969 112
virtual water 8–9, 31
agricultural subsidies, and 66–7
food supplies, and
state responsibilities 66, 70–71
water scarcity 58
water security *vs.* food security 68–9, 74–5
green *vs.* blue water 61–2, 65–6

international flows 9
international institutional role 68–9
international trade, interaction with 64
policies
   procedural fairness *vs.* equity 72
   production efficiency and transfer 62–3
   purpose 60–63, 71–2
   research focus 64
   territorial perspectives 63, 71–2
theory
   circularity challenges 72–3
   limitations 74–5
   principles 58–60, 69
   questions and problems for 60
   water basin approaches, and 72–4
   water cost influences 64–5
   water pricing, and 64–7
water-intensive goods
   production drivers 62–3, 65–6, 68
   subsidies, and 66–7
   trade in 60, 66
Vivendi v. Argentina case 161–2

Ward, Philip J. 285
Warner, Jeroen 65
wastewater
   charges, international comparison 309–11
   collection systems 87–8
   management challenges 122–5
   re-use policies 295
   standards implementation 124
   treatment processes 88–9
   untreated discharges 122–4
wastewater services
   see also sanitation services
   collection systems 87–8
   functions 87–9
   treatment processes 88–9
water acidification 137–8
water basin resources management 73–4
water charges
   see also Ireland
   affordability, and approaches 312–18
   challenges 311–14
   cost recovery model 303
   demand management, and 301–2
   environmental sustainability, and 301
   financial sustainability, and 303
   fixed tariff model 302–5
   international comparison 309–11
   marginal pricing model 302–3
   policy trade-offs and challenges 302–5
   polluter pays principle 301, 303–4, 312
   RPI-X formula 307–9
   social justice, and 299–300
   Universal Free Allowances 299–300, 312
   variable tariff model 302–5
   wastewater charges, international comparison 309–11
   water funding model 300–301
water credits 314–15
Water Dialogues 232–3
water footprint
   climatic and environmental influences on 61
   definition 31, 59
   food and beverage industry 222–4
   industrial 221–2
   international standards 31
   operational footprint 221–2
   variability 61–2
   virtual water, and 61–2, 73–4
   WTO rules, and 31–2
water, generally
   see also right to water
   classification as right 168–70
   clean water, definition 203
   definition 117–18, 127–8
   forms 105
   interconnectedness 106–7, 111–12
   Law of the Sea, and 127–8
   privatization, as violation of rights 168–9
   threats to 105–6
water management
   see also sanitation services; water services
   integrated resources management 73–4
   models 7
   private sector role 6

Index 409

residential consumption, successful methods for 301–2
Sustainable Development Goals, and 114–15, 123, 257
trends 1
unsustainable policies 23
volumatic pricing, and 301–2
wastewater management challenges 122–5
water pollution
  algal blooms 122–3
  trends 122–3
water poverty
  food shortages, and 58
water prices
  see also water charges
  affordability, and 154–5
  concession agreements 85–6
  full cost recovery paradigm 218
  green water vs. blue water 65–6
  human rights paradigm 218
  marginal cost pricing 302
  marginal pricing model 302–3
  per-unit area fees 65
  price caps 307
  RPI-X formula 307–9
  virtual water 64–7
  volumetric pricing 301
  water-intensive production, relationship between 64–7
water purification systems 86–7
water quality
  acidification 137–8
  clean water distribution systems 86–7
  definition 86
  influences on 106
  international law protections 116–17, 173
  trends 124
water regulation
  challenges 141–2
  fragmentation 141–2
  institutional influences 4–6
  Law of the Sea approach 127–41
    common heritage of mankind, and 130–32
    development 125–9
    ecological policy developments 132–6

political and economic influences on 129–31
need for 70
relevance of 91–2
state host right to regulate
  conventional law 98–9
  disputes 99–103
  indirect right 98–9
  investment agreement conflicts 91–3, 148–52
  public interest protection 92–3
terrestrial approach 107–25
  environmental protection, and 115–25
  no harm principle 116–20
  right to water, and 108–15
  wastewater management 122–5
water resources
  agricultural use trends 116
  climate change, and 106, 284–5, 321–2
  communal sharing, attitudes to 198
  energy resources, and 4, 107
  food security, and 4, 68–9, 74–5
  global resource, as 4–5
  industry water use trends 3, 221–5
  interconnectedness 106–7
  management trends 1
  population growth, and 1, 275–80
  shared resources, definition 5
  supply and demand trends 1–2
  threats to 105–6
water salinity 134–5, 295–6
water scarcity
  food and beverage industry, and 223–4
  food shortages, and 58
  ground water contamination, and 223
  trends 13, 106, 108, 123–4, 167
  water value, and 167
water security
  hydro-politics, concept development 56
  international food trade, and 58
  international institutional role 68–9
  research trends 55–7
  vs. food security 4, 58, 68–9, 74–5

water services
  see also sanitation services; water utility companies
  access trends 270–74
  affordability, and
    approaches 312–18
    challenges 311–14
  arbitration
    fair and equitable treatment, and 160–64
    investment treaty conflicts 99–103, 160–61, 163
    right to regulate 99–103, 160–61, 163
    trends 79, 82, 89
  classification
    ambiguity and clarity 46–7, 51–4
    under GATT and GATS 45–7, 51–2, 54
    subcategories 45–6, 54
  competition 279
  conflicts 298
  developmental assistance 287
  expansion trends 77
  full cost recovery paradigm 218
  globalisation 226–8
  human rights paradigm 218
  international regulation 78
  investment
    competition for 322
    concession agreements 83–6, 226
    cost estimates 264–8, 275–84, 294–5
    foreign 80–82
    infrastructure, needs of 78, 259–61
    needs 78, 259–68
    purposes 81–2
    total costs of 264–8
    trends 77
  multinational companies
    efficiency, and 324–9, 332–42
    role 215
  per unit costs 263
  private participation in infrastructure (PPI) schemes 287–91
  privatisation 38–9
    concession agreements 83–6, 226
    influences on 77–8
  productivity and efficiency, and 306–7
  multinational company role 324–9, 332–42
  public funding for 286–7
  purpose 298
  regulation, role of 307–8
  responsibility, trends 38–9, 70–71
  water charges, international comparison 309–11
water services regulation
  efficiency glide path 307–9
  host state rights and obligations 91–104
    activities in breach 161–3
    investment agreement conflicts 91–3, 148–52
  limitations of 307
water stress 23, 108
water utility companies
  corporate social responsibility, and 231–2
  efficiency, and 324–9
    studies 332–42
  global stakeholders 226–8, 322–3
  investment, and
    concession agreements 83–6, 226
    disputes 228–31
    international investment regime 228–31
    privatization 225–6
    public-private partnerships 225, 287–91
    strategies 225–6
  multinational company presence
    benefits 324–6
    trends 323, 325
  privatization
    opponents 226
    right to water, and 92, 101–2, 168, 218, 228–31, 229
    trends 226–8
  reform in 225
water wars 108
  Bolivia 92, 101–2, 168, 218, 229
  research trends 55–6
watercourses
  non-navigational uses 174
weapon, water as 1–2

Index 411

West Pakistan Salt Miners Labour Union (CBA) Khewra, Jhelum v. Director, Industries and Mineral Development, Punjab, Lahore case 187
World Trade Organisation (WTO)
  Agreement on Agriculture 24–7
  agricultural subsidies 24–30
  Enabling Clauses 32–4
  GATS
    background 36, 39–40
    definitions 39–40
    differences from GATT 41–2
    human, animal or plant life and health 44
    key provisions 40–41
    limitations, exceptions and exclusions 40, 43–4
    market access 41–2
    most favoured nation treatment 41–2
    national treatment obligations 41
    product classification 36–7
    public morals and public order 44
    transparency 41
  GATT
    Agreement on Technical Barriers to Trade, and 32
    agricultural subsidies 25
    differences from GATS 41–2
    general exceptions 34
    general exceptions to Article XX 34
    process and production distinctions 32–4
    product definition criteria 32
    services classification 45
  most favoured-nation treatment 33–4, 41–2
  national treatment obligations 34, 41
  Services Sectorial Classification List 45
  Subsidies and Counterveiling Measures Agreement 24, 27–8
  Technical Barriers to Trade Agreement 32–4
  Trade Facilitation Agreement 29–30
  water footprint, and 31–2
  water services classification 45–7, 51–4
Wouters, Patricia 73

Yepes, Tito 259–62, 264, 276